Xibei Yang
Jingyu Yang

Incomplete Information System and Rough Set Theory

Models and Attribute Reductions

Xibei Yang
Jingyu Yang

Incomplete Information System and Rough Set Theory

Models and Attribute Reductions

Science Press
Beijing

Springer

Authors

Xibei Yang
School of Computer Science and Engineering, Jiangsu University of Science and Technology, Zhenjiang Jiangsu 212003, P.R. China
E-mail: yangxibei@hotmail.com

Jingyu Yang
School of Computer Science and Technology, Nanjing University of Science and Technology, Nanjing Jiangsu 210094, P.R. China
E-mail: yangjy@mail.njust.edu.cn

ISBN 978-7-03-032476-4
Science Press Beijing

ISBN 978-3-642-25934-0 ISBN 978-3-642-25935-7 (eBook)
Springer Heidelberg Dordrecht London New York

Library of Congress Control Namber: 2011942907

Printed on acid-free paper

Springer is part of Springer Science+Business Media (www.springer.com)

Foreword

It is a challenging task to preface the book of Dr. Xibei Yang and Prof. Jingyu Yang which is because that their works have a wide range of knowledge and wide scope. In the last decade, Dr. Yang and Prof. Yang have done excellent job in rough set theory. The book "*Incomplete Information System and Rough Set Theory: Models and Attribute Reductions*" is a masterpiece. It involves some of the most basic contents in incomplete data processing, and has the authority, systematicness, width, profundity, and strong explanatory. This book has rich example, commentary and references, which makes people feel that reading this book is enjoyable. It is devoted to scholarly research on knowledge reduction, rule extraction, uncertainty reasoning, decision evaluation, and granular computing etc.

Rough set theory, proposed by Pawlak, has become a well-established mechanism for uncertainty management in a wide variety of applications related to artificial intelligence. Pawlak's rough set model was constructed on the basis of an indiscernibility relation, it then cannot be used to deal with the incomplete information system with unknown values. Presently, though many results about rough sets in incomplete information system have been obtained in many journal articles, there is not a systematic book, which can summarize these researching results. Dr. Yang and Prof. Yang have successfully done such work in this book. Furthermore, what I want to point out is that this book not only includes many widely known results, but also introduces many unpublished idea such as neighborhood system and multigranulation approaches into incomplete information system. From this point of view, this book is a landmark work to rough set theory.

This book may be considered as the material for master candidate and doctor candidate in institutes of higher learning, it can also be referred by researchers in the fields of artificial intelligence, machine learning and so on. I do hope that this book will receive a global acceptance on the merit of its quality and contribution to the development of rough set theory.

Prof. Yuhua Qian
Email: jinchengqyh@126.com
School of Computer and Information
Technology, Shanxi University,
Taiyuan, Shanxi, 030006, P. R. China
September 2011

Preface

Incomplete Information System and Rough Set Theory: Models And Attribute Reductions provides evidence of present growth in the rough set approach to the incomplete information system. The topics discussed in this book have received significant attentions in recent years because researchers can apply new tools for problem solving. This book reflects a number of approaches those were either directly or indirectly begun by the seminal work on rough set by Zdzisław Pawlak.

It is well-know that the knowledge representation system or the so-called information system plays a crucial role in Pawlak's rough set theory. Evidence of the growth of various rough set-based research streams can be found in the rough set database[1]. However, in many practical applications, since the difficulties of acquisitions of knowledge, incomplete instead of the complete information systems can be seen everywhere. Therefore, how to employ the rough set approach to deal with the incomplete information systems is very important to the development of the rough set theory.

This book includes several important approaches. These approaches have significant implications in a number of research areas such as hierarchical structure, attribute reduction, dominance-based rough set, decision rules and multigranulation rough set.

The book contains four parts.

1. In Part I, the indiscernibility relation based rough sets such as Pawlak's rough set, variable precision rough set and multigranulation rough set are presented. Especially, it should be noticed that multigranulation rough set approach provides a new direction to the development of the rough set theory.
2. In Part II, the tolerance relation, maximal consistent block, descriptor, similarity relation, different relation, neighborhood system techniques are used to deal with the incomplete information system, respectively. Most importantly, as the first model for granular computing, neighborhood system approach provides a new view on the model of rough set.
3. In Part III, the concept of the dominance-based rough set is introduced into the incomplete environment and then several expanded dominance-based rough sets have been explored.
4. In Part IV, the multigranulation rough set approach is introduced into the incomplete environment, which provides a new direction for the developing of the multigranulation rough set theory.

The primary target audience of this book are those who are interested in the fields of machine learning, data mining and knowledge discovery, especially for those who

[1] http://rsds.univ.rzeszow.pl. Accessed 7 April 2011.

are working in rough set theory, and granular computing. This book could also be a useful reference for university courses in both fields of applied computational mathematics and computer sciences.

Xibei Yang
Jingyu Yang
May 2011

Acknowledgements

This work is supported by the Natural Science Foundation of China (Nos.61100116, 60632050, 60903110), Natural Science Feundation of Jiangsu Province of China (No. BK2011492), Natural Science Foundation of Jiangsu Higher Education Institution of China (No.11KJB 520004), Postdoctoral Science Foundation of China (No.20100481149), and Postdoctoral Science Foundation of Jiangsu Province of China (No. 1101137C).

We extend our thanks to the following researchers: Prof. Tsau Young Lin, Prof. Weizhi Wu, Prof. Jusheng Mi, Dr. Yuhua Qian, Prof. Jian Yang, Prof. Dongjun Yu, Prof. Chen Wu, Dr. Xiaoning Song, Prof. Gongzhi Luo, Prof. Yunsong Qi, Dr. Wei Wei, Dr. Yanhong She, Dr. Yuecheng Yu, Dr. Dakuan Wei, Dr. Ming Zhang, Dr. Lijuan Wang, M.S. Huili Dou, M.S. Junyi Zhou.

Special thanks go to Haina Zhang, our production editor, who kept us on schedule.

Contents

Part III Dominance-based Rough Sets and Incomplete Information Systems

Part IV Incomplete Information Systems and Multigranulation Rough Sets

Part I
Indiscernibility Relation Based Rough Sets

Chapter 1

Indiscernibility Relation, Rough Sets and Information System

Abstract Pawlak's rough set model, was firstly constructed on the basis of an indiscernibility relation. Such an indiscernibility relation is an intersection of some equivalence relations in knowledge base and then it is also an equivalence relation. This chapter introduced the basic concepts of Pawlak's rough set, Ziarko's variable precision rough set and Qian's multigranulation rough sets. These models were all proposed on the basis of indiscernibility relation. Variable precision rough set generalizes classical rough approximation by introducing a threshold β. Such β value represents a bound on the conditional probability of an equivalence class, which are classified into the target concept. Multigranulation rough set uses a family of the indiscernibility relation instead of a single one to construct rough approximation. In multigranulation rough set approach, the optimistic and pessimistic multigranulation rough sets are two basic models.

1.1 Pawlak's Rough Approximation

1.1.1 Rough Set

The theory of rough set [1–10] was firstly proposed by Pawlak in 1980s for the study of intelligent systems characterized by inexact, uncertain or vague information. Presently, with the rapid development of the rough set, such theory has been demonstrated to be useful in the fields such as pattern recognition [11–17], decision support system [18–22], medical analysis [23–26], data mining [27–35] and so on. It has become one of the flash points in the research area of information science.

From the viewpoint of the rough set, since we have no enough knowledge in the universe of discourse, the set, or the so-called concept / target then cannot be clearly defined. However, we can use a pair of approximations, i.e. lower and upper approximations to describe the target roughly. Therefore, the lower and upper approximations are two key notions in Pawlak's rough set theory, they were firstly constructed on the basis of an indiscernibility relation (reflexive, symmetric and transitive, i.e. equivalence relation). The main difference between rough set and other mathematical tools for dealing with uncertain problems is that rough set does not need any prior information beyond the problem itself. The main goal of the rough set analysis is to synthesize approximations of concepts from the acquired data. To know what is rough set, we need the following notions.

1. Let $U \neq \emptyset$ be the universe of discourse, $\mathbf{R}$ is a family of the equivalence relations on U, the pair $K = (U, \mathbf{R})$ is referred to as a knowledge base [1];
2. If $P \subseteq \mathbf{R}$ and $P \neq \emptyset$, then $\cap P$ (intersection of all equivalence relations in P) is also an equivalence relation, and will be denoted by $IND(P)$, it is referred to as an indiscernibility relation over P in Pawlak's rough set theory.

$\forall R \in \mathbf{R}$, we use U/R to represent the family of all equivalence classes in terms of the equivalence relation R (or classification of U in terms of R). Such family of equivalence classes is also referred to as categories or concepts of R. Therefore, $\forall x \in U$, $[x]_R$ is used to denotes a category (equivalence class), which contains x in terms of R.

Suppose that $P \subseteq \mathbf{R}$, then $IND(P)$ is also an equivalence relation, $U/IND(P)$ is then the family of all equivalence classes in terms of the set of equivalence relations P, each element in $U/IND(P)$ is referred to as a P-basic knowledge, $[x]_P = \{y \in U : (x, y) \in IND(P)\}$ is the equivalence class of P, which contains x. $\forall X \subseteq U$, if X is the union of some P-basic knowledge, then X is P-definable; otherwise X is P-undefinable. To describe the P-undefinable set more clearly, Pawlak proposed his rough set model as Definition 1.1 shows.

Definition 1.1. [1] Let $K = (U, \mathbf{R})$ be a knowledge base, $P \subseteq \mathbf{R}$, then $\forall X \subseteq U$, the lower approximation and upper approximation of X are denoted by $\underline{P}(X)$ and $\overline{P}(X)$, respectively, where

$$\underline{P}(X) = \{x \in U : [x]_P \subseteq X\} \tag{1.1}$$

$$= \cup\{Y \in U/IND(P) : Y \subseteq X\} \tag{1.2}$$

$$\overline{P}(X) = \{x \in U : [x]_P \cap X \neq \emptyset\} \tag{1.3}$$

$$= \cup\{Y \in U/IND(P) : Y \cap X \neq \emptyset\} \tag{1.4}$$

The pair $[\underline{P}(X), \overline{P}(X)]$ is referred to as the rough set of X in terms of the indiscernibility relation $IND(P)$.

Remark 1.1. By Definition 1.1, we can see that two different approaches have been employed for the constructing of lower and upper approximations. The first one is referred to as "point-based" approach (see Eq. (1.1) and Eq. (1.3)) while the second one is referred to as "set-based" approach (see Eq. (1.2) and Eq. (1.4)).

- In "point-based" approach, the lower approximation (Eq. (1.1)) is the set of elements, which can be certainly classified as elements of X; the upper approximation (Eq. (1.3)) is the set of the elements, which can be possibly classified as elements of X.

- In "set-based" approach, the lower approximation (Eq. (1.2)) is the union of some equivalence classes, which can be certainly classified as part of X; the upper approximation (Eq. (1.4)) is the union of some equivalence classes, which can be possibly classified as part of X.

Obviously, since indiscernibility relation (equivalence relation) is used in Pawlak's rough set model, the result of "point-based" rough approximation is same to the "set-based" rough approximation. However, it should be noticed that in some more general binary relations based rough set models, the rough approximations in terms of the above two approaches are not necessarily equivalent.

Following Definition 1.1, we shall also employ the following denotations [1]:

- P-positive region of X: $POS_P(X) = \underline{P}(X)$;
- P-negative region of X: $NEG_P(X) = U - \overline{P}(X)$;
- P-boundary region of X: $BN_P(X) = \overline{P}(X) - \underline{P}(X)$.

The positive region $POS_P(X)$ is the collection of objects, which can be classified with full certainty as members of the set X by using P. The negative region $NEG_P(X)$ is the collection of objects, which can be determined without any ambiguity, employing P, that they do not belong the the set X. The boundary region is the undecidable are, that is, none of the objects belonging to the boundary can be classified with certainty into X or the complement of X, i.e. $\sim X$ as far as knowledge P is concerned [1].

Proposition 1.1. [1] *Let* $K = (U, \mathbf{R})$ *be a knowledge base,* $P \subseteq \mathbf{R}$*, then we have following properties about the lower and upper approximations*:

1. $\underline{P}(U) = \overline{P}(U) = U$;
2. $\underline{P}(\varnothing) = \overline{P}(\varnothing) = \varnothing$;
3. $\underline{P}(X) \subseteq X$;
4. $X \subseteq \overline{P}(X)$;
5. $\underline{P}(X \cap Y) = \underline{P}(X) \cap \underline{P}(Y)$;
6. $\overline{P}(X \cup Y) = \overline{P}(X) \cup \overline{P}(Y)$;
7. $\underline{P}(X \cup Y) \supseteq \underline{P}(X) \cup \underline{P}(Y)$;
8. $\overline{P}(X \cap Y) \subseteq \overline{P}(X) \cap \overline{P}(Y)$;
9. $\underline{P}(\underline{P}(X)) = \underline{P}(X)$;
10. $\overline{P}(\overline{P}(X)) = \overline{P}(X)$;
11. $\underline{P}(\sim X) = \sim \overline{P}(X)$;
12. $\overline{P}(\sim X) = \sim \underline{P}(X)$;
13. $X \subseteq Y \rightarrow \underline{P}(X) \subseteq \underline{P}(Y)$;
14. $X \subseteq Y \rightarrow \overline{P}(X) \subseteq \overline{P}(Y)$.

By the rough set model showed in Definition 1.1, Pawlak [1] introduced the following approximate equality of sets.

Definition 1.2. [1] Let $K = (U, \mathbf{R})$ be a knowledge base, $P \subseteq \mathbf{R}$, $\forall X, Y \subseteq U$, then

1. X and Y are referred to as lower equivalent in terms of P if and only if $\underline{P}(X) = \underline{P}(Y)$;
2. X and Y are referred to as upper equivalent in terms of P if and only if $\overline{P}(X) = \overline{P}(Y)$;
3. X and Y are referred to as rough equivalent in terms of P if and only if $\underline{P}(X) = \underline{P}(Y)$ and $\overline{P}(X) = \overline{P}(Y)$.

By Definition 1.2, we can see that

1. if X and Y are lower equivalent in terms of P, then the positive regions of X and Y are the same;
2. if X and Y are upper equivalent in terms of P, then the negative regions of X and Y are the same;
3. if X and Y are rough equivalent in terms of P, then both positive and negative regions of X and Y are the same, i.e. the boundary regions of X and Y are the same.

Definition 1.3. [1] Let $K = (U, \mathbf{R})$ be a knowledge base, $P \subseteq \mathbf{R}$, $\forall X \subseteq U$, $\forall x \in U$, the rough membership of x in X is denoted by $\mu_X^P(x)$ such that

$$\mu_X^P(x) = \frac{|[x]_P \cap X|}{|[x]_P|} \tag{1.5}$$

where $|X|$ is the cardinal number of set X.

Obviously, $0 \leqslant \mu_X^P(x) \leqslant 1$ holds. By the concept of rough membership, we can re-define the lower approximation, upper approximation and boundary region such that

$$\underline{P}(X) = \{x \in U : \mu_X^P(x) = 1\}$$
$$\overline{P}(X) = \{x \in U : \mu_X^P(x) > 0\}$$
$$BN_P(X) = \{x \in U : 0 < \mu_X^P(x) < 1\}$$

Proposition 1.2. *Let $K = (U, \mathbf{R})$ be a knowledge base, $P \subseteq \mathbf{R}$, $X, Y \subseteq U$, we have following properties about rough membership:*

1. $\mu_U^P(x) = 1$;
2. $\mu_\varnothing^P(x) = 0$;
3. $x \in X \Rightarrow \mu_X^P(x) \neq 0$;
4. $x \notin X \Rightarrow \mu_X^P(x) \neq 1$;
5. $\mu_X^P(x) = 1 \Leftrightarrow x \in \underline{P}(X)$;
6. $\mu_X^P(x) = 0 \Leftrightarrow x \notin \overline{P}(X)$;
7. $0 < \mu_X^P(x) < 1 \Leftrightarrow x \in BN_P(X)$;
8. *if* $(x, y) \in IND(P)$, *then* $\mu_X^P(x) = \mu_X^P(y)$;
9. $\mu_{U-X}^P(x) = 1 - \mu_X^P(x)$;
10. $X \subseteq Y \Rightarrow \mu_X^P(x) \leqslant \mu_Y^P(x)$;
11. $\mu_{X \cup Y}^P(x) = \mu_X^P(x) + \mu_Y^P(x) - \mu_{X \cap Y}^P(x)$;
12. $X \cap Y = \varnothing \Rightarrow \mu_{X \cup Y}^P(x) = \mu_X^P(x) + \mu_Y^P(x)$;

13. $\mu^P_{X\cup Y}(x) \geqslant \max(\mu^P_X(x), \mu^P_Y(x))$;
14. $\mu^P_{X\cap Y}(x) \leqslant \min(\mu^P_X(x), \mu^P_Y(x))$;
15. *if* $\mathbf{X} = \{X_1, X_2, \cdots\}$ *is a family of sets, which are disjoint to each other, then*

$$\mu^P_{\cup X_i}(x) = \sum_{X_i \in \mathbf{X}} \mu^P_{X_i}(x) \tag{1.6}$$

where $X_i \in \mathbf{X}$.

1.1.2 Uncertainty Measurements and Knowledge Granulation

1.1.2.1 Accuracy of Approximation and Rough Accuracy

Inexactness of a set is due to the existence of a boundary region [1]. The greater the boundary region of a set, the lower the accuracy of such set. To express this idea more precisely, Pawlak introduced the concept of accuracy of approximation [1] such that

$$\alpha_P(X) = \frac{|\underline{P}(X)|}{|\overline{P}(X)|} \tag{1.7}$$

The accuracy measure $\alpha_P(X)$ is intended to capture the degree of completeness of our knowledge about the set X. By the basic property of rough set, we know $0 \leqslant \alpha_P(X) \leqslant 1$ holds for each $P \subseteq \mathbf{R}$ and $X \subseteq U$. If $\alpha_P(X) = 1$, then the boundary region of X is empty, X is P-definable; if $\alpha_P(X) < 1$, then X has some non-empty boundary region and consequently it is P-undefinable.

On the other hand, in Ref. [36] Yao said that the accuracy measure can also be interpreted by using the well-known Marczewski-Steinhaus metric, or MZ metric for short. By introducing the MZ metric into the lower and upper approximations, we have

$$\begin{aligned}\rho_P(X) &= 1 - \frac{|\underline{P}(X) \cap \overline{P}(X)|}{|\underline{P}(X) \cup \overline{P}(X)|}\\ &= 1 - \frac{|\underline{P}(X)|}{|\overline{P}|(X)}\\ &= 1 - \alpha_P(X)\end{aligned}$$

$\rho_R(X)$ is referred to as the rough accuracy.

By the above equation, the accuracy of approximation may be viewed as an inverse of MZ metric when applied to lower and upper approximations. In other words, the distance between the lower and upper approximations determines the accuracy of approximation. The greater the rough accuracy is, the smaller the accuracy of approximation we have.

Moreover, to investigate the accuracy of approximation and rough accuracy under different granulation, a partial relation [37] is defined such that: given two families of the equivalence relations P_1 and P_2, i.e. $P_1, P_2 \subseteq \mathbf{R}$, $U/IND(P_1) \preceq U/IND(P_2)$ (or

$U/IND(P_2) \succeq U/IND(P_1)$) if and only if, for each $Y_i \in U/IND(P_1)$, there exists $X_j \in U/IND(P_2)$ such that $Y_i \subseteq X_j$ where $U/IND(P_1) = \{Y_1, Y_2, \cdots\}$ and $U/IND(P_2) = \{X_1, X_2, \cdots\}$ are partitions induced by indiscernibility relations $IND(P_1)$ and $IND(P_2)$, respectively. In this case, we say that P_2 is coarser than P_1, or P_1 is finer than P_2. If $U/IND(P_1) \preceq U/IND(P_2)$ and $U/IND(P_1) \neq U/IND(P_2)$, we say P_2 is strictly coarser than P_1 (or P_1 is strictly finer than P_2), denoted by $U/IND(P_1) \prec U/IND(P_2)$ (or $U/IND(P_2) \succ U/IND(P_1)$).

Theorem 1.1. *Let $K = (U, \mathbf{R})$ be a knowledge base, $P_1, P_2 \subseteq \mathbf{R}$, if $U/IND(P_1) \preceq U/IND(P_2)$, then $\forall X \subseteq U$, we have*

$$\alpha_{P_1}(X) \geqslant \alpha_{P_2}(X), \quad \rho_{P_1}(X) \leqslant \rho_{P_2}(X)$$

Proof. It can be derived directly from the definition of the accuracy of approximation. and rough accuracy. □

The above theorem tells us that by the accuracy of approximation and rough accuracy, we can compare the uncertainties between rough sets. However, such two measurements have their own limitations, let us use an example to illustrate it.

Example 1.1. Suppose that $U = \{x_1, x_2, \cdots, x_{10}\}$ is the universe of discourse, $IND(P_1)$, $IND(P_2)$ and $IND(P_3)$ are three indiscernibility relations on U such that

$$U/IND(P_1) = \Big\{\{x_1, x_2, x_3, x_4\}, \{x_5, x_6, x_7\}, \{x_8, x_9, x_{10}\}\Big\}$$
$$U/IND(P_2) = \Big\{\{x_1, x_2\}, \{x_3, x_4\}, \{x_5, x_6, x_7\}, \{x_8, x_9\}, \{x_{10}\}\Big\}$$
$$U/IND(P_3) = \Big\{\{x_1\}, \{x_2\}, \{x_3\}, \{x_4\}, \{x_5, x_6, x_7\}, \{x_8\}, \{x_9\}, \{x_{10}\}\Big\}$$

If $X = \{x_1, x_2, x_3, x_4, x_6, x_7\}$, then the rough approximations of X by using the above three indiscernibility relations are:

$$\underline{P_1}(X) = \underline{P_2}(X) = \underline{P_3}(X) = \{x_1, x_2, x_3, x_4\}$$
$$\overline{P_1}(X) = \overline{P_2}(X) = \overline{P_3}(X) = \{x_1, x_2, x_3, x_4, x_5, x_6, x_7\}$$

from which we can conclude that no matter what indiscernibility relation is selected, the rough approximations are equivalent. Thus, the accuracies of approximations and rough accuracies are:

$$\alpha_{P_1}(X) = \alpha_{P_2}(X) = \alpha_{P_3}(X) = 4/7$$

$$\rho_{P_1}(X) = \rho_{P_2}(X) = \rho_{P_3}(X) = 3/7$$

In Example 1.1, we can see that $U/IND(P_3) \preceq U/IND(P_2) \preceq U/IND(P_1)$, i.e. there is a partial relation among three different indiscernibility relations, nevertheless, the same accuracies of approximations and rough accuracies are obtained for these three rough sets of X. Therefore, it is necessary to introduce more effective measurements. To solve such problem, Liang introduced the concept of knowledge granulation [38–41] into uncertainty measurements of the rough set.

1.1.2.2 Knowledge Granulation

Definition 1.4. [37, 38] Let $K = (U, \mathbf{R})$ be a knowledge base, $\forall X \subseteq U$, GK is a mapping from the power set of X to the set of real numbers, GK is referred to as a knowledge granulation in knowledge base K if and only if GK satisfies the following conditions:

1. non-negativity: $GK(P) \geqslant 0$ for each $P \subseteq \mathbf{R}$;
2. invariability: $GK(P_1) = GK(P_2)$ for each $P_1, P_2 \subseteq \mathbf{R}$, if there is a bijective mapping function $f : F(P_1) \to F(P_2)$ such that $|[x]_{P_1}| = |f([x]_{P_2})| (\forall x \in U)$, where $F(P_1) = \{[x]_{P_1} : x \in U\}$ and $F(P_2) = \{[x]_{P_2} : x \in U\}$;
3. monotonicity: $GK(P_1) < GK(P_2)$ for each $P_1, P_2 \subseteq \mathbf{R}$ with $P_1 \preceq P_2$.

Definition 1.4 is the axiomatic definition of knowledge granulation. A widely used mathematical form of knowledge granulation is defined in Definition 1.5.

Definition 1.5. [39, 40] Let $K = (U, \mathbf{R})$ be a knowledge base, $P \subseteq \mathbf{R}$, $U/IND(P) = \{Y_1, Y_2, \cdots\}$ is the partition determined by the indiscernibility relation $IND(P)$, a knowledge granulation in terms of P is defined by

$$GK(P) = \frac{1}{|U|^2} \sum_{Y_i \in U/IND(P)} |Y_i|^2 \tag{1.8}$$

Obviously, $GK(P)$ in Definition 1.5 satisfies with all the conditions showed in Definition 1.4, thus, it is a knowledge granulation.

Theorem 1.2. [37] *Let $K = (U, \mathbf{R})$ be a knowledge base, $P \subseteq \mathbf{R}$,*

1. *if $U/IND(P) = \omega = \{\{x\} : x \in U\}$, then $GK(P)$ achieves minimum value, i.e.* $\frac{1}{|U|}$;
2. *if $U/IND(P) = \delta = \{U\}$, then $GK(P)$ achieves maximal value, i.e.* 1.

Definition 1.6. [39] Let $K = (U, \mathbf{R})$ be a knowledge base, $P \subseteq \mathbf{R}$, $\forall X \subseteq U$, the roughness of X with respect to P is defined as follows:

$$Roughness_P(X) = \rho_P(X) \cdot \mathrm{GK}(P) \tag{1.9}$$

the accuracy of X with respect to P is defined as follows:

$$Accuracy_P(X) = 1 - \rho_P(X) \cdot \mathrm{GK}(P) \tag{1.10}$$

Theorem 1.3. [41] *Let $K = (U, \mathbf{R})$ be a knowledge base, $P_1, P_2 \subseteq \mathbf{R}$, $\forall X \subseteq U$, if $U/IND(P_1) \preceq U/IND(P_2)$, then we have*

$$\begin{aligned} Roughness_{P_1}(X) &\leqslant Roughness_{P_2}(X) \\ Accuracy_{P_1}(X) &\geqslant Accuracy_{P_2}(X) \end{aligned} \tag{1.11}$$

The above theorem states that the roughness of X with respect to P decreases with P becoming finer, the accuracy of X with respect to P increases with P becoming finer.

Example 1.2. Following Example 1.1, compute the roughnesses and accuracies of X with respect to three different indiscernibility relations, respectively.

In Example 1.1, we have

$$GK(P_1) = \frac{1}{|U|^2} \sum_{Y_i \in U/IND(P_1)} |Y_i|^2 = \frac{1}{100}(16 + 9 + 9) = \frac{17}{50}$$

$$GK(P_2) = \frac{1}{|U|^2} \sum_{Y_i \in U/IND(P_2)} |Y_i|^2 = \frac{1}{100}(4 + 4 + 9 + 4 + 1) = \frac{11}{50}$$

$$GK(P_3) = \frac{1}{|U|^2} \sum_{Y_i \in U/IND(P_3)} |Y_i|^2 = \frac{1}{100}(1 + 1 + 1 + 1 + 9 + 1 + 1 + 1) = \frac{8}{50}$$

Therefore, the roughnesses of X with respect to $IND(P_1)$, $IND(P_2)$ and $IND(P_3)$ are:

$$Roughness_{P_1}(X) = \frac{4}{7} \cdot \frac{17}{50} = 0.19$$

$$Roughness_{P_2}(X) = \frac{4}{7} \cdot \frac{11}{50} = 0.13$$

$$Roughness_{P_3}(X) = \frac{4}{7} \cdot \frac{8}{50} = 0.09$$

the accuracies of X with respect to $IND(P_1)$, $IND(P_2)$ and $IND(P_3)$ are:

$$Accuracy_{P_1}(X) = 1 - \rho_{R_1}(X) \cdot GK(R_1) = 0.81$$

$$Accuracy_{P_2}(X) = 1 - \rho_{R_2}(X) \cdot GK(R_2) = 0.89$$

$$Accuracy_{P_3}(X) = 1 - \rho_{R_3}(X) \cdot GK(R_3) = 0.91$$

By comparing with Example 1.1, in Example 1.2, we can see that different roughnesses and accuracies are obtained by three different indiscernibility relations. From this point of view, roughness and accuracy are better than Pawlak's accuracy of approximation and Yao's rough accuracy, respectively, in the measuring of rough uncertainty.

1.1.3 Knowledge Reductions

Given a knowledge base $K = (U, \mathbf{R})$, one fundamental aspect of rough set theory involves the search for particular subsets of the equivalence relations, which provide the same information for classification or some purposes as all of the equivalence relations in $\mathbf{R}$. Such subsets are called reducts. The approach to find such subsets is referred to as knowledge reduction while the obtained subset is referred to as reduct. Obviously, different requirements may induce different reducts. In the following, we briefly presented the reduct and relative reduct, which were widely used in Pawlak's rough set theory.

1.1.3.1 Reduct of Indiscernibility Relation

Definition 1.7. [1] Let $K = (U, \mathbf{R})$ be a knowledge base, $\forall R \in \mathbf{R}$,

1. if $IND(\mathbf{R}) = IND(\mathbf{R} - \{R\})$, then R is referred to as dispensable in $\mathbf{R}$, otherwise R is indispensable in $\mathbf{R}$;
2. if each $R \in \mathbf{R}$ is indispensable in $\mathbf{R}$, then the family of the equivalence relations $\mathbf{R}$ is referred to as independent, otherwise $\mathbf{R}$ is dependent.

Definition 1.8. [1] Let $K = (U, \mathbf{R})$ be a knowledge base, $Q \subseteq P \subseteq \mathbf{R}$,

1. if $IND(Q) = IND(P)$ and Q is independent, then Q is referred to as a reduct of P;
2. the set of all indispensable relations in P will be referred to as the core of P, and will be denoted by $CORE(P)$.

Theorem 1.4. [1] *Let $K = (U, \mathbf{R})$ be a knowledge base, $P \subseteq \mathbf{R}$,*

$$CORE(P) = \cap RED(P) \tag{1.12}$$

where $\cap RED(P)$ is the intersection of all reducts of P.

The use of the concept of the core is twofold:

1. firstly, it can be considered as a basis for computation of reducts because the core is included in each reduct, and its computation is straightforward;
2. secondly, the core can be interpreted as the set of most characteristic part of knowledge, which can not be deleted when reducing the knowledge.

1.1.3.2 Relative Reduct of Positive Region

In Definition 1.8, a reduct of P is actually a minimal subset of the equivalence relations in P, which can be used to preserve the indiscernibility relation $IND(P)$. Moreover, the concept of reduct can also be introduced into the rough set model. For example, Pawlak's proposed his relative reduct as Definition 1.9 shows.

Definition 1.9. [1] Let $K = (U, \mathbf{R})$ be a knowledge base, $Q \subseteq \mathbf{R}$, $S \subseteq P \subseteq \mathbf{R}$, if $U/IND(Q) = \{X_1, X_2, \cdots, X_m\}$, then the positive region of Q in terms of P is

$$POS_P(Q) = \bigcup_{i=1,2,\cdots,m} \underline{P}(X)$$

1. $\forall R \in P$, if $POS_P(Q) = POS_{P-\{R\}}(Q)$, then R is referred to as Q-dispensable in P, otherwise R is Q-indispensable in P;
2. if each R in P is Q-indispensable in P, then P is referred to as Q-independent;
3. if $POS_P(Q) = POS_S(Q)$ and S is Q-independent, then $S \subseteq P$ is referred to as a Q-reduct of P;
4. the set of all Q-indispensable relations in P will be called the Q-core of P, and will be denoted by $CORE_Q(P)$.

Theorem 1.5. [1] *Let $K = (U, \mathbf{R})$ be a knowledge base, then*

$$CORE_Q(P) = \cap RED_Q(P) \tag{1.13}$$

where $\cap RED_Q(P)$ is is the intersection of all Q-reducts of P.

1.1.4 Knowledge Dependency

Definition 1.10. [1] Let $K = (U, \mathbf{R})$ be a knowledge base, $P, Q \subseteq \mathbf{R}$, then

1. knowledge Q depends on knowledge P if and only if $IND(P) \subseteq IND(Q)$, it is denoted by $P \Rightarrow Q$;
2. knowledge P and Q are equivalent, denoted as $P \equiv Q$, if and only if both $IND(P) \subseteq IND(Q)$ and $IND(Q) \subseteq IND(P)$ hold;
3. knowledge P and Q are independent, denoted as $P \not\equiv Q$, if and only if neither $P \Rightarrow Q$ nor $Q \Rightarrow P$ holds.

Theorem 1.6. [1] *Let $K = (U, \mathbf{R})$ be a knowledge base, $P, Q \subseteq \mathbf{R}$, then the following conditions are equivalent*:

1. $P \Rightarrow Q$;
2. $IND(P \cup Q) = IND(P)$;
3. $POS_P(Q) = U$;
4. $\underline{P}(X) = X$ *for each* $X \in U/IND(Q)$.

Proposition 1.3. [1] *Let $K = (U, \mathbf{R})$ be a knowledge base, $P, Q, R, S \subseteq \mathbf{R}$, then*

1. *if $P \Rightarrow Q$ and $P' \supset P$, then $P' \Rightarrow Q$;*
2. *if $P \Rightarrow Q$ and $Q' \subset Q$, then $P \Rightarrow Q'$;*
3. *$P \Rightarrow Q$ and $Q \Rightarrow R$ imply $P \Rightarrow R$;*
4. *$P \Rightarrow R$ and $Q \Rightarrow R$ imply $P \cup Q \Rightarrow R$;*
5. *$P \Rightarrow R \cup Q$ imply $P \Rightarrow R$ and $P \Rightarrow Q$;*
6. *$P \Rightarrow Q$ and $Q \cup R \Rightarrow S$ imply $P \cup R \Rightarrow S$;*
7. *$P \Rightarrow Q$ and $R \Rightarrow S$ imply $P \cup R \Rightarrow Q \cup S$.*

The derivation (dependency) can also be partial, which means that only part of knowledge Q is derivable from knowledge P. The partial derivability can be defined by using the notion of the positive region of knowledge.

Definition 1.11. [1] Let $K = (U, \mathbf{R})$ be a knowledge base, $P, Q \subseteq \mathbf{R}$, we say that knowledge Q depends in a degree k $(0 \leqslant k \leqslant 1)$ from knowledge P, symbolically $P \Rightarrow_k Q$, where

$$k = \gamma_P(Q) = \frac{|POS_P(Q)|}{|U|} \tag{1.14}$$

If $k = 1$, we will say that Q totally depends from P, if $0 < k < 1$, we say that Q partially depends from P, and if $k = 0$, we say that Q is totally independent from P.

The above described ideas can also be interpreted as an ability to classify objects. More precisely, if $k = 1$, then all elements of the universe can be classified into elementary categories of $U/IND(Q)$ by using knowledge P; if $k \neq 1$, only those elements of the universe which belong to the positive region can be classified into elementary categories of knowledge Q, employing knowledge P. In particular, if $k = 0$, none of the elements of the universe can be classified using knowledge P to elementary categories of Q.

1.2 Variable Precision Rough Set

1.2.1 Inclusion Error and Variable Precision Rough Set

In Pawlak's rough set model, the classification analysis must be completely correct or certain. In practice, however, it seems that admitting some level of uncertainty in the classification process may lead to a deeper understanding and a better utilization of properties of the data being analyzed. To overcome such limitation in traditional rough set model, partial classification was taken into account by Ziarko in Ref. [42]. He extended rough approximation by introducing a threshold β. The β value represents a bound on the conditional probability of a proportion of objects in a condition class, which are classified into the same decision class. Ziarko's approach is referred to as the Variable Precision Rough Set (VPRS) model [42].

Given a universe U of discourse, let $U/IND(P) = \{X_1, X_2, \cdots, X_m\}$ be the partition on U by indiscernibility relation $IND(P)$, $\forall X \subseteq U$ and $\forall X_i \in U/IND(P)$, we say that the set X_i is included in the set X with an admissible error β if and only if:

$$X_i \subseteq^{\beta} X \Leftrightarrow e(X_i, X) \leqslant \beta \tag{1.15}$$

where

$$e(X_i, X) = 1 - \frac{|X_i \cap X|}{|X_i|} \tag{1.16}$$

The quantity $e(X_i, X)$ is referred to as the inclusion error of X_i in X. The value of β should be limited: $0 \leqslant \beta < 0.5$. By the concept of inclusion error, the variable precision lower approximation and upper approximation of X are denoted by $\underline{P}^{\beta}(X)$ and $\overline{P}^{\beta}(X)$, respectively, where

$$\underline{P}^{\beta}(X) = \cup\{X_i \in U/IND(P) : e(X_i, X) \leqslant \beta\} \tag{1.17}$$

$$\overline{P}^{\beta}(X) = \cup\{X_i \in U/IND(P) : e(X_i, X) < 1 - \beta\} \tag{1.18}$$

The pair $[\underline{P}^{\beta}(X), \overline{P}^{\beta}(X)]$ is referred to as a variable precision rough set of X in terms of the indiscernibility relation $IND(P)$. By the variable precision lower approximation and variable precision upper approximation, the variable precision boundary region is denoted by $BN_P^{\beta}(X)$ where

$$BN_P^{\beta}(X) = \overline{P}^{\beta}(X) - \underline{P}^{\beta}(X). \tag{1.19}$$

Proposition 1.4. *Let $K = (U, \mathbf{R})$ be a knowledge base, $P \subseteq \mathbf{R}$, then we have following properties about the variable precision lower and upper approximation*:

1. $\underline{R}^{\beta}(U) = \overline{R}^{\beta}(U) = U$;
2. $\underline{R}^{\beta}(\varnothing) = \overline{R}^{\beta}(\varnothing) = \varnothing$;
3. $\underline{R}^{\beta}(X) \subseteq \overline{R}^{\beta}(X)$;
4. $\underline{R}^{\beta}(X \cap Y) \subseteq \underline{R}^{\beta}(X) \cap \underline{R}^{\beta}(Y)$;
5. $\overline{R}^{\beta}(X \cup Y) \supseteq \overline{R}^{\beta}(X) \cup \overline{R}^{\beta}(Y)$;

6. $\underline{R}^{\beta}(X \cup Y) \supseteq \underline{R}^{\beta}(X) \cup \underline{R}^{\beta}(Y)$;
7. $\overline{R}^{\beta}(X \cap Y) \subseteq \overline{R}^{\beta}(X) \cap \overline{R}^{\beta}(Y)$;
8. $\beta_1 \geqslant \beta_2 \Rightarrow \underline{R}^{\beta_1}(X) \supseteq \underline{R}^{\beta_2}(X), \overline{R}^{\beta_1}(X) \subseteq \overline{R}^{\beta_2}(X)$;
9. $\underline{R}^{\beta}(\sim X) = \sim \overline{R}^{\beta}(X)$;
10. $\overline{R}^{\beta}(\sim X) = \sim \underline{R}^{\beta}(X)$.

Remark 1.2. It should be noticed that in the variable precision rough set model, $\underline{P}^{\beta}(X) \subseteq X$ and $X \subseteq \overline{P}^{\beta}(X)$ do not always hold. This is different from the property in Pawlak's rough set model.

1.2.2 Several Reducts in Variable Precision Rough Set

Similar to the reduct in classical rough set model, the reduct in variable precision rough set is still an interesting issue to be addressed. Based on Zirako's variable precision rough set model, Beynon [43] proposed the concept of β-reduct while Mi [44] proposed the approximate distribution reduct.

Definition 1.12. Let $K = (U, \mathbf{R})$ be a knowledge base, $S \subseteq P \subseteq \mathbf{R}$, $U/IND(Q) = \{X_1, X_2, \cdots, X_m\}$, we denote

$$\delta_P^{\beta} = \frac{\Sigma\{|\underline{P}^{\beta}(X_i)| : i \leqslant m\}}{|U|} \tag{1.20}$$

$$\lambda_P^{\beta} = \frac{\Sigma\{|\overline{P}^{\beta}(X_i)| : i \leqslant m\}}{|U|} \tag{1.21}$$

$$L_P^{\beta} = \{\underline{P}^{\beta}(X_1), \underline{P}^{\beta}(X_2), \cdots, \underline{P}^{\beta}(X_m)\} \tag{1.22}$$

$$H_P^{\beta} = \{\overline{P}^{\beta}(X_1), \overline{P}^{\beta}(X_2), \cdots, \overline{P}^{\beta}(X_m)\} \tag{1.23}$$

$$BN_P^{\beta} = \{BN_P^{\beta}(X_1), BN_P^{\beta}(X_2), \cdots, BN_P^{\beta}(X_m)\} \tag{1.24}$$

1. If $\delta_S^{\beta} = \delta_P^{\beta}$, then S is referred to as a β lower approximate consistent set of P; if S is a β lower approximate consistent set of P and $\forall S' \subset S$, S' is not the β lower approximate consistent set of P, then S is referred to as a β lower approximate reduct of P(in Ref. [43], it is called the β reduct or approximate reduct).
2. If $\lambda_S^{\beta} = \lambda_P^{\beta}$, then S is referred to as a β upper approximate consistent set of P; if S is a β upper approximate consistent set of P and $\forall S' \subset S$, S' is not the β upper approximate consistent set of P, then S is referred to as a β upper approximate reduct of P.
3. If $L_S^{\beta} = L_P^{\beta}$, then S is referred to as a β lower approximate distribution consistent set of P; if S is a β lower approximate distribution consistent set of P and $\forall S' \subset S$, S' is not the β lower approximate distribution consistent set of P, then S is referred to as a β lower approximate distribution reduct of P.
4. If $H_S^{\beta} = H_P^{\beta}$, then S is referred to as a β upper approximate distribution consistent set of P; if S is a β upper approximate distribution consistent set of P and $\forall S' \subset S$, S' is not the β upper approximate distribution consistent set of P, then S is referred to as a β upper approximate distribution reduct of P.

5. If $BN_S^\beta = BN_P^\beta$, then S is referred to as a β boundary region distribution consistent set of P; if S is a β boundary region distribution consistent set of P and $\forall S' \subset S$, S' is not the β boundary region distribution consistent set of P, then S is referred to as a β boundary region distribution reduct of P.

By Definition 1.12, we can see that

- The β lower approximate consistent set of P is a subset of P, which preserves the sum of cardinal numbers of variable precision lower approximations of equivalence classes induced by Q; the β reduct of P is a minimal subset of P, which preserves the sum of cardinal numbers of variable precision lower approximations of equivalence classes induced by Q.
- The β upper approximate consistent set of P is a subset of P, which preserves the sum of cardinal numbers of variable precision upper approximations of equivalence classes induced by Q; the β upper approximate reduct of P is a minimal subset of P, which preserves the sum of cardinal numbers of variable precision upper approximations of equivalence classes induced by Q.
- The β lower approximate distribution consistent set of P is a subset of P, which preserves the variable precision lower approximations of the equivalence classes induced by Q; the β lower approximate distribution reduct of P is a minimal subset of P, which preserves the variable precision lower approximations of the equivalence classes induced by Q.
- The β upper approximate distribution consistent set of P is a subset of P, which preserves the variable precision upper approximations of the equivalence classes induced by Q; the β upper approximate distribution reduct of P is a minimal subset of P, which preserves the variable precision upper approximations of the equivalence classes induced by Q.
- The β boundary region distribution consistent set of P is a subset of P, which preserves the variable precision boundary regions of the equivalence classes induced by Q; the β boundary region distribution reduct of P is a minimal subset of P, which preserves the variable precision boundary regions of the equivalence classes induced by Q.

By analyzing the above different types of reducts, Inuiguchi presented the following property of the variable precision rough set in Ref. [45]:

$$\frac{1}{m} \geqslant \beta \Rightarrow U = \bigcup_{i=1}^{m} \overline{P}^{\beta}(X_i). \tag{1.25}$$

Eq. (1.25) tells us that the union of variable precision upper approximations does not always equal to U but when $1/m \geqslant \beta$. For such reason, Inuiguchi defined the β unpredictable region such that

$$UN_P^\beta(Q) = U - \bigcup_{i=1}^{m} \overline{P}^{\beta}(X_i). \tag{1.26}$$

By the β unpredictable region in variable precision rough set, Inuiguchi proposed the following definition of reduct.

Definition 1.13. [45] Let $K = (U, \mathbf{R})$ be a knowledge base, $S \subseteq P \subseteq \mathbf{R}$, $U/IND(Q) = \{X_1, X_2, \cdots, X_m\}$,

if $UN_S^{\beta}(Q) = UN_P^{\beta}(Q)$, then S is referred to as a β unpredictable region consistent set of P; if S is a β unpredictable region distribution consistent set of P and $\forall S' \subset S$, S' is not the β unpredictable region distribution consistent set of P, then S is referred to as a β unpredictable region reduct of P.

By Definition 1.13, we can see that the β unpredictable region consistent set of P is a subset of P, which preserves the unpredictable region in the variable precision rough set model; a β unpredictable region reduct of P is the minimal subset of P, which preserves the unpredictable region in the variable precision rough set model.

1.3 Multigranulation Rough Set

In Pawlak's rough set model, the target is approximated by using one and only one indiscernibility relation (equivalence relation). However, Qian et al. claimed that we often need to describe a target concept through multi binary relations on the universe according to a user's requirements or targets of problem solving [46]. They presented several practical cases to illustrate such restrictions.

Case 1: In some data analysis issues, for the same object, there is a contradiction or inconsistent relationship between its values under one indiscernibility relation $IND(P)$ and those under another indiscernibility relation $IND(Q)$. In other words, we cannot perform the intersection operation between $IND(P)$ and $IND(Q)$, and then the target concept cannot be approximated by using $U/IND(P \cup Q)$.

Case 2: In the process of some decision making, the decision or the view of each of decision makers may be independent for the same project (or a sample, object and element) in the universe. In such case, the intersection operation between any two decision makers will be redundant for decision making.

Case 3: To extract decision rules from distributive information systems and groups of intelligent agents through rough set approach, knowledge representation and rough approximation should be investigated. For the reduction of the time complexity of rule extractions, it is unnecessary for us to perform the intersection operation among all the sites in the context of distributive information systems.

For the above reason, Qian et al. proposed the concept of Multigranulation Rough Set (MGRS) [47–49], in which the target is approximate by two or more equivalence relations. In Qian's multigranulation rough set theory, two different multigranulation cases, i.e. optimistic and pessimistic multigranulation rough sets have been considered.

1.3.1 Optimistic Multigranulation Rough Set

Definition 1.14. [49] Let $K = (U, \mathbf{R})$ be a knowledge base, $\{R_1, R_2, \cdots, R_m\} \subseteq \mathbf{R}$ is a set of the equivalence relations, $\forall X \subseteq U$, the optimistic multigranulation lower approximation and upper approximation of X are denoted by $\underline{\sum_{i=1}^{m} R_i}^{OPT}(X)$ and $\overline{\sum_{i=1}^{m} R_i}^{OPT}(X)$, respectively, where

$$\underline{\sum_{i=1}^{m} R_i}^{OPT}(X) = \{x \in U : [x]_{R_1} \subseteq X \vee [x]_{R_2} \subseteq X \vee \cdots \vee [x]_{R_m} \subseteq X\} \tag{1.27}$$

$$\overline{\sum_{i=1}^{m} R_i}^{OPT}(X) = \sim \left(\underline{\sum_{i=1}^{m} R_i}^{OPT}(\sim X) \right) \tag{1.28}$$

The pair $\left[\underline{\sum_{i=1}^{m} R_i}^{OPT}(X), \overline{\sum_{i=1}^{m} R_i}^{OPT}(X)\right]$ is referred to as an optimistic multigranulation rough set of X in terms of the set of the equivalence relations $\{R_1, R_2, \cdots, R_m\}$. By Definition 1.14, it can be seen that the lower approximation in optimistic multigranulation rough set is defined through the equivalence classes induced by multi independent equivalence relations, whereas standard rough lower approximation is represented via those derived by only one indiscernibility relation though it is the intersection of some equivalence relations. The upper approximation in optimistic multigranulation rough set can be characterized by the complementary set of the lower approximation of the target concept.

Theorem 1.7. *Let* $K = (U, \mathbf{R})$ *be a knowledge base,* $\{R_1, R_2, \cdots, R_m\} \subseteq \mathbf{R}$, $\forall X \subseteq U$, *we have*

$$\overline{\sum_{i=1}^{m} R_i}^{OPT}(X) = \{x \in U : [x]_{R_1} \cap X \neq \varnothing \wedge [x]_{R_2} \cap X \neq \varnothing \wedge \cdots \wedge [x]_{R_m} \cap X \neq \varnothing\} \tag{1.29}$$

Proof. By Definition 1.14, we have $\overline{\sum_{i=1}^{m} R_i}^{OPT}(X) = \sim \left(\underline{\sum_{i=1}^{m} R_i}^{OPT}(\sim X) \right)$, thus

$$\begin{aligned}\overline{\sum_{i=1}^{m} R_i}^{OPT}(X) &= \sim \{x \in U : [x]_{R_1} \subseteq (\sim X) \vee [x]_{R_2} \subseteq (\sim X) \vee \cdots \vee [x]_{R_m} \subseteq (\sim X)\} \\ &= \sim \{x \in U : [x]_{R_1} \cap X = \varnothing \vee [x]_{R_2} \cap X = \varnothing \vee \cdots \vee [x]_{R_m} \cap X = \varnothing\} \\ &= \{x \in U : [x]_{R_1} \cap X \neq \varnothing \wedge [x]_{R_2} \cap X \neq \varnothing \wedge \cdots \wedge [x]_{R_m} \cap X \neq \varnothing\}. \quad \square\end{aligned}$$

Theorem 1.8. [49] *Let* $K = (U, \mathbf{R})$ *be a knowledge base,* $\{R_1, R_2, \cdots, R_m\} \subseteq \mathbf{R}$, *suppose that* $U/R_1 \preceq U/R_2 \preceq \cdots \preceq U/R_m$, *then we have*

$$\underline{\sum_{i=1}^{m} R_i}^{OPT}(X) = \underline{R_1}(X) \tag{1.30}$$

$$\overline{\sum_{i=1}^{m} R_i}^{OPT}(X) = \overline{R_1}(X) \tag{1.31}$$

Theorem 1.8 tells us that if there is a partial relation among the partitions, which are determined by the set of equivalence relations, then the optimistic multigranulation rough set is equivalent to the rough set in terms of the *finest* partition.

Proposition 1.5. [49] *Let $K = (U, \mathbf{R})$ be a knowledge base, $\{R_1, R_2, \cdots, R_m\} \subseteq \mathbf{R}$, $\forall X \subseteq U$, then we have following properties about the optimistic multigranulation lower and upper approximations:*

1. $\underline{\sum_{i=1}^{m} R_i}^{OPT}(X) \subseteq X \subseteq \overline{\sum_{i=1}^{m} R_i}^{OPT}(X)$;
2. $\underline{\sum_{i=1}^{m} R_i}^{OPT}(\varnothing) = \overline{\sum_{i=1}^{m} R_i}^{OPT}(\varnothing) = \varnothing$,
   $\underline{\sum_{i=1}^{m} R_i}^{OPT}(U) = \overline{\sum_{i=1}^{m} R_i}^{OPT}(U) = U$;
3. $\underline{\sum_{i=1}^{m} R_i}^{OPT}(X) = \bigcup_{i=1}^{m} \underline{R_i}(X)$,
   $\overline{\sum_{i=1}^{m} R_i}^{OPT}(X) = \bigcap_{i=1}^{m} \overline{R_i}(X)$;
4. $\underline{\sum_{i=1}^{m} R_i}^{OPT}\left(\underline{\sum_{i=1}^{m} R_i}^{OPT}(X)\right) = \underline{\sum_{i=1}^{m} R_i}^{OPT}(X)$,
   $\overline{\sum_{i=1}^{m} R_i}^{OPT}\left(\overline{\sum_{i=1}^{m} R_i}^{OPT}(X)\right) = \overline{\sum_{i=1}^{m} R_i}^{OPT}(X)$;
5. $\underline{\sum_{i=1}^{m} R_i}^{OPT}(\sim X) = \sim\left(\overline{\sum_{i=1}^{m} R_i}^{OPT}(X)\right)$,
   $\overline{\sum_{i=1}^{m} R_i}^{OPT}(\sim X) = \sim\left(\underline{\sum_{i=1}^{m} R_i}^{OPT}(X)\right)$.

Proposition 1.5 shows the basic properties of the optimistic multigranulation rough set. For example, 1 says that the optimistic multigranulation lower approximation is included into the target concept, and the optimistic multigranulation upper approximation includes the target concept; 2 shows the normality and conormality of the optimistic multigranulation rough set; 3 says that the optimistic multigranulation lower approximation is the union of all the single equivalence relation based lower approximations, while the optimistic multigranulation upper approximation is the intersection of all the single equivalence relation based upper approximations; 4 shows the idempotency of the optimistic multigranulation rough set; 5 shows the complement of the optimistic multigranulation rough set.

Proposition 1.6. [49] *Let* $K = (U, \mathbf{R})$ *be a knowledge base,* $\{R_1, R_2, \cdots, R_m\} \subseteq \mathbf{R}$*, then* $\forall X_1, X_2, \cdots, X_n \subseteq U$*, then we have following properties about the optimistic multigranulation lower and upper approximations:*

1. $$\underline{\sum_{i=1}^{m} R_i}^{OPT}\left(\bigcap_{j=1}^{n} X_j\right) = \bigcup_{i=1}^{m}\left(\bigcap_{j=1}^{n} \underline{R_i}(X_j)\right),$$
$$\overline{\sum_{i=1}^{m} R_i}^{OPT}\left(\bigcup_{j=1}^{n} X_j\right) = \bigcap_{i=1}^{m}\left(\bigcup_{j=1}^{n} \overline{R_i}(X_j)\right);$$
2. $$\underline{\sum_{i=1}^{m} R_i}^{OPT}\left(\bigcap_{j=1}^{n} X_j\right) \subseteq \bigcap_{j=1}^{n}\left(\underline{\sum_{i=1}^{m} R_i}^{OPT}(X_j)\right),$$
$$\overline{\sum_{i=1}^{m} R_i}^{OPT}\left(\bigcup_{j=1}^{n} X_j\right) \supseteq \bigcup_{j=1}^{n}\left(\overline{\sum_{i=1}^{m} R_i}^{OPT}(X_j)\right);$$
3. $$\underline{\sum_{i=1}^{m} R_i}^{OPT}\left(\bigcup_{j=1}^{n} X_j\right) \supseteq \bigcup_{j=1}^{n}\left(\underline{\sum_{i=1}^{m} R_i}^{OPT}(X_j)\right),$$
$$\overline{\sum_{i=1}^{m} R_i}^{OPT}\left(\bigcap_{j=1}^{n} X_j\right) \subseteq \bigcap_{j=1}^{n}\left(\overline{\sum_{i=1}^{m} R_i}^{OPT}(X_j)\right).$$

Proposition 1.6 shows the relationships between the approximations of a single set and that of multi-sets through using the optimistic multigranulation rough set approach.

Theorem 1.9. [49] *Let* $K = (U, \mathbf{R})$ *be a knowledge base, if* $\{R_1, R_2, \cdots, R_m\} = \mathbf{R}$*, then* $\forall X \subseteq U$*, we have*

$$\underline{\sum_{i=1}^{m} R_i}^{OPT}(X) \subseteq \underline{\mathbf{R}}(X) \tag{1.32}$$

$$\overline{\sum_{i=1}^{m} R_i}^{OPT}(X) \supseteq \overline{\mathbf{R}}(X) \tag{1.33}$$

Theorem 1.9 tells us that by comparing with Pawlak's rough set, optimistic multigranulation rough set can induce smaller lower approximation and greater upper approximation.

Definition 1.15. [49] Let $K = (U, \mathbf{R})$ be a knowledge base, $\{R_1, R_2, \cdots, R_m\} \subseteq \mathbf{R}$, $\forall X \subseteq U$, the accuracy of optimistic multigranulation approximation is denoted by

$\alpha_{R_1,R_2,\cdots,R_m}(X)$ where

$$\alpha^{OPT}_{R_1,R_2,\cdots,R_m}(X) = \frac{\left|\underline{\sum_{i=1}^{m} R_i}^{OPT}(X)\right|}{\left|\overline{\sum_{i=1}^{m} R_i}^{OPT}(X)\right|} \tag{1.34}$$

Theorem 1.10. [49] *Let $K=(U,\mathbf{R})$ be a knowledge base, $\forall X\subseteq U$, if $\{R_1,R_2,\cdots,R_n\}\subseteq \{R_1,R_2,\cdots,R_m\}\subseteq \mathbf{R}$, then*

$$\alpha^{OPT}_{R_1,R_2,\cdots,R_m}(X) \geqslant \alpha^{OPT}_{R_1,R_2,\cdots,R_n}(X)$$

1.3.2 Pessimistic Multigranulation Rough Set

Definition 1.16. [48] Let $K = (U,\mathbf{R})$ be a knowledge base, $\{R_1,R_2,\cdots,R_m\}\subseteq \mathbf{R}$ is a set of the equivalence relations, $\forall X\subseteq U$, the pessimistic multigranulation lower approximation and upper approximation of X are denoted by $\underline{\sum_{i=1}^{m} R_i}^{PES}(X)$ and $\overline{\sum_{i=1}^{m} R_i}^{PES}(X)$, respectively, where

$$\underline{\sum_{i=1}^{m} R_i}^{PES}(X)=\{x\in U : [x]_{R_1}\subseteq X\wedge [x]_{R_2}\subseteq X\wedge\cdots\wedge [x]_{R_m}\subseteq X\} \tag{1.35}$$

$$\overline{\sum_{i=1}^{m} R_i}^{PES}(X)=\sim\left(\underline{\sum_{i=1}^{m} R_i}^{PES}(\sim X)\right) \tag{1.36}$$

The pair $\left[\underline{\sum_{i=1}^{m} R_i}^{PES}(X), \overline{\sum_{i=1}^{m} R_i}^{PES}(X)\right]$ is referred to as a pessimistic multigranulation rough set of X in terms of the set of the equivalence relations $\{R_1,R_2,\cdots,R_m\}$. By comparing with Definition 1.14, i.e. optimistic multigranulation rough set, the lower approximation of pessimistic multigranulation rough set needs the inclusion between equivalence class and target on each equivalence relation. Both optimistic and pessimistic multigranulation upper approximations are defined on the complements of the optimistic and pessimistic multigranulation lower approximations, respectively.

Theorem 1.11. *Let $K = (U,\mathbf{R})$ be a knowledge base, $\{R_1,R_2,\cdots,R_m\}\subseteq \mathbf{R}$, $\forall X\subseteq U$, we have*

$$\overline{\sum_{i=1}^{m} R_i}^{PES}(X) = \{x\in U : [x]_{R_1}\cap X\neq\varnothing\vee [x]_{R_2}\cap X\neq\varnothing\vee\cdots\vee [x]_{R_m}\cap X\neq\varnothing\} \tag{1.37}$$

Proof. By Definition 1.16, we have $\overline{\sum_{i=1}^{m} R_i}^{PES}(X) = \sim\left(\underline{\sum_{i=1}^{m} R_i}^{PES}(\sim X)\right)$, thus

$$\begin{aligned}\overline{\sum_{i=1}^{m} R_i}^{PES}(X) &= \sim \{x \in U : [x]_{R_1} \subseteq \sim X \wedge [x]_{R_2} \subseteq \sim X \wedge \cdots \wedge [x]_{R_m} \subseteq \sim X\}\\ &= \sim \{x \in U : [x]_{R_1} \cap X = \varnothing \wedge [x]_{R_2} \cap X = \varnothing \wedge \cdots \wedge [x]_{R_m} \cap X = \varnothing\}\\ &= \{x \in U : [x]_{R_1} \cap X \neq \varnothing \vee [x]_{R_2} \cap X \neq \varnothing \vee \cdots \vee [x]_{R_m} \cap X \neq \varnothing\}. \quad \square\end{aligned}$$

Theorem 1.12. [48] *Let $K = (U, \mathbf{R})$ be a knowledge base, $\{R_1, R_2, \cdots, R_m\} \subseteq \mathbf{R}$, suppose that $U/IND(R_1) \preceq U/IND(R_2) \preceq \cdots \preceq U/IND(R_m)$, then we have*

$$\underline{\sum_{i=1}^{m} R_i}^{PES}(X) = \underline{R_m}(X) \tag{1.38}$$

$$\overline{\sum_{i=1}^{m} R_i}^{PES}(X) = \overline{R_m}(X) \tag{1.39}$$

The above theorem tells us that if there is a partial relation among the partitions which are determined by the set of equivalence relations, then the pessimistic multigranulation rough set is equivalent to the rough set in terms of the *coarsest* partition.

Proposition 1.7. [48] *Let $K = (U, \mathbf{R})$ be a knowledge base, $\{R_1, R_2, \cdots, R_m\} \subseteq \mathbf{R}$, $\forall X \subseteq U$, then we have following properties about the pessimistic multigranulation lower and upper approximations:*

1. $\underline{\sum_{i=1}^{m} R_i}^{PES}(X) \subseteq X \subseteq \overline{\sum_{i=1}^{m} R_i}^{PES}(X)$;
2. $\underline{\sum_{i=1}^{m} R_i}^{PES}(\varnothing) = \overline{\sum_{i=1}^{m} R_i}^{PES}(\varnothing) = \varnothing$,
   $\underline{\sum_{i=1}^{m} R_i}^{PES}(U) = \overline{\sum_{i=1}^{m} R_i}^{PES}(U) = U$;
3. $\underline{\sum_{i=1}^{m} R_i}^{PES}(X) = \bigcap_{i=1}^{m} \underline{R_i}(X)$,
   $\overline{\sum_{i=1}^{m} R_i}^{PES}(X) = \bigcup_{i=1}^{m} \overline{R_i}(X)$;
4. $\underline{\sum_{i=1}^{m} R_i}^{PES}\left(\underline{\sum_{i=1}^{m} R_i}^{PES}(X)\right) = \underline{\sum_{i=1}^{m} R_i}^{PES}(X)$,

$$\overline{\sum_{i=1}^{m} R_i}^{PES}\left(\overline{\sum_{i=1}^{m} R_i}^{PES}(X)\right) = \overline{\sum_{i=1}^{m} R_i}^{PES}(X);$$

5. $$\underline{\sum_{i=1}^{m} R_i}^{PES}(\sim X) = \sim\left(\overline{\sum_{i=1}^{m} R_i}^{PES}(X)\right),$$
$$\overline{\sum_{i=1}^{m} R_i}^{PES}(\sim X) = \sim\left(\underline{\sum_{i=1}^{m} R_i}^{PES}(X)\right).$$

Proposition 1.7 shows the basic properties of the pessimistic multigranulation rough set. The semantic explanation of Proposition 1.7 is similar to that of Proposition 1.5.

Proposition 1.8. [48] *Let* $K = (U, \mathbf{R})$ *be the knowledge base,* $\{R_1, R_2, \cdots, R_m\} \subseteq \mathbf{R}$, *then* $\forall X_1, X_2, \cdots, X_n \subseteq U$, *then we have following properties about the pessimistic multigranulation lower and upper approximations*:

1. $$\underline{\sum_{i=1}^{m} R_i}^{PES}\left(\bigcap_{j=1}^{n} X_j\right) = \bigcap_{i=1}^{m}\left(\bigcap_{j=1}^{n} \underline{R_i}(X_j)\right),$$
$$\overline{\sum_{i=1}^{m} R_i}^{PES}\left(\bigcup_{j=1}^{n} X_j\right) = \bigcup_{i=1}^{m}\left(\bigcup_{j=1}^{n} \overline{R_i}(X_j)\right);$$

2. $$\underline{\sum_{i=1}^{m} R_i}^{PES}\left(\bigcap_{j=1}^{n} X_j\right) = \bigcap_{j=1}^{n}\left(\underline{\sum_{i=1}^{m} R_i}^{PES}(X_j)\right),$$
$$\overline{\sum_{i=1}^{m} R_i}^{PES}\left(\bigcup_{j=1}^{n} X_j\right) = \bigcup_{j=1}^{n}\left(\overline{\sum_{i=1}^{m} R_i}^{PES}(X_j)\right);$$

3. $$\underline{\sum_{i=1}^{m} R_i}^{PES}\left(\bigcup_{j=1}^{n} X_j\right) \supseteq \bigcup_{j=1}^{n}\left(\underline{\sum_{i=1}^{m} R_i}^{PES}(X_j)\right),$$
$$\overline{\sum_{i=1}^{m} R_i}^{PES}\left(\bigcap_{j=1}^{n} X_j\right) \subseteq \bigcap_{j=1}^{n}\left(\overline{\sum_{i=1}^{m} R_i}^{PES}(X_j)\right).$$

Proposition 1.8 shows the relationship between the approximations of a single set and that of multi-sets through using the pessimistic multigranulation rough set approach.

Theorem 1.13. [48] *Let $K = (U, \mathbf{R})$ be a knowledge base, if $\{R_1, R_2, \cdots, R_m\} = \mathbf{R}$, then $\forall X \subseteq U$, we have*

$$\underline{\sum_{i=1}^{m} R_i}^{PES}(X) \subseteq \underline{\mathbf{R}}(X) \tag{1.40}$$

$$\overline{\sum_{i=1}^{m} R_i}^{PES}(X) \supseteq \overline{\mathbf{R}}(X) \tag{1.41}$$

Theorem 1.13 tells us that by comparing with Pawlak's rough set, pessimistic multigranulation rough set can also induce smaller lower approximation and greater upper approximation.

Theorem 1.14. [48] *Let $K = (U, \mathbf{R})$ be the knowledge base, if $\{R_1, R_2, \cdots, R_m\} = \mathbf{R}$, then $\forall X \subseteq U$, we have*

$$\underline{\sum_{i=1}^{m} R_i}^{PES}(X) \subseteq \underline{\sum_{i=1}^{m} R_i}^{OPT}(X) \tag{1.42}$$

$$\overline{\sum_{i=1}^{m} R_i}^{OPT}(X) \supseteq \overline{\sum_{i=1}^{m} R_i}^{PES}(X) \tag{1.43}$$

Theorem 1.14 shows the relationships between optimistic and pessimistic rough approximations.

Definition 1.17. [48] Let $K = (U, \mathbf{R})$ be a knowledge base, $\{R_1, R_2, \cdots, R_m\} \subseteq \mathbf{R}$, $\forall X \subseteq U$, the accuracy of pessimistic multigranulation approximation is denoted by $\alpha^{PES}_{R_1, R_2, \cdots, R_m}(X)$ where

$$\alpha^{PES}_{R_1, R_2, \cdots, R_m}(X) = \frac{\left|\underline{\sum_{i=1}^{m} R_i}^{PES}(X)\right|}{\left|\overline{\sum_{i=1}^{m} R_i}^{PES}(X)\right|} \tag{1.44}$$

Theorem 1.15. [48] *Let $K = (U, \mathbf{R})$ be a knowledge base, $\forall X \subseteq U$, if $\{R_1, R_2, \cdots, R_n\} \subseteq \{R_1, R_2, \cdots, R_m\} \subseteq \mathbf{R}$, then*

$$\alpha^{PES}_{R_1, R_2, \cdots, R_m}(X) \leqslant \alpha^{PES}_{R_1, R_2, \cdots, R_n}(X)$$

1.3.3 Multigranulation Rough Memberships

In Pawlak's rough set model, there is a direct relationship between rough approximation and the membership such that

$$\mu^P_X(x) = 1 \Leftrightarrow x \in \underline{P}(X) \tag{1.45}$$

$$0 < \mu^P_X(x) \leqslant 1 \Leftrightarrow x \in \overline{P}(X) \tag{1.46}$$

It should be noticed that since more than one equivalence relations are used in multigranulation rough set approach, the re-definition of the rough membership has become a necessity.

Definition 1.18. Let $K = (U, \mathbf{R})$ be a knowledge base, $\{R_1, R_2, \cdots, R_m\} \subseteq \mathbf{R}$, $\forall X \subseteq U$, the maximal and minimal rough memberships of x in X are denoted by $\mu_{X_{MAX}}^{\sum_{i=1}^m R_i}(x)$ and $\mu_{X_{MIN}}^{\sum_{i=1}^m R_i}(x)$, respectively, where

$$\mu_{X_{MAX}}^{\sum_{i=1}^m R_i}(x) = \vee_{i=1}^m \mu_X^{R_i}(x) \tag{1.47}$$

$$\mu_{X_{MIN}}^{\sum_{i=1}^m R_i}(x) = \wedge_{i=1}^m \mu_X^{R_i}(x) \tag{1.48}$$

Theorem 1.16. *Let $K = (U, \boldsymbol{R})$ be a knowledge base, $\{R_1, R_2, \cdots, R_m\} \subseteq \boldsymbol{R}$, $\forall X \subseteq U$, we have*

1. $\mu_{X_{MAX}}^{\sum_{i=1}^m R_i}(x) = 1 \Leftrightarrow x \in \underline{\sum_{i=1}^m R_i}^{OPT}(X)$;
2. $0 < \mu_{X_{MIN}}^{\sum_{i=1}^m R_i}(x) \leqslant 1 \Leftrightarrow x \in \overline{\sum_{i=1}^m R_i}^{OPT}(X)$;
3. $\mu_{X_{MIN}}^{\sum_{i=1}^m R_i}(x) = 1 \Leftrightarrow x \in \underline{\sum_{i=1}^m R_i}^{PES}(X)$;
4. $0 < \mu_{X_{MAX}}^{\sum_{i=1}^m R_i}(x) \leqslant 1 \Leftrightarrow x \in \overline{\sum_{i=1}^m R_i}^{PES}(X)$.

Proof. We only prove 1, others can be proved analogously.

$$\begin{aligned}\mu_{X_{MAX}}^{\sum_{i=1}^m R_i}(x) = 1 &\Leftrightarrow \vee_{i=1}^m \mu_X^{R_i}(x) = 1 \\ &\Leftrightarrow \exists i \in \{1, 2, \cdots, m\} \text{ s.t. } [x]_{R_i} \subseteq X \\ &\Leftrightarrow x \in \underline{\sum_{i=1}^m R_i}^{OPT}(X)\end{aligned}$$ □

1.4 Hierarchical Structures on Multigranulation Spaces

Hierarchical structure plays a fundamental role in the development of the granular computing. Generally speaking, we can use a binary relation to represent the hierarchical structure on granulation spaces. However, it should be noticed that most of the proposed binary relations can only be used to judge whether a single-granulation space is finer or coarser than another single-granulation space, i.e. these binary relations are used to compare different single-granulation spaces. For example, if $U/IND(P_1) \preceq U/IND(P_2)$, then the partition based granulation space $U/IND(P_1)$ is finer than the partition based granulation space $U/IND(P_2)$.

In the above section, we have mentioned the rough sets in multigranulation space and then, it is an interesting issue to discuss the hierarchical structures on the multigranulation spaces.

Given a knowledge base in which $R \in \mathbf{R}$, since R is an equivalence relation, then we can induce a partition based granulation space such that U/R. Each element in such granulation space is an equivalence class. Moreover, suppose that $\mathscr{R} \subseteq \mathbf{R}$, i.e. $\mathscr{R}$ is a family of the equivalence relations. Since each equivalence relation in $\mathscr{R}$ can induce a granulation space, then the integration of all these granulation spaces forms a multigranulation space. Formally, a partitions based multigranulation space is denoted by $MS_{\mathscr{R}}$ such that

$$MS_{\mathscr{R}} = \{U/R : R \in \mathscr{R}\} \tag{1.49}$$

In Eq. (1.49), the multigranulation space is a family of the partitions on the universe of discourse. In the following, we will propose three different hierarchical structures to investigate the finer or coarser relationships between different partitions based multigranulation spaces.

1.4.1 Definitions of Three Hierarchical Structures

Definition 1.19. (The first hierarchical structure) Let $K = (U, \mathbf{R})$ be a knowledge base in which $\mathscr{R}_1, \mathscr{R}_2 \subseteq \mathbf{R}$, if $\forall [x]_{R_{2j}} (R_{2j} \in \mathscr{R}_2)$, there must be $[x]_{R_{1k}} (R_{1k} \in \mathscr{R}_1)$ such that $[x]_{R_{1k}} \subseteq [x]_{R_{2j}}$, then we say that the partitions based multigranulation space $MS_{\mathscr{R}_1}$ is finer than the partitions based multigranulation space $MS_{\mathscr{R}_2}$ or $MS_{\mathscr{R}_2}$ is coarser than $MS_{\mathscr{R}_1}$, it is denoted by $MS_{\mathscr{R}_1} \preceq_1 MS_{\mathscr{R}_2}$ or $MS_{\mathscr{R}_2} \succeq_1 MS_{\mathscr{R}_1}$; if $MS_{\mathscr{R}_1} \preceq_1 MS_{\mathscr{R}_2}$ and $MS_{\mathscr{R}_1} \neq MS_{\mathscr{R}_2}$, then we say that the partitions based multigranulation space $MS_{\mathscr{R}_1}$ is strictly finer than the partitions based multigranulation space $MS_{\mathscr{R}_2}$, it is denoted by $MS_{\mathscr{R}_1} \prec_1 MS_{\mathscr{R}_2}$.

In the first hierarchical structure, if for each equivalence class ($[x]_{R_{2j}}$) in the multigranulation space $MS_{\mathscr{R}_2}$, there must be an equivalence class ($[x]_{R_{1k}}$) in the multigranulation space $MS_{\mathscr{R}_1}$ such that $[x]_{R_{1k}} \subseteq [x]_{R_{2j}}$, then the multigranulation space $MS_{\mathscr{R}_1}$ is considered as to be finer than the multigranulation space $MS_{\mathscr{R}_2}$.

Definition 1.20. (The second hierarchical structure) Let $K = (U, \mathbf{R})$ be a knowledge base in which $\mathscr{R}_1, \mathscr{R}_2 \subseteq \mathbf{R}$, if $\forall [x]_{R_{1k}} (R_{1k} \in \mathscr{R}_1)$, there must be $[x]_{R_{2j}} (R_{2j} \in \mathscr{R}_2)$ such that $[x]_{R_{1k}} \subseteq [x]_{R_{2j}}$, then we say that the partitions based multigranulation space $MS_{\mathscr{R}_1}$ is finer than the partitions based multigranulation space $MS_{\mathscr{R}_2}$ or $MS_{\mathscr{R}_2}$ is coarser than $MS_{\mathscr{R}_1}$, it is denoted by $MS_{\mathscr{R}_1} \preceq_2 MS_{\mathscr{R}_2}$ or $MS_{\mathscr{R}_2} \succeq_2 MS_{\mathscr{R}_1}$; if $MS_{\mathscr{R}_1} \preceq_2 MS_{\mathscr{R}_2}$ and $MS_{\mathscr{R}_1} \neq MS_{\mathscr{R}_2}$, then we say that the partitions based multigranulation space $MS_{\mathscr{R}_1}$ is strictly finer than the partitions based multigranulation space $MS_{\mathscr{R}_2}$, it is denoted by $MS_{\mathscr{R}_1} \prec_2 MS_{\mathscr{R}_2}$.

In the second hierarchical structure, if for each equivalence class ($[x]_{R_{1k}}$) in the multigranulation space $MS_{\mathscr{R}_1}$, there must be an equivalence class ($[x]_{R_{2j}}$) in the multigranulation space $MS_{\mathscr{R}_2}$ such that $[x]_{R_{1k}} \subseteq [x]_{R_{2j}}$, then the multigranulation space $MS_{\mathscr{R}_1}$ is considered as to be finer than the multigranulation space $MS_{\mathscr{R}_2}$.

Definition 1.19 and Definition 1.20 are two different hierarchical structures, which are proposed on the partitions based multigranulation spaces. The following example will show that there is not a necessary causality between these two hierarchical structures.

Example 1.3. Suppose that $U = \{x_1, x_2, x_3, x_4, x_5, x_6, x_7, x_8\}$ is the universe, three families of the equivalence relations are given by $\mathscr{R}_1 = \{R_{11}, R_{12}\}$, $\mathscr{R}_2 = \{R_{21}, R_{22}\}$ and $\mathscr{R}_3 = \{R_{31}, R_{32}\}$ such that

$$U/R_{11} = \{\{x_1, x_2\}, \{x_3, x_4\}, \{x_5, x_6\}, \{x_7, x_8\}\}$$
$$U/R_{12} = \{\{x_1, x_4, x_5\}, \{x_2, x_3\}, \{x_6, x_7, x_8\}\}$$
$$U/R_{21} = \{\{x_1, x_2, x_3, x_4, x_5, x_6\}, \{x_7, x_8\}\}$$
$$U/R_{22} = \{\{x_1, x_2, x_3, x_4\}, \{x_5, x_6\}, \{x_7, x_8\}\}$$
$$U/R_{31} = \{\{x_1, x_2, x_3, x_4, x_5, x_6, x_7, x_8\}\}$$
$$U/R_{32} = \{\{x_1, x_2, x_3, x_7, x_8\}, \{x_4, x_6\}, \{x_5\}\}$$

By Definition 1.19, we can see that $MS_{\mathscr{R}_1} \preceq_1 MS_{\mathscr{R}_2}$ holds obviously. However, $MS_{\mathscr{R}_1} \preceq_2 MS_{\mathscr{R}_2}$ does not hold since $[x_6]_{R_{12}} \nsubseteq [x_6]_{R_{21}}$ and $[x_6]_{R_{12}} \nsubseteq [x_6]_{R_{22}}$.

By Definition 1.20, we can see that $MS_{\mathscr{R}_1} \preceq_2 MS_{\mathscr{R}_3}$ holds obviously. However, $MS_{\mathscr{R}_1} \preceq_1 MS_{\mathscr{R}_3}$ does not hold since $[x_4]_{R_{32}} \nsubseteq [x_4]_{R_{11}}$ and $[x_4]_{R_{32}} \nsubseteq [x_4]_{R_{12}}$.

Remark 1.3. Example 1.3 tells us that there is not a necessary causality from the fist hierarchical structure to the second hierarchical structure; conversely, there is also not a necessary causality from the second hierarchical structure to the first hierarchical structure.

Theorem 1.17. *Let $K = (U, \mathbf{R})$ be a knowledge base, $\preceq_1$ is reflexive and transitive.*

Proof.

1. Suppose that $\mathscr{R} \subseteq \mathbf{R}$, then $\forall R \in \mathscr{R}$, there must be $R' \in \mathscr{R}$ such that $R = R'$. Thus, $\forall [x]_R (R \in \mathscr{R})$, there must be $[x]_{R'} (R' \in \mathscr{R})$ such that $[x]_{R'} = [x]_R$, i.e. $[x]_{R'} \subseteq [x]_R$, by Definition 1.19, $MS_{\mathscr{R}} \preceq_1 MS_{\mathscr{R}}$ holds, from which we can conclude that $\preceq_1$ is reflexive.
2. Suppose that $\mathscr{R}_1, \mathscr{R}_2, \mathscr{R}_3 \subseteq \mathbf{R}$, $MS_{\mathscr{R}_1} \preceq_1 MS_{\mathscr{R}_2}$ and $MS_{\mathscr{R}_2} \preceq_1 MS_{\mathscr{R}_3}$. By condition $MS_{\mathscr{R}_1} \preceq_1 MS_{\mathscr{R}_2}$ we know that $\forall [x]_{R_{2j}} (R_{2j} \in \mathscr{R}_2)$, there must be $[x]_{R_{1k}} (R_{1k} \in \mathscr{R}_1)$ such that $[x]_{R_{1k}} \subseteq [x]_{R_{2j}}$; by $MS_{\mathscr{R}_2} \preceq_1 MS_{\mathscr{R}_3}$ we know that $\forall [x]_{R_{3l}} (R_{3l} \in \mathscr{R}_3)$, there must be $[x]_{R_{2j}} (R_{2j} \in \mathscr{R}_2)$ such that $[x]_{R_{2j}} \subseteq [x]_{R_{3l}}$. Therefore, we can conclude that $\forall [x]_{R_{3l}} (R_{3l} \in \mathscr{R}_3)$, there must be $[x]_{R_{1k}} (R_{1k} \in \mathscr{R}_1)$ such that $[x]_{R_{1k}} \subseteq [x]_{R_{3l}}$, i.e. $MS_{\mathscr{R}_1} \preceq_1 MS_{\mathscr{R}_3}$, it follows that $\preceq_1$ is transitive. □

Theorem 1.18. *Let $K = (U, \mathbf{R})$ be a knowledge base, $\preceq_2$ is reflexive and transitive.*

Proof. The proof of Theorem 1.18 is similar to the proof of Theorem 1.17. □

The following Example 1.4 will show that $\preceq_1$ and $\preceq_2$ are not necessarily antisymmetric.

Example 1.4. Suppose that $U = \{x_1, x_2, x_3, x_4, x_5, x_6, x_7, x_8\}$ is the universe, three families of the equivalence relations are given by $\mathscr{R}_1 = \{R_{11}, R_{12}\}$, $\mathscr{R}_2 = \{R_{21}, R_{22}\}$ and $\mathscr{R}_3 = \{R_{31}, R_{32}\}$ such that

$$U/R_{11} = \{\{x_1, x_2, x_3\}, \{x_4, x_5, x_6\}, \{x_7, x_8\}\}$$
$$U/R_{12} = \{\{x_1\}, \{x_2, x_3\}, \{x_4, x_5\}, \{x_6\}, \{x_7, x_8\}\}$$
$$U/R_{21} = \{\{x_1, x_2, x_3\}, \{x_4, x_5\}, \{x_6\}, \{x_7, x_8\}\}$$
$$U/R_{22} = \{\{x_1\}, \{x_2, x_3\}, \{x_4, x_5\}, \{x_6\}, \{x_7, x_8\}\}$$
$$U/R_{31} = \{\{x_1\}, \{x_2\}, \{x_3\}, \{x_4, x_5\}, \{x_6\}, \{x_7\}, \{x_8\}\}$$
$$U/R_{32} = \{\{x_1, x_2, x_3\}, \{x_4, x_5, x_6\}, \{x_7, x_8\}\}$$

By Definition 1.19 and Definition 1.20, we can see that $MS_{\mathscr{R}_1} \preceq_1 MS_{\mathscr{R}_2}$ and $MS_{\mathscr{R}_2} \preceq_1 MS_{\mathscr{R}_1}$. However, $MS_{\mathscr{R}_1} \neq MS_{\mathscr{R}_2}$, from which we can see that $\preceq_1$ is a binary relation without the condition of the antisymmetric.

By Definition 1.19 and Definition 1.20, we can see that $MS_{\mathscr{R}_1} \preceq_2 MS_{\mathscr{R}_3}$ and $MS_{\mathscr{R}_3} \preceq_2 MS_{\mathscr{R}_1}$. However, $MS_{\mathscr{R}_1} \neq MS_{\mathscr{R}_3}$, from which we can see that $\preceq_2$ is also a binary relation without the condition of the antisymmetric.

By the investigations of the above two different hierarchical structures, the third hierarchical structure on partitions based multigranulation spaces will be proposed as Definition 1.21 shows.

Definition 1.21. (The third hierarchical structure) Let $K = (U, \mathbf{R})$ be a knowledge base in which $\mathscr{R}_1, \mathscr{R}_2 \subseteq \mathbf{R}$, if $\forall [x]_{R_{1k}} (R_{1k} \in \mathscr{R}_1)$ and $\forall [x]_{R_{2j}} (R_{2j} \in \mathscr{R}_2)$, we have $[x]_{R_{1k}} \subseteq [x]_{R_{2j}}$, then we say that the multigranulation space $MS_{\mathscr{R}_1}$ is finer than the multigranulation space $MS_{\mathscr{R}_2}$ or $MS_{\mathscr{R}_2}$ is coarser than $MS_{\mathscr{R}_1}$, it is denoted by $MS_{\mathscr{R}_1} \preceq_3 MS_{\mathscr{R}_2}$ or $MS_{\mathscr{R}_2} \succeq_3 MS_{\mathscr{R}_1}$; if $MS_{\mathscr{R}_1} \preceq_3 MS_{\mathscr{R}_2}$ and $MS_{\mathscr{R}_1} \neq MS_{\mathscr{R}_2}$, then we say that $MS_{\mathscr{R}_1}$ is strictly finer than $MS_{\mathscr{R}_2}$, it is denoted by $MS_{\mathscr{R}_1} \prec_3 MS_{\mathscr{R}_2}$.

In the third hierarchical structure, if for each equivalence class ($[x]_{R_{1k}}$) in the multigranulation space $MS_{\mathscr{R}_1}$ and for each equivalence class ($[x]_{R_{2j}}$) in the multigranulation space $MS_{\mathscr{R}_2}$, $[x]_{R_{1k}} \subseteq [x]_{R_{2j}}$ holds, then the multigranulation space $MS_{\mathscr{R}_1}$ is considered as to be finer than the multigranulation space $MS_{\mathscr{R}_2}$.

Theorem 1.19. *Let $K = (U, \mathbf{R})$ be a knowledge base, $\preceq_3$ is antisymmetric and transitive.*

Proof.

1. Suppose that $\mathscr{R}_1, \mathscr{R}_2 \subseteq \mathbf{R}$ and $MS_{\mathscr{R}_1} \neq MS_{\mathscr{R}_2}$, then there must be $[x]_{R_{1k}}$ ($R_{1k} \in \mathscr{R}_1$) and $[x]_{R_{2j}} (R_{2j} \in \mathscr{R}_2)$ such that $[x]_{R_{1k}} \neq [x]_{R_{2j}}$, it follows that $[x]_{R_{1k}} \nsubseteq [x]_{R_{2j}}$ or $[x]_{R_{1k}} \nsupseteq [x]_{R_{2j}}$. In other words, $\exists [x]_{R_{1k}} (R_{1k} \in \mathscr{R}_1)$ and $\exists [x]_{R_{2j}} (R_{2j} \in \mathscr{R}_2)$ such that $[x]_{R_{1k}} \nsubseteq [x]_{R_{2j}}$ or $[x]_{R_{1k}} \nsupseteq [x]_{R_{2j}}$, from which we can conclude that $MS_{\mathscr{R}_1} \npreceq_3 MS_{\mathscr{R}_2}$ or $MS_{\mathscr{R}_2} \npreceq_3 MS_{\mathscr{R}_1}$ and then $\preceq_3$ is antisymmetric.
2. Suppose that $\mathscr{R}_1, \mathscr{R}_2, \mathscr{R}_3 \subseteq \mathbf{R}$, $MS_{\mathscr{R}_1} \preceq_3 MS_{\mathscr{R}_2}$ and $MS_{\mathscr{R}_2} \preceq_3 MS_{\mathscr{R}_3}$. By condition $MS_{\mathscr{R}_1} \preceq_3 MS_{\mathscr{R}_2}$ we know that $\forall [x]_{R_{1k}} (R_{1k} \in \mathscr{R}_1)$ and $\forall [x]_{R_{2j}} (R_{2j} \in \mathscr{R}_2)$,

$[x]_{R_{1k}} \subseteq [x]_{R_{2j}}$ holds; by $MS_{\mathscr{R}_2} \preceq_3 MS_{\mathscr{R}_3}$ we know that $\forall [x]_{R_{2j}}(R_{2j} \in \mathscr{R}_2)$ and $\forall [x]_{R_{3l}}(R_{3l} \in \mathscr{R}_3)$, $[x]_{R_{2j}} \subseteq [x]_{R_{3l}}$ holds. Therefore, we can conclude that $\forall [x]_{R_{1k}}(R_{1k} \in \mathscr{R}_1)$ and $\forall [x]_{R_{3l}}(R_{3l} \in \mathscr{R}_3)$, $[x]_{R_{1k}} \subseteq [x]_{R_{3l}}$ holds, i.e. $MS_{\mathscr{R}_1} \preceq_3 MS_{\mathscr{R}_3}$, $\preceq_3$ is transitive. □

Theorem 1.20. *Let $K = (U, \mathbf{R})$ be a knowledge base in which $\mathscr{R}_1, \mathscr{R}_2 \subseteq \mathbf{R}$, if $MS_{\mathscr{R}_1} \preceq_3 MS_{\mathscr{R}_2}$, then $MS_{\mathscr{R}_1} \preceq_1 MS_{\mathscr{R}_2}$ and $MS_{\mathscr{R}_1} \preceq_2 MS_{\mathscr{R}_2}$.*

Proof. By the above definitions of three hierarchical structures, it is a trivial to prove this theorem. □

The following example will show that the inverse of Theorem 1.20 does not holds.

Example 1.5. Suppose that $U = \{x_1, x_2, x_3, x_4, x_5, x_6, x_7, x_8\}$ is the universe, two families of the equivalence relations are given by $\mathscr{R}_1 = \{R_{11}, R_{12}\}$ and $\mathscr{R}_2 = \{R_{21}, R_{22}\}$ such that

$$U/R_{11} = \{\{x_1, x_2, x_3\}, \{x_4, x_5, x_6\}, \{x_7, x_8\}\}$$
$$U/R_{12} = \{\{x_1\}, \{x_2, x_3\}, \{x_4, x_5\}, \{x_6\}, \{x_7, x_8\}\}$$
$$U/R_{21} = \{\{x_1, x_2, x_3, x_4, x_5, x_6, x_7, x_8\}\}$$
$$U/R_{22} = \{\{x_1\}, \{x_2, x_3\}, \{x_4, x_5, x_6\}, \{x_7, x_8\}\}$$

By Definition 1.19 and Definition 1.20, $MS_{\mathscr{R}_1} \preceq_1 MS_{\mathscr{R}_2}$ and $MS_{\mathscr{R}_1} \preceq_2 MS_{\mathscr{R}_2}$ hold obviously. However, $MS_{\mathscr{R}_1} \preceq_3 MS_{\mathscr{R}_2}$ does not hold since $[x_1]_{R_{11}} \nsubseteq [x_1]_{R_{22}}$.

Remark 1.4. Example 1.5 tells us that there is not a necessary causality from the integration of the fist and second hierarchical structures to the third hierarchical structure.

1.4.2 Relationships Between Hierarchical Structures and Multigranulation Rough Sets

In Section 1.3, the optimistic and pessimistic multigranulation rough sets have been briefly introduced. In Section 1.4.1, three different hierarchical structures have been proposed. In the following, we will investigate the relationships between these hierarchical structures and the varieties of the multigranulation rough sets.

Given two different multigranulation spaces such that $MS_{\mathscr{R}_1} = \{R_1, R_2, \cdots, R_m\}$ and $MS_{\mathscr{R}_2} = \{R_1, R_2, \cdots, R_{m'}\}$, the optimistic multigranulation rough sets, which are constructed on such two multigranulation spaces are denoted by $\left[\underline{\sum_{i=1}^{m} R_i}^{OPT}(X), \overline{\sum_{i=1}^{m} R_i}^{OPT}(X)\right]$ and $\left[\underline{\sum_{i=1}^{m'} R_i}^{OPT}(X), \overline{\sum_{i=1}^{m'} R_i}^{OPT}(X)\right]$, respectively, the pessimistic multigranulation rough sets, which are constructed on such two multigranulation spaces are denoted by $\left[\underline{\sum_{i=1}^{m} R_i}^{PES}(X), \overline{\sum_{i=1}^{m} R_i}^{PES}(X)\right]$ and $\left[\underline{\sum_{i=1}^{m'} R_i}^{PES}(X), \overline{\sum_{i=1}^{m'} R_i}^{PES}(X)\right]$, respectively.

Theorem 1.21. *Let $K = (U, \mathbf{R})$ be a knowledge base in which $\mathscr{R}_1, \mathscr{R}_2 \subseteq \mathbf{R}$, $MS_{\mathscr{R}_1} = \{U/R_1, U/R_2, \cdots, U/R_m\}$ and $MS_{\mathscr{R}_2} = \{U/R_1, U/R_2, \cdots, U/R_{m'}\}$, if $MS_{\mathscr{R}_1} \preceq_1 MS_{\mathscr{R}_2}$, then $\underline{\sum_{i=1}^{m} R_i}^{OPT}(X) \supseteq \underline{\sum_{i=1}^{m'} R_i}^{OPT}(X)$ for each $X \subseteq U$.*

Proof. $\forall x \in \underline{\sum_{i=1}^{m'} R_i}^{OPT}(X)$, then by Definition 1.14, there must be $R_{2j} \in \mathscr{R}_2$ such that $[x]_{R_{2j}} \subseteq X$. By condition we know that $MS_{\mathscr{R}_1} \preceq_1 MS_{\mathscr{R}_2}$, then by Definition1.19, there must be $[x]_{R_{1k}} (R_{1k} \in \mathscr{R}_1)$ such that $[x]_{R_{1k}} \subseteq [x]_{R_{2j}}$, it follows that $[x]_{R_{1k}} \subseteq X$, i.e. $x \in \underline{\sum_{i=1}^{m} R_i}^{OPT}(X)$. □

Theorem 1.21 tells us that by using the first hierarchical structure, if the partitions based multigranulation space is finer, then the greater optimistic multigranulation lower approximation will be obtained.

Theorem 1.22. *Let $K = (U, \mathbf{R})$ be a knowledge base in which $\mathscr{R}_1, \mathscr{R}_2 \subseteq \mathbf{R}$, $MS_{\mathscr{R}_1} = \{U/R_1, U/R_2, \cdots, U/R_m\}$ and $MS_{\mathscr{R}_2} = \{U/R_1, U/R_2, \cdots, U/R_{m'}\}$, if $MS_{\mathscr{R}_1} \preceq_1 MS_{\mathscr{R}_2}$, then $\overline{\sum_{i=1}^{m} R_i}^{OPT}(X) \subseteq \overline{\sum_{i=1}^{m'} R_i}^{OPT}(X)$ for each $X \subseteq U$.*

Proof. Since $MS_{\mathscr{R}_1} \preceq_1 MS_{\mathscr{R}_2}$, then by Theorem 1.21, we have $\underline{\sum_{i=1}^{m} R_i}^{OPT}(\sim X) \supseteq \underline{\sum_{i=1}^{m'} R_i}^{OPT}(\sim X)$ where $\sim X \subseteq U$. By Definition 1.14, we have $\overline{\sum_{i=1}^{m} R_i}^{OPT}(X) = \sim \underline{\sum_{i=1}^{m} R_i}^{OPT}(\sim X)$ and $\overline{\sum_{i=1}^{m'} R_i}^{OPT}(X) = \sim \underline{\sum_{i=1}^{m'} R_i}^{OPT}(\sim X)$, then we obtain $\overline{\sum_{i=1}^{m} R_i}^{OPT}(X) \subseteq \overline{\sum_{i=1}^{m'} R_i}^{OPT}(X)$ obviously. □

Theorem 1.22 tells us that by using the first hierarchical structure, if the partitions based multigranulation space is finer, then the smaller optimistic multigranulation upper approximation will be obtained.

Theorem 1.23. *Let $K = (U, \mathbf{R})$ be a knowledge base in which $\mathscr{R}_1, \mathscr{R}_2 \subseteq \mathbf{R}$, $MS_{\mathscr{R}_1} = \{U/R_1, U/R_2, \cdots, U/R_m\}$ and $MS_{\mathscr{R}_2} = \{U/R_1, U/R_2, \cdots, U/R_{m'}\}$, if $\underline{\sum_{i=1}^{m} R_i}^{OPT}(X) \supseteq \underline{\sum_{i=1}^{m'} R_i}^{OPT}(X)$ for each $X \subseteq U$, then $MS_{\mathscr{R}_1} \preceq_1 MS_{\mathscr{R}_2}$.*

Proof. Suppose that $MS_{\mathscr{R}_1} \not\preceq_1 MS_{\mathscr{R}_2}$, then by Definition 1.19, $\exists [x]_{R_{2j}}(R_{2j} \in \mathscr{R}_2)$ such that $[x]_{R_{1k}} \nsubseteq [x]_{R_{2j}}$ for each $[x]_{R_{1k}}(R_{1k} \in \mathscr{R}_1)$. $x \in \underline{\sum_{i=1}^{m'} R_i}^{OPT}([x]_{R_{2j}})$ holds obviously since $[x]_{R_{2j}} \subseteq [x]_{R_{2j}}$. However, $[x]_{R_{1k}} \nsubseteq [x]_{R_{2j}}$ holds for each $[x]_{R_{1k}}(R_{1k} \in \mathscr{R}_1)$, it follows that $x \notin \underline{\sum_{i=1}^{m} R_i}^{OPT}([x]_{R_{2j}})$. From discussions above, we know that $\exists Y \subseteq U$ such that $\underline{\sum_{i=1}^{m} R_i}^{OPT}(Y) \nsupseteq \underline{\sum_{i=1}^{m'} R_i}^{OPT}(Y)$. □

Theorem 1.23 is the inverse of Theorem 1.21, i.e. in the partitions based multigranulation spaces, if the optimistic multigranulation lower approximation is greater, then the multigranulation space is finer in terms of the first hierarchical structure.

Theorem 1.24. *Let $K = (U, \mathbf{R})$ be a knowledge base in which $\mathscr{R}_1, \mathscr{R}_2 \subseteq \mathbf{R}$, $MS_{\mathscr{R}_1} = \{U/R_1, U/R_2, \cdots, U/R_m\}$ and $MS_{\mathscr{R}_2} = \{U/R_1, U/R_2, \cdots, U/R_{m'}\}$, if $\overline{\sum_{i=1}^{m} R_i}^{OPT}(X) \subseteq \overline{\sum_{i=1}^{m'} R_i}^{OPT}(X)$ for each $X \subseteq U$, then $MS_{\mathscr{R}_1} \preceq_1 MS_{\mathscr{R}_2}$.*

Proof. Since $\overline{\sum_{i=1}^{m} R_i}^{OPT}(X) \subseteq \overline{\sum_{i=1}^{m'} R_i}^{OPT}(X)$ for each $X \subseteq U$, then by Definition 1.14, we know that $\underline{\sum_{i=1}^{m} R_i}^{OPT}(X) \supseteq \underline{\sum_{i=1}^{m'} R_i}^{OPT}(X)$ for each $X \subseteq U$. By Theorem 1.23, $MS_{\mathscr{R}_1} \preceq_1 MS_{\mathscr{R}_2}$ holds. □

Theorem 1.24 is the inverse of Theorem 1.22, i.e. in the partitions based multigranulation spaces, if the optimistic multigranulation upper approximation is smaller, then the multigranulation space is finer in terms of the first hierarchical structure.

Theorem 1.25. *Let $K = (U, \mathbf{R})$ be a knowledge base in which $\mathscr{R}_1, \mathscr{R}_2 \subseteq \mathbf{R}$, $MS_{\mathscr{R}_1} = \{U/R_1, U/R_2, \cdots, U/R_m\}$ and $MS_{\mathscr{R}_2} = \{U/R_1, U/R_2, \cdots, U/R_{m'}\}$, if $MS_{\mathscr{R}_1} \preceq_2 MS_{\mathscr{R}_2}$, then $\underline{\sum_{i=1}^{m} R_i}^{PES}(X) \supseteq \underline{\sum_{i=1}^{m'} R_i}^{PES}(X)$ for each $X \subseteq U$.*

Proof. $\forall x \in \underline{\sum_{i=1}^{m'} R_i}^{PES}(X)$, by Definition 1.16, we know that $[x]_{R_{2j}} \subseteq X$ holds for each $R_{2j} \in \mathscr{R}_2$. By condition, $MS_{\mathscr{R}_1} \preceq_2 MS_{\mathscr{R}_2}$, then by Definition 1.20, $\forall [x]_{R_{1k}}(R_{1k} \in \mathscr{R}_1)$, there must be $[x]_{R_{2j}}(R_{2j} \in \mathscr{R}_2)$ such that $[x]_{R_{1k}} \subseteq [x]_{R_{2j}}$. Since $[x]_{R_{2j}} \subseteq X$ for each $R_{2j} \in$

$\mathscr{R}_2$, then we can conclude that $[x]_{R_{1k}} \subseteq X$ for each $R_{1k} \in \mathscr{R}_1$, i.e. $x \in \underline{\sum_{i=1}^{m} R_i}^{PES}(X)$. □

Theorem 1.25 tells us that by using the second hierarchical structure, if the partitions based multigranulation space is finer, then the greater pessimistic multigranulation lower approximation will be obtained.

Theorem 1.26. *Let $K = (U, \mathbf{R})$ be a knowledge base in which $\mathscr{R}_1, \mathscr{R}_2 \subseteq \mathbf{R}$, $MS_{\mathscr{R}_1} = \{U/R_1, U/R_2, \cdots, U/R_m\}$ and $MS_{\mathscr{R}_2} = \{U/R_1, U/R_2, \cdots, U/R_{m'}\}$, if $MS_{\mathscr{R}_1} \preceq_2^P MS_{\mathscr{R}_2}$, then $\overline{\sum_{i=1}^{m} R_i}^{PES}(X) \subseteq \overline{\sum_{i=1}^{m'} R_i}^{PES}(X)$ for each $X \subseteq U$.*

Proof. Since $MS_{\mathscr{R}_1} \preceq_2 MS_{\mathscr{R}_2}$, then by Theorem 1.25, we have $\underline{\sum_{i=1}^{m} R_i}^{PES}(\sim X) \supseteq \underline{\sum_{i=1}^{m'} R_i}^{PES}(\sim X)$ where $\sim X \subseteq U$. By Definition 1.16, we have $\overline{\sum_{i=1}^{m} R_i}^{PES}(X) = \sim \underline{\sum_{i=1}^{m} R_i}^{PES}(\sim X)$ and $\overline{\sum_{i=1}^{m'} R_i}^{PES}(X) = \sim \underline{\sum_{i=1}^{m'} R_i}^{PES}(\sim X)$, then we obtain $\overline{\sum_{i=1}^{m} R_i}^{PES}(X) \subseteq \overline{\sum_{i=1}^{m'} R_i}^{PES}(X)$ obviously. □

Theorem 1.26 tells us that by using the second hierarchical structure, if the partitions based multigranulation space is finer, then the smaller pessimistic multigranulation upper approximation will be obtained.

Theorem 1.27. *Let $K = (U, \mathbf{R})$ be a knowledge base in which $\mathscr{R}_1, \mathscr{R}_2 \subseteq \mathbf{R}$, $MS_{\mathscr{R}_1} = \{U/R_1, U/R_2, \cdots, U/R_m\}$ and $MS_{\mathscr{R}_2} = \{U/R_1, U/R_2, \cdots, U/R_{m'}\}$, if $\underline{\sum_{i=1}^{m} R_i}^{PES}(X) \supseteq \underline{\sum_{i=1}^{m'} R_i}^{PES}(X)$ for each $X \subseteq U$, then $MS_{\mathscr{R}_1} \preceq_2 MS_{\mathscr{R}_2}$.*

Proof. Suppose that $MS_{\mathscr{R}_1} \npreceq_2 MS_{\mathscr{R}_2}$, then by Definition 1.20, $\exists [x]_{R_{1k}} (R_{1k} \in \mathscr{R}_1)$ such that $[x]_{R_{1k}} \nsubseteq [x]_{R_{2j}}$ for each $[x]_{R_{2j}} (R_{2j} \in \mathscr{R}_2)$. Let $Y = \cup\{[x]_{R_{2j}} : \forall R_{2j} \in \mathscr{R}_2\}$, by Definition 1.16, we know that $x \in \underline{\sum_{i=1}^{m'} R_i}^{PES}(Y)$ since $[x]_{R_{2j}} \subseteq Y$ for each $R_{2j} \in \mathscr{R}_2$. However, since $[x]_{R_{1k}} \nsubseteq [x]_{R_{2j}}$ for each $[x]_{R_{2j}} (R_{2j} \in \mathscr{R}_2)$, then $[x]_{R_{1k}} \nsubseteq Y$, it follows that $x \notin \underline{\sum_{i=1}^{m} R_i}^{PES}(Y)$. From discussions above, we know that $\exists Y \subseteq U$ such that

$$\underline{\sum_{i=1}^{m} R_i}^{PES}(Y) \nsupseteq \underline{\sum_{i=1}^{m'} R_i}^{PES}(Y).$$ □

Theorem 1.27 is the inverse of Theorem 1.25, i.e. in the partitions based multigranulation spaces, if the pessimistic multigranulation lower approximation is greater, then the multigranulation space is finer in terms of the second hierarchical structure.

Theorem 1.28. *Let $K = (U, \mathbf{R})$ be a knowledge base in which $\mathscr{R}_1, \mathscr{R}_2 \subseteq \mathbf{R}$, $MS_{\mathscr{R}_1} = \{U/R_1, U/R_2, \cdots, U/R_m\}$ and $MS_{\mathscr{R}_2} = \{U/R_1, U/R_2, \cdots, U/R_{m'}\}$, if $\overline{\sum_{i=1}^{m} R_i}^{PES}(X) \subseteq \overline{\sum_{i=1}^{m'} R_i}^{PES}(X)$ for each $X \subseteq U$, then $MS_{\mathscr{R}_1} \preceq_2 MS_{\mathscr{R}_2}$.*

Proof. Since $\overline{\sum_{i=1}^{m} R_i}^{PES}(X) \subseteq \overline{\sum_{i=1}^{m'} R_i}^{PES}(X)$, for each $X \subseteq U$, then by Definition 1.16, we know that $\underline{\sum_{i=1}^{m} R_i}^{PES}(X) \supseteq \underline{\sum_{i=1}^{m'} R_i}^{PES}(X)$ for each $X \subseteq U$. By Theorem 1.27, $MS_{\mathscr{R}_1} \preceq_2 MS_{\mathscr{R}_2}$ holds. □

Theorem 1.28 is the inverse of Theorem 1.26, i.e. in the partitions based multigranulation spaces, if the pessimistic multigranulation upper approximation is smaller, then the multigranulation space is finer in terms of the second hierarchical structure.

Since the third hierarchical structure can generate the first and the second hierarchical structures (see Theorem 1.20), then by the above theorems, it is not difficult to obtain the following corollaries.

Corollary 1.1. *Let $K = (U, \mathbf{R})$ be a knowledge base in which $\mathscr{R}_1, \mathscr{R}_2 \subseteq \mathbf{R}$, $MS_{\mathscr{R}_1} = \{U/R_1, U/R_2, \cdots, U/R_m\}$ and $MS_{\mathscr{R}_2} = \{U/R_1, U/R_2, \cdots, U/R_{m'}\}$, if $MS_{\mathscr{R}_1} \preceq_3 MS_{\mathscr{R}_2}$, then $\underline{\sum_{i=1}^{m} R_i}^{OPT}(X) \supseteq \underline{\sum_{i=1}^{m'} R_i}^{OPT}(X)$ for each $X \subseteq U$.*

Corollary 1.2. *Let $K = (U, \mathbf{R})$ be a knowledge base in which $\mathscr{R}_1, \mathscr{R}_2 \subseteq \mathbf{R}$, $MS_{\mathscr{R}_1} = \{U/R_1, U/R_2, \cdots, U/R_m\}$ and $MS_{\mathscr{R}_2} = \{U/R_1, U/R_2, \cdots, U/R_{m'}\}$, if $MS_{\mathscr{R}_1} \preceq_3 MS_{\mathscr{R}_2}$, then $\overline{\sum_{i=1}^{m} R_i}^{OPT}(X) \subseteq \overline{\sum_{i=1}^{m'} R_i}^{OPT}(X)$ for each $X \subseteq U$.*

Corollary 1.3. *Let $K = (U, \mathbf{R})$ be a knowledge base in which $\mathscr{R}_1, \mathscr{R}_2 \subseteq \mathbf{R}$, $MS_{\mathscr{R}_1} = \{U/R_1, U/R_2, \cdots, U/R_m\}$ and $MS_{\mathscr{R}_2} = \{U/R_1, U/R_2, \cdots, U/R_{m'}\}$, if $MS_{\mathscr{R}_1} \preceq_3 MS_{\mathscr{R}_2}$, then $\underline{\sum_{i=1}^{m} R_i}^{PES}(X) \supseteq \underline{\sum_{i=1}^{m'} R_i}^{PES}(X)$ for each $X \subseteq U$.*

Corollary 1.4. *Let* $K = (U, \mathbf{R})$ *be a knowledge base in which* $\mathscr{R}_1, \mathscr{R}_2 \subseteq \mathbf{R}$, $MS_{\mathscr{R}_1} = \{U/R_1, U/R_2, \cdots, U/R_m\}$ *and* $MS_{\mathscr{R}_2} = \{U/R_1, U/R_2, \cdots, U/R_{m'}\}$, *if* $MS_{\mathscr{R}_1} \preceq_3 MS_{\mathscr{R}_2}$, *then* $\overline{\sum_{i=1}^{m} R_i}^{PES}(X) \subseteq \overline{\sum_{i=1}^{m'} R_i}^{PES}(X)$ *for each* $X \subseteq U$.

1.5 Information System

1.5.1 Information System and Rough Set

The issue of knowledge representation is of primary importance in current research in the field of cognitive science and artificial intelligence. In cognitive science, knowledge representation is mainly focusing on storing and processing files while in the field of artificial intelligence, the main objective of knowledge representation is to store knowledge, which can be processed by the computer.

In Pawlak's rough set theory, since knowledge is understood as partition (classification),which can be viewed as a semantic definition of knowledge, then he employed a tabular representation of knowledge, which can be viewed as special kind of "formal knowledge" to represent equivalence relations (or partitions) in symbolic form for computer processing. Such a table can also be called Knowledge Representation System / Information System / Attribute-value System. To unify our discussion, the table is referred to as information system in this book.

Actually, an information system is a data table, in which columns are labeled by attributes while rows are labeled by objects. Formally, an information system is a quadruple such that $I = (U, AT, V, f)$, in which

- U is the universe of discourse;
- AT is the set of attributes;
- $\forall a \in AT$, V_a is the domain of a and then V is the domain of all attributes, i.e. $V = \bigcup_{a \in AT} V_a$
- f is the information function such that $f(x, a) \in V_a$ for each $x \in U$ and $a \in AT$.

Obviously, the information system I can be used to express the knowledge in knowledge base $K = (U, \mathbf{R})$. Each attribute in I is corresponding to an equivalence relation in $(U, \mathbf{R})$ while $\forall R \in \mathbf{R}$, R is corresponding to an attribute in I. The information system converts the knowledge in knowledge base to the data structure, which can be accepted by computer. For an information system I, one can describe relationships between objects through their attributes values. With respect to a subset of attributes such that $A \subseteq AT$, an indiscernibility relation $IND(A)$ may be defined as

$$IND(A) = \{(x, y) \in U^2 : f(x, a) = f(y, a), \forall a \in A\} \tag{1.50}$$

The indiscernibility relation $IND(A)$ can be used to define the lower and upper approximations as Definition 1.1 shows, i.e.

$$\underline{A}(X) = \{x \in U : [x]_A \subseteq X\} = \cup\{Y \in U/IND(A) : Y \subseteq X\} \tag{1.51}$$

$$\overline{A}(X) = \{x \in U : [x]_A \cap X \neq \emptyset\} = \cup\{Y \in U/IND(A) : Y \cap X \neq \emptyset\} \tag{1.52}$$

where $[x]_A = \{y \in U : (x, y) \in IND(A)\}$, $U/IND(A)$ is the partition induced by set of attributes A.

A decision system is a special form of the information system such that $I = (U, AT \cup D, V, f)$, in which AT is referred to as the set of condition attributes while D is the set of decision attributes and $AT \cap D = \emptyset$. Generally speaking, the decision system can be analyzed by rough set for the discovering of decision rules.

Definition 1.22. [1] Let $I = (U, AT \cup D, V, f)$ be a decision system, if $IND(AT) \subseteq IND(D)$, then I is referred to as consistent, otherwise, it is referred to as inconsistent.

1.5.2 Rough Sets in Multiple-source Information Systems

Pawlak's rough set can be used to deal with one and only one information system. In Ref. [50], Khan et al. presented a rough set approach to deal with multiple-source information systems. A multiple-source information systems is a family of information systems such that $\mathscr{I} = \{I_1, I_2, \cdots\}$, for each information system $I_i \in \mathscr{I}$, we have an indiscernibility relation such that $IND(AT_i)$ in which AT_i is the set of attributes in I_i.

Definition 1.23. [50] Let $\mathscr{I}$ be a multiple-source information systems, then $\forall X \subseteq U$, the strong lower approximation $\underline{\mathscr{I}}_S(X)$, weak lower approximation $\underline{\mathscr{I}}_W(X)$, strong upper approximation $\overline{\mathscr{I}}_S(X)$, and weak upper approximation $\overline{\mathscr{I}}_W(X)$ of X, respectively, are defined as follows:

$$\underline{\mathscr{I}}_S(X) = \bigcap_{I_i \in \mathscr{I}} \underline{AT_i}(X) \tag{1.53}$$

$$\underline{\mathscr{I}}_W(X) = \bigcup_{I_i \in \mathscr{I}} \underline{AT_i}(X) \tag{1.54}$$

$$\overline{\mathscr{I}}_S(X) = \bigcap_{I_i \in \mathscr{I}} \overline{AT_i}(X) \tag{1.55}$$

$$\overline{\mathscr{I}}_W(X) = \bigcup_{I_i \in \mathscr{I}} \overline{AT_i}(X) \tag{1.56}$$

in which $\underline{AT_i}(X)$ and $\overline{AT_i}(X)$ are lower and upper approximations of X in the information system I_i respectively.

From the viewpoint of granular computing, each partition in terms of an information system indicates a family of information granules, i.e. granulation space. Obviously, these information granules are all equivalence classes in the above rough approximations. If we use each subset of attributes in multigranulation rough sets to construct an information system, then it is interesting to explore the relationships between rough sets in multiple-source information systems and multigranulation rough sets.

Theorem 1.29. *Let I be an information system system in which $A_1, A_2, \cdots, A_m \subseteq AT$, $\forall A_i$ $(1 \leqslant i \leqslant m)$, we can construct an information system $I_i = (U, A_i)$ and then the multiple-source information systems is $\mathscr{I} = \{I_1, I_2, \cdots, I_m\}$, we have*

$$\underline{\sum_{i=1}^{m} A_i}^{O}(X) = \underline{\mathscr{I}}_W(X) \tag{1.57}$$

$$\overline{\sum_{i=1}^{m} A_i}^{O}(X) = \overline{\mathscr{I}}_S(X) \tag{1.58}$$

$$\underline{\sum_{i=1}^{m} A_i}^{P}(X) = \underline{\mathscr{I}}_S(X) \tag{1.59}$$

$$\overline{\sum_{i=1}^{m} A_i}^{P}(X) = \overline{\mathscr{I}}_W(X) \tag{1.60}$$

Proof. We only prove Eq. (1.57), others can be proved analogously.

$\forall x \in U$,

$$\begin{aligned} x \in \underline{\sum_{i=1}^{m} A_i}^{O}(X) &\Leftrightarrow \exists A_i (1 \leqslant i \leqslant m) \text{ s.t. } [x]_{A_i} \subseteq X \\ &\Leftrightarrow \exists A_i (1 \leqslant i \leqslant m) \text{ s.t. } x \in \underline{A_i}(X) \\ &\Leftrightarrow x \in \bigcup_{I_i \in \mathscr{I}} \underline{A_i}(X) \\ &\Leftrightarrow x \in \underline{\mathscr{I}}_W(X) \end{aligned}$$

□

Theorem 1.29 shows that the multigranulation rough set can be considered as rough sets in multiple-source information systems and verse vice. The main difference between these two approaches is that the multigranulation rough set is defined by using a family of the binary relations, the rough sets in multiple-source information systems is constructed on the basis of a family of the information systems.

1.5.3 Several Reducts in Decision System

The concept of knowledge reduction we mentioned in Section 1.1.3, can also be introduced into information system. Since equivalence relations in the knowledge base are corresponding to the attributes in the information system, then given an information system, a reduct is a subset of the attributes instead of a subset of the equivalence relations, which can preserve the same information for classification or some other purpose.

In an inconsistent decision system (see Definition 1.22), Kryszkiewicz [51] proposed five types of attribute reductions as Definition 1.24 shows, and then presented some static relationships among these reducts.

Definition 1.24. Let I be an inconsistent decision system in which $A \subseteq AT$, let us denote the generalized decision value of x such that

$$\partial_A(x) = \{f(y, d) : y \in [x]_A, d \in D\} \tag{1.61}$$

then define the following:

1. If $[x]_A \subseteq \overline{AT}([x]_D)$, then A is referred to as a possible consistent attributes set of I for $x \in U$; if A is a possible consistent attributes set of I for $x \in U$ and $\forall B \subset A$, B is not the possible consistent attributes set of I for $x \in U$, then A is referred to as a possible reduct of I for $x \in U$.
2. If $\forall y \in [x]_A$, $\partial_A(y) = \partial_{AT}(y) = \partial_{AT}(x)$, then A is referred to as an approximate consistent attributes set of I for $x \in U$; if A is an approximate consistent attributes set of I for $x \in U$ and $\forall B \subset A$, B is not the approximate consistent attributes set of I for $x \in U$, then A is referred to as an approximate reduct I for $x \in U$.
3. If $\partial_A(x) = \partial_{AT}(x)$, then A is referred to as a generalized decision consistent attributes set of I for $x \in U$; if A is a generalized decision consistent attributes set of I for $x \in U$ and $\forall B \subset A$, B is not the generalized decision consistent attributes set of I for $x \in U$, then A is referred to as an generalized decision reduct of I for $x \in U$.
4. If $\mu^A_{X_i}(x) = \mu^{AT}_{X_i}(x)$ for each $X_i \in U/IND(D)$, then A is referred to as a μ-decision consistent attributes set of I for x; if A is a μ-decision consistent attributes set of I for x and $\forall B \subset A$, B is not the μ-decision consistent attributes set of I for x, then A is referred to as a μ-decision reduct of I for $x \in U$.
5. If $\forall y \in [x]_A$, $\mu^A_{[x]_D}(y) = \mu^{AT}_{[x]_D}(y) = \mu^{AT}_{[x]_D}(x)$, then A is referred to as a μ-consistent attributes set of I for $x \in U$; if A is a μ-consistent attributes of I for $x \in U$ and $\forall B \subset A$, B is not the μ-consistent attributes of I for $x \in U$, then A is referred to as a μ-reduct of I for $x \in U$.

Theorem 1.30. [51] *Let I be an inconsistent decision system in which $A \subseteq AT$, then*

1. *If A is a possible consistent attributes set of I for $x \in U$, then A is a generalized decision consistent attributes set of I for $x \in U$, i.e.*

$$[[x]_A \subseteq \overline{AT}([x]_D)] \Rightarrow [\partial_A(x) = \partial_{AT}(x)]$$

2. *If A is an approximate consistent attributes set of I for $x \in U$, then A is a possible consistent attributes set of I for $x \in U$, i.e.*

$$[\forall y \in [x]_A, \partial_A(y) = \partial_{AT}(y) = \partial_{AT}(x)] \Rightarrow [[x]_A \subseteq \overline{AT}(X_{d(x)})]$$

3. *If A is a μ-consistent attributes of I for $x \in U$, then A is a possible consistent attributes set of I for $x \in U$, i.e.*

$$[\forall y \in [x]_A, \mu^A_{[x]_D}(y) = \mu^{AT}_{[x]_D}(y) = \mu^{AT}_{[x]_D}(x)] \Rightarrow [[x]_A \subseteq \overline{AT}(X_{d(x)})]$$

4. *If A is a μ-decision consistent attributes of I for x, then A is an approximate consistent attributes set of I for $x \in U$, i.e.*

$$[\forall X_i \in U/IND(D), \mu^A_{X_i}(x) = \mu^{AT}_{X_i}(x)] \Rightarrow [\forall y \in [x]_A, \partial_A(y) = \partial_{AT}(y) = \partial_{AT}(x)]$$

5. *If A is a μ-decision consistent attributes of I for x, then A is a μ-consistent attributes of I for $x \in U$, i.e.*

$$[\forall X_i \in U/IND(D), \mu^A_{X_i}(x) = \mu^{AT}_{X_i}(x)] \Rightarrow [\forall y \in [x]_A, \mu^A_{[x]_D}(y) = \mu^{AT}_{[x]_D}(y) = \mu^{AT}_{[x]_D}(x)]$$

By Definition 1.24, it is not difficult to observe that Kryszkiewicz's five different reducts are all defined for a nominated object x in the decision system. Following Kryszkiewicz's work, Zhang et al. [52] further proposed the concept of maximum distribution reduct and possible distribution reduct. In Ref. [52], the reducts are defined for the whole decision system instead of a single object any more.

Definition 1.25. Let I be a decision system in which $A \subseteq AT$, $U/IND(D) = \{X_1, X_2, \cdots, X_m\}$ is the partition determined by the set of decision attributes D, let us denote

$$\mu_A(x) = \{\mu^A_{X_1}(x), \mu^A_{X_2}(x), \cdots, \mu^A_{X_m}(x)\} \tag{1.62}$$

$$\gamma_A(x) = \{X_i : \mu^A_{X_i}(x) = max\{\mu^A_{X_i}(x) : X_i \in IND(D)\}\} \tag{1.63}$$

$$\delta_A(x) = \{X_i : [x]_A \cap X_i \neq \emptyset, X_i \in IND(D)\} \tag{1.64}$$

$$\eta_A(x) = \frac{\sum\{|\overline{P}(X_i)| : X_i \in U/IND(D)\}}{U} \tag{1.65}$$

then define the following:

1. If $\mu_A(x) = \mu_{AT}(x)$ for each $x \in U$, then A is referred to as a distribution consistent attributes set of I; if A is a distribution consistent attributes set of I and $\forall B \subset A$, B is not the distribution consistent attributes set of I, then A is referred to as a distribution reduct of I.
2. If $\gamma_A(x) = \gamma_{AT}(x)$ for each $x \in U$, then A is referred to as a maximum distribution consistent attributes set of I; if A is a maximum distribution consistent attributes set of I and no proper subset of A is the maximum distribution consistent attributes set of I, then A is referred to as a maximum distribution reduct of I.
3. If $\delta_A(x) = \delta_{AT}(x)$ for each $x \in U$, then A is referred to as an assignment consistent attributes set of I; if A is an assignment consistent attributes set of I and no proper subset of A is the assignment consistent attributes set of I, then A is referred to as an assignment reduct of I.
4. If $\eta_A(x) = \eta_{AT}(x)$ for each $x \in U$, then A is referred to as a possible distribution consistent attributes set of I; if A is a possible distribution consistent attributes set of I and no proper subset of A is the possible distribution consistent attributes set of I, then A is referred to as a possible distribution reduct of I.

Theorem 1.31. [52] *Let I be an decision system in which $A \subseteq AT$, then*

1. *If A is a a distribution consistent attributes set of I, then A is a possible distribution consistent attributes set of I, i.e.*

$$[\forall x \in U, \mu_A(x) = \mu_{AT}(x)] \Rightarrow [\forall x \in U, \eta_A(x) = \eta_{AT}(x)]$$

2. *If A is a a distribution consistent attributes set of I, then A is a maximum distribution consistent attributes set of I, i.e.*

$$[\forall x \in U, \mu_A(x) = \mu_{AT}(x)] \Rightarrow [\forall x \in U, \gamma_A(x) = \gamma_{AT}(x)]$$

3. *If A is a distribution consistent attributes set of I, then A is an assignment consistent attributes set of I, i.e.*

$$[\forall x \in U, \mu_A(x) = \mu_{AT}(x)] \Rightarrow [\forall x \in U, \eta_A(x) = \eta_{AT}(x)]$$

4. *A is an assignment consistent attributes set of I if and only if A is a possible distribution consistent attributes set of I, i.e.*

$$[\forall x \in U, \delta_A(x) = \delta_{AT}(x)] \Leftrightarrow [\forall x \in U, \delta_A(x) = \delta_{AT}(x)]$$

Remark 1.5. *It should be noticed that the converse of* 1 *in Theorem* 1.31 *is not always true, i.e., a possible distribution consistent attributes set may not be a distribution consistent attributes set.*

Remark 1.6. *It should be noticed that the converse of* 2 *in Theorem* 1.31 *is not always true, i.e., a maximum distribution consistent attributes set may not be a distribution consistent attributes set.*

Remark 1.7. *There is no static relationship between maximum distribution consistent attributes sets and possible consistent attributes set.*

1.6 Conclusions

Pawlak's rough set model, was proposed through an indiscernibility relation. Such indiscernibility relation is an equivalence relation since it is the intersection of the equivalence relations in knowledge base.

In this chapter, we introduced two important expanded rough set models, which are also constructed on the basis of the equivalence relation, one is the variable precision rough set and the other is the multigranulation rough set:

1. The variable precision rough set is a robust rough set model since it can be used to handle the problems with misclassification by incorporating a controlled degree of misclassification into knowledge representation of rough set formalism. It should be noticed that though the variable precision rough set is proposed by using an indiscernibility relation, it can also be generalized. For example, the variable precision fuzzy rough set [53–55], the variable precision dominans-based rough set [56–58], the adapted variable precision rough set [59], probabilistic rough set [60–66], Bayesian rough set [67–70] and so on have been studied by many researchers in recent years.
2. In multigranulation rough set approach, the target is approximated by a family of the binary relations instead of a single binary relation. It should be noticed that multigranulation rough set has two different forms, one is the optimistic case and the other is the pessimistic case. Since the multigranulation rough set was firstly proposed in 2007 and then it has a wider space to be generalized and applied.

Finally, information system is an important knowledge representation method in rough set theory. By using information system, knowledge in knowledge base may be expressed and then processed by computers. One attribute in the information system is corresponding to an equivalence relation in the knowledge base, one attribute value in the information system is then corresponding to an equivalence class, which is generated from an equivalence relation in the knowledge base, such the attribute value is actually the "name" of the equivalence class.

References

[1] Pawlak, Z.: Rough sets-theoretical aspects of reasoning about data. Kluwer Academic Publishers, (1991)

[2] Pawlak, Z.: Rough set theory and its applications to data analysis. Cybernet. Syst. **29**, 661–688 (1998)

[3] Pawlak, Z.: AI and intelligent industrial applications: the rough set perspective. Cybernet. Syst. **31**, 227–252 (2000)

[4] Pawlak, Z.: Drawing conclusions from data-the rough set way. Int. J. Intell. Syst. **16**, 3–11 (2001)

[5] Pawlak, Z.: Rough sets and intelligent data analysis. Inform. Sci. **147**, 1–12 (2002)

[6] Pawlak, Z.: A rough set view on Bayes' theorem. Int. J. Intell. Syst. **18**, 487–498 (2003)

[7] Pawlak, Z.: Some remarks on conflict analysis. Eur. J. Oper. Res. **166**, 649–654 (2005)

[8] Pawlak, Z., Skowron, A.: Rudiments of rough sets. Inform. Sci. **177**, 3–27 (2007)

[9] Pawlak, Z., Skowron, A.: Rough sets: some extensions. Inform. Sci. **177**, 28–40 (2007)

[10] Pawlak, Z., Skowron, A.: Rough sets and boolean reasoning. Inform. Sci. **177**, 41–73 (2007)

[11] Bhatt, R. B., Gopal, M.: On fuzzy-rough sets approach to feature selection. Pattern Recogn. Lett. **26**, 965–975 (2005)

[12] Hu, Q. H., Liu, J. F., Yu D. R.: Mixed feature selection based on granulation and approximation. Knowl.-Based Syst. **21**, 294–304 (2008)

[13] Kim, D.: Data classification based on tolerant rough set. Pattern Recognit. **34**, 1613–1624 (2001)

[14] Parthaláin, N. M., Shen Q.: Exploring the boundary region of tolerance rough sets for feature selection. Pattern Recognit. **42**, 655–667 (2009)

[15] Shen, Q., Jensen, R.: Selecting informative features with fuzzy–rough sets and its application for complex systems monitoring. Pattern Recognit. **37**, 1351–1363 (2004)

[16] Swiniarski, R. W., Skowron, A.: Rough set method in feature selection and recognition. Pattern Recogn. Lett. **24**, 833–849 (2003)

[17] Wang, J. Y., Zhou, J.: Research of reduct features in the variable precision rough set model. Neurocomput. **72**, 2643–2648 (2009)

[18] Kulak, O.: A decision support system for fuzzy multi-attribute selection of material handling equipments. Expert Syst. Appl. **29**, 310–319 (2005)

[19] Li, G. D., Yamaguchi, D., Nagai, M.: A grey-based rough decision-making approach to supplier selection. Int. J. Adv. Manuf. Techn. **36**, 1032–1040 (2008)

[20] Liou J. J. H.: A novel decision rules approach for customer relationship management of the airline market. Expert Syst. Appl. **36**, 4374–4381 (2009)

[21] Shen, L. X., Loh, H. T.: Applying rough sets to market timing decisions. Decis. Support Syst. **37**, 583–597 (2004)

[22] Xie, G., Zhang, J., Lai, K. K., Yu, L.: Variable precision rough set for group decision–making: an application. Int. J. Approx. Reason. **49**, 331–343 (2008)

[23] Li, H. L., Chen, M. H.: Induction of multiple criteria optimal classification rules for biological and medical data. Comput. Biol. Med. **38**, 42–52 (2008)

[24] Tsumoto, S.: Automated extraction of medical expert system rules from clinical databases based on rough set theory. Inform. Sci. **112**, 67–84 (1998)

[25] Tsumoto, S.: Automated discovery of positive and negative knowledge in clinical databases. IEEE Eng. Med. Biol. **19**, 56–62 (2000)

[26] Zhang, Z. W., Shi, Y., Gao, G. X.: A rough set-based multiple criteria linear programming approach for the medical diagnosis and prognosis. Expert Syst. Appl. **36**, 8932–8937 (2009)

[27] Ananthanarayana, V. S., Narasimha, M. M., Subramanian, D. K.: Tree structure for efficient data mining using rough sets. Pattern Recogn. Lett. **24**, 851–862 (2003)

[28] Chan, C. C.: A rough set approach to attribute generalization in data mining. Inform. Sci. **107**, 177–194 (1998)

[29] Chen, Y. L., Weng, C. H.: Mining fuzzy association rules from questionnaire data. Knowl.-Based Syst. **22**, 46–56 (2009)

[30] Hulse, J. V., Khoshgoftaar, T.: Knowledge discovery from imbalanced and noisy data. Data Knowl. Eng. **68**, 1513–1542 (2009)

[31] Li, T. R., Ruan, D., Geert, W., Song, J., Xu, Y.: A rough sets based characteristic relation approach for dynamic attribute generalization in data mining. Knowl.-Based Syst. **20**, 485–494 (2007)

[32] Lingras, P. J., Yao, Y. Y.: Data mining using extensions of the rough set model. J. Am. Soc. Inf. Sci. Tec. **49**, 415–422 (1998)

[33] Liu, J. F., Hu, Q. H., Yu, D. R.: A comparative study on rough set based class imbalance learning. Knowl.-Based Syst. **21**, 753–763 (2008)

[34] Su, C. T., Chen, L. S., Yih, Y.: Knowledge acquisition through information granulation for imbalanced data. Expert Syst. Appl. **31**, 531–541 (2006)

[35] Zhai, L. Y., Khoo, L. P., Zhong, Z.W.: A dominance-based rough set approach to Kansei Engineering in product development. Expert Syst. Appl. **36**, 393–402 (2009)

[36] Yao, Y. Y.: Information granulation and rough set approximation. Int. J. Intell. Syst. **16**, 87–104 (2001)

[37] Qian, Y. H., Liang, J. Y., Dang, C. Y.: Knowledge structure, knowledge granulation and knowledge distance in a knowledge base. Int. J. Approx. Reason. **50**, 174–188 (2009)

[38] Liang, J. Y., Qian, Y. H.: Axiomatic approach of knowledge granulation in information system. In: 19th Australian Joint Conference on Artificial Intelligence, pp. 1074–1078 (2006)

[39] Liang, J. Y., Dang, C. Y., Chin, K. S., Yam Richard C. M.: A new method for measuring uncertainty and fuzziness in rough set theory. Int. J. Gen. Syst. **31**, 331–342 (2002)

[40] Liang, J. Y., Shi, Z. Z.: The information entropy, rough entropy and knowledge granulation in rough set theory. Int. J. Uncertain. Fuzz. **12**, 37–46 (2004)

[41] Liang, J. Y., Wang, J. H., Qian, Y. H.: A new measure of uncertainty based on knowledge granulation for rough sets. Inform. Sci. **179**, 458–470 (2009)

[42] Ziarko, W.: Variable precision rough set model. J. Comput. Syst. Sci. **46**, 39–59 (1993)

[43] Beynon, M.: Reducts within the variable precision rough sets model: a further investigation. Eur. J. Oper. Res. **134**, 592–605 (2001)

[44] Mi, J. S., Wu, W. Z., Zhang, W. X.: Approaches to knowledge reduction based on variable precision rough set model. Inform. Sci. **159**, 255–272 (2004)

[45] Inuiguchi, M.: Attribute reduction in variable precision rough set model. Int. J. Uncertain. Fuzz. **14**, 461–479 (2006)

[46] Qian, Y. H., Dang, C. Y., Liang, J. Y.: MGRS in incomplete information systems. In: 2007 IEEE International Conference on Granular Computing, pp. 163–168 (2007)

[47] Qian, Y. H., Liang, J. Y., Dang, C. Y.: Incomplete multigranulation rough set, IEEE T. Syst. Man Cy. B. **20**, 420–431 (2010)

[48] Qian, Y. H., Liang, J. Y., Wei, W. Pessimistic rough decision. In: Second International Workshop on Rough Sets Theory, pp. 440–449 (2010)

[49] Qian, Y. H., Liang, J. Y., Yao, Y. Y., Dang, C. Y.: MGRS: A multi-granulation rough set. Inform. Sci. **180**, 949–970 (2010)

[50] Khan, M. A., Banerjee, M.: Formal reasoning with rough sets in multiple-source approximation systems. Int. J. Approx. Reason. **49**, 466–477 (2008)

[51] Kryszkiewicz, M.: Comparative study of alternative types of knowledge reduction in inconsistent systems. Int. J. Intell. Syst. **16**, 105–120 (2001)

[52] Zhang, W. X., Mi, J. S., Wu, W. Z.: Approaches to knowledge reductions in inconsistent systems. Int. J. Intell. Syst. **18**, 989–1000 (2003)

[53] Mieszkowicz-Rolka, A., Rolka, L.: Variable precision fuzzy rough sets. Transactions on Rough Sets I, LNCS **3100**, 144–160 (2004)

[54] Mieszkowicz-Rolka, A., Rolka, L.: Fuzzy rough approximations of process data. Int. J. Approx. Reason. **49**, 301–315 (2008)

[55] Zhao, S. Y., Tsang, E. C. C., Chen D. G.: The model of fuzzy variable precision rough sets. IEEE. T. Fuzzy Syst. **17**, 451–476 (2009)

[56] Hu, Q. H., Yu, D. R.: Variable precision dominance based rough set model and reduction algorithm for preference-ordered data. In: Proceedings of the Third International Conference on Machine Learning and Cybernetics, pp. 2279–2284 (2004)

[57] Inuiguchi, M., Yoshioka, Y.: Variable-precision dominance-based rough set approach. In: The Fifth International Conference on Rough Sets and Current Trends in Computing, pp. 203–212 (2006)

[58] Inuiguchi, M., Yoshioka, Y., Kusunoki, Y.: Variable-precision dominance-based rough set approach and attribute reduction. Int. J. Approx. Reason. **50**, 1199–1214 (2009)

[59] Ningler, M., Stockmanns, G., Schneider, G., Kochs, H. D., Kochs E.: Adapted variable precision rough set approach for EEG analysis. Artif. Intell. Med. **47**, 239–261 (2009)

[60] Yao, Y. Y., Zhao, Y.: Attribute reduction in decision-teoretic rough set models. Inform. Sci. **178**, 3356–3373 (2008)

[61] Ziarko, W.: Probablistic decision tables in the variable precision rough set model. Comput. Intell. **17**, 593–603 (2001)

[62] Ziarko, W.: Probabilistic approach to rough sets. Int. J. Approx. Reason. **49**, 272–284 (2008)

[63] Yao, Y. Y.: Probabilistic approaches to rough sets. Expert Syst. **20**, 287–297 (2003)

[64] Yao, Y. Y.: Probabilistic rough set approximations. Int. J. Approx. Reason. **49** 255–271 (2008)

[65] Yao, Y. Y.: Three-way decisions with probabilistic rough sets. Inform. Sci. **180**, 341–353 (2010)

[66] Yao, Y. Y.: The superiority of three-way decisions in probabilistic rough set models. Inform. Sci. **181**, 1080–1096 (2011)

[67] Ślęzak, D.: The rough Bayesian model for distributed decision systems. In: 4th International Conference on Rough Sets and Current Trends in Computing, pp. 384–393 (2004)

[68] Ślęzak, D., Ziarko, W.: Bayesian rough set model. In: Proceeding of the International Workshop on Foundation of Data Mining, pp. 131–135 (2002)

[69] Ślęzak, D., Ziarko, W.: Attribute reduction in the Bayesian version of variable precision rough set model. In: Proceeding of the International Workshop on Rough Sets and Knowledge Discovery, pp. 263–273 (2003)

[70] Ślęzak, D., Ziarko, W.: The investigation of the Bayesian rough set model. Int. J. Approx. Reason. **40**, 81–91 (2005)

Part II
Incomplete Information Systems and Rough Sets

Chapter 2

Expansions of Rough Sets in Incomplete Information Systems

Abstract Generally speaking, an incomplete information system indicates a system with unknown values. In this chapter, several expanded rough sets approaches to incomplete information system have been introduced. Firstly, by assuming that the unknown values can be compared with any values in the domains of the corresponding attributes, the tolerance relation, valued tolerance relation, maximal consistent block, descriptor can be used to construct rough approximations, respectively. Secondly, by assuming that the unknown values cannot be compared with any values in the domains of the corresponding attributes, the similarity relation, difference relation can be used to construct rough approximations, respectively. Finally, by considering the above two different semantic explanations of the unknown values, the characteristic relation can be used to construct rough approximation.

Since Pawlak's rough set model is constructed on the basis of an indiscernibility relation, it can only be used to deal with information system without unknown values, i.e. complete information system. A complete information system indicates that all objects have deterministic values on every attribute. Therefore, given a knowledge base $K = (U, \mathbf{R})$, one can construct a complete information system, in which each column is corresponding to an equivalence relation in such knowledge base. The indiscernibility relation is then the classification analysis of all the attributes in such complete information system.

Gore, in his influential book "Earth in the Balance" [1], notes that "We must acknowledge that we never have complete information. Yet we have to make decisions anyway $\cdots$" This quote illustrates not only the difficulty of making decisions about environmental issues, but also the fact that making such decisions with partial information is ultimately inevitable [2]. Therefore, how to employ the rough set technique to the incomplete information systems[3–19] plays a crucial role in the development of the rough set theory.

Presently, two strategies have been used in rough set theory to process incomplete information systems [20, 21]:

1. data reparation: it is an indirect method that transforms an incomplete information system into a complete information system, where we can acquire knowledge with

the classical rough set theory;

2. model extension: it is a direct method that extends basic concepts of the classical rough set theory in incomplete information systems by relaxing the requirement of indiscernibility relation of reflexivity, symmetry and transitivity.

It should be noticed that the first strategy may change the original information of incomplete information systems and then all what will be discussed in this book are based on the second strategy, i.e. model extension. From the viewpoint of the direct method, the unknown values in an incomplete information system may have two different semantic explanations[22–25] such that:

1. in the first case, all unknown values are "do not care" conditions;
2. in the second case, all unknown values are lost.

The "do not care" unknown value is the "everything is possible" value. Such an interpretation of the unknown value corresponds to the idea that the unknown value is just "missing", but it does exist. On the other hand, if the unknown value is regarded as lost, we think that objects may be described incompletely not only because of our imperfect knowledge, but also because it may be definitely impossible to describe them with all of the attributes. Thus, lost unknown value is a non-existing one and it is not comparable with any other values.

Without loss of generality, an incomplete information system will still denoted without confusion by $I = (U, AT, V, f)$. Here, $\forall x \in U$ and $\forall a \in AT$:

1. if "$f(x, a) = *$", then we assume that the unknown value x holds on a is "do not care" [23, 24];
2. if "$f(x, a) = ?$", then we assume that the unknown value x holds on a is lost [23, 24].

An incomplete decision system is an incomplete information system such that $I = (U, AT \cup D, V, f)$ where $AT \cap D = \emptyset$. To simplify our discussions, we only consider one and only one decision attribute in the incomplete decision system, i.e. $I = (U, AT \cup \{d\}, V, f)$. Additionally, for the rest of the book we will assume that all decision values are specified, i.e., they are known.

2.1 Tolerance Relation Based Rough Set Approach

Given an incomplete information system I, if all unknown values are "do not care" in I, then "*" is referred to as equivalent to any other values in the domain of the corresponding attribute. From this point of view, Kryszkiewicz [6] proposed her tolerance relation as Definition 2.1 shows.

2.1.1 Tolerance Relation and Its Reducts

Definition 2.1. [6]Let I be an incomplete information system in which $A \subseteq AT$, then the tolerance relation in terms of A is denoted by $TOL(A)$ where

$$TOL(A) = \{(x, y) \in U^2 : f(x, a) = f(y, a) \vee f(x, a) = * \vee f(y, a) = *\} \tag{2.1}$$

Obviously, tolerance relation $TOL(A)$ is reflexive and symmetric, but it is not necessarily transitive.

Given an incomplete information system I, $\forall x \in U$, the tolerance class of x in terms of $TOL(A)$ is denoted by $TOL_A(x)$ [6] such that

$$TOL_A(x) = \{y \in U : (x,y) \in TOL(A)\} \tag{2.2}$$

$TOL_A(x)$ is the maximal set of objects, which are *possibly* indiscernible by A with x.

Proposition 2.1. *Let I be an incomplete information system in which $A \subseteq AT$, we have*

$$TOL(A) = \bigcap_{a \in A} TOL(\{a\}) \tag{2.3}$$

Example 2.1. Table 2.1 shows a car incomplete information system, which was analyzed by Kryszkiewicz in Ref. [6].

In Table 2.1,

- $U = \{x_1, x_2, x_3, x_4, x_5, x_6\}$ is the universe of discourse;
- *Price*, *Mileage*, *Size* and *Max-Speed* are four attributes (In the sequel, *P*, *M*, *S* and *MX* will stand for *Price*, *Mileage*, *Size* and *Max-Speed*, respectively); the domains of these attributes are:

$$V_P = \{\text{high}, \text{low}\}$$
$$V_M = \{\text{high}, \text{low}\}$$
$$V_S = \{\text{full}, \text{compact}\}$$
$$V_{MX} = \{\text{high}, \text{low}\}$$

Table 2.1 Car incomplete information system

Car	*Price*	*Mileage*	*Size*	*Max-Speed*
x_1	high	low	full	low
x_2	low	*	full	low
x_3	*	*	compact	low
x_4	high	*	full	high
x_5	*	*	full	high
x_6	low	high	full	*

By Definition 2.1, we obtain the following tolerance classes:

$$TOL_{AT}(x_1) = \{x_1\}$$
$$TOL_{AT}(x_2) = \{x_2, x_6\}$$
$$TOL_{AT}(x_3) = \{x_3\}$$
$$TOL_{AT}(x_4) = \{x_4, x_5\}$$
$$TOL_{AT}(x_5) = \{x_4, x_5, x_6\}$$
$$TOL_{AT}(x_6) = \{x_2, x_5, x_6\}$$

Definition 2.2. Let I be an incomplete information system in which $A \subseteq AT$:

1. if $TOL(A) = TOL(AT)$, then A is referred to as a tolerance relation consistent attributes set in I;
2. if A is a tolerance relation consistent attributes set in I and $\forall B \subset A$, B is not the tolerance relation consistent attributes set in I, then B is referred to as a tolerance relation reduct in I.

By Definition 2.2, we can see that a tolerance relation consistent attributes set in I is a subset of the attributes, which preserves the original tolerance relation in I; a tolerance relation reduct in I is a minimal subset of the attributes, which preserves the original tolerance relation in I. To present the practical approach to compute such tolerance relation reduct, we need the following notions.

$$DIS^{\mathrm{T}}(x,y) = \begin{cases} \{a \in AT : (x,y) \notin TOL(\{a\})\}, & (x,y) \notin TOL(AT) \\ \varnothing, & \text{otherwise} \end{cases}$$

$DIS^{\mathrm{T}}(x,y)$ is referred to as the tolerance relation discernibility attributes set in I w.r.p. to the pair of objects (x,y). Moreover, the matrix

$$\boldsymbol{M}^{\mathrm{T}} = \{DIS^{\mathrm{T}}(x,y) : (x,y) \notin TOL(AT)\} \tag{2.4}$$

is referred to as the tolerance relation discernibility matrix in I.

Theorem 2.1. *Let I be an incomplete information system in which $A \subseteq AT$, then we have*

A is a tolerance relation consistent attributes set in $I \Leftrightarrow A \cap DIS^{\mathrm{T}}(x,y) \neq \varnothing$ for each $DIS^{\mathrm{T}}(x,y) \in \boldsymbol{M}^{\mathrm{T}}$.

Proof. $\Rightarrow$: Suppose $\exists DIS^{\mathrm{T}}(x,y) \in \boldsymbol{M}^{\mathrm{T}}$ such that $A \cap DIS^{\mathrm{T}}(x,y) = \varnothing$. Then, by the definition of $DIS^{\mathrm{T}}(x,y)$, we have $(x,y) \in TOL(A)$. Moreover, since A is a tolerance relation consistent attributes set in I, then $TOL(A) = TOL(AT)$, from which we can conclude that $(x,y) \in TOL(AT)$, which is contradictive to the assumption because $DIS^{\mathrm{T}}(x,y) \in \boldsymbol{M}^{\mathrm{T}} \Rightarrow (x,y) \notin TOL(AT)$.

$\Leftarrow$: By Proposition 2.1, we have $A \subseteq AT \Rightarrow TOL(AT) \subseteq TOL(A)$. Thus, it must be proved that $TOL(A) \subseteq TOL(AT)$. Since for each $DIS^{\mathrm{T}}(x,y) \in \boldsymbol{M}^{\mathrm{T}}$ ($DIS^{\mathrm{T}}(x,y) \in \boldsymbol{M}^{\mathrm{T}} \Rightarrow (x,y) \notin TOL(AT)$), we have $A \cap DIS^{\mathrm{T}}(x,y) \neq \varnothing$, then $(x,y) \notin TOL(A)$, from which we can conclude that $(x,y) \notin TOL(AT) \Rightarrow (x,y) \notin TOL(A)$, i.e. $TOL(A) \subseteq TOL(AT)$. □

Definition 2.3. Let I be an incomplete information system, define

$$\Delta^{\mathrm{T}} = \bigwedge_{DIS^{\mathrm{T}}(x,y) \in \boldsymbol{M}^{\mathrm{T}}} \vee DIS^{\mathrm{T}}(x,y) \tag{2.5}$$

Δ^{T} is referred to as the tolerance relation discernibility function in I.

By Boolean Reasoning technique, it is not difficult to obtain the following theorem by Theorem 2.1.

Theorem 2.2. *Let I be an incomplete information system in which $A \subseteq AT$, then*

A is a tolerance relation reduct in I if and only if $\wedge A$ is the prime implicant of the tolerance relation discernibility function in I, i.e. Δ^{T}.

Example 2.2. Take for instance Table 2.1, following Example 2.1, we obtain the tolerance relation discernibility matrix $\boldsymbol{M}^{\mathrm{T}}$ in Table 2.1 as Table 2.2 shows.

Table 2.2 Tolerance relation discernibility matrix $\boldsymbol{M}^{\mathrm{T}}$ in Table 2.1

	x_1	x_2	x_3	x_4	x_5	x_6
x_1		P	S, MX	MX	MX	P
x_2	P		S, MX	P, MX	MX	
x_3	S, MX	S, MX		S	S	S
x_4	MX	P, MX	S			P
x_5	MX	MX	S			
x_6	P		S	P		

By Definition 2.3, the tolerance relation discernibility function is

$$\begin{aligned}\Delta^{\mathrm{T}} &= \bigwedge_{DIS^{\mathrm{T}}(x,y)\in \boldsymbol{M}^{\mathrm{T}}} \vee DIS^{\mathrm{T}}(x,y)\\ &= P \wedge MX \wedge S \wedge (S \vee MX) \wedge (P \vee MX)\\ &= P \wedge S \wedge MX\end{aligned}$$

Then by Theorem 2.2, we know that the set of attributes $\{P, S, MX\}$ is the tolerance relation reduct in Table 2.1.

2.1.2 Tolerance Relation Based Rough Set and Generalized Decision Reduct

Similar to the classical rough set approach, Kryszkiewicz [6] also presented her rough set model in the incomplete information system by using the tolerance relation. It should be noticed that Kryszkiewicz's expanded rough set model is constructed through the "point-based" approach (see Remark 1.1).

Definition 2.4. [6] Let I be an incomplete information system in which $A \subseteq AT$, $\forall X \subseteq U$, the lower approximation and upper approximation of X in terms of the tolerance relation $TOL(A)$ are denoted by $\underline{TOL_A}(X)$ and $\overline{TOL_A}(X)$, respectively, where

$$\underline{TOL_A}(X) = \{x \in U : TOL_A(x) \subseteq X\} \tag{2.6}$$

$$\overline{TOL_A}(X) = \{x \in U : TOL_A(x) \cap X \neq \varnothing\} \tag{2.7}$$

The pair $\left[\underline{TOL_A}(X), \overline{TOL_A}(X)\right]$ is referred to as a rough set of X in terms of the tolerance relation $TOL(A)$.

Proposition 2.2. [6] *Let I be an incomplete information system in which $A \subseteq AT$, $\forall X \subseteq U$, we have*

$$\underline{TOL_A}(X) \subseteq \underline{TOL_{AT}}(X); \tag{2.8}$$

$$\overline{TOL_A}(X) \supseteq \overline{TOL_{AT}}(X). \tag{2.9}$$

Proposition 2.2 shows that the tolerance relation based lower and upper approximations are monotonous with respect to the variety of the attributes.

Based on the tolerance relation based rough set, Kryszkiewicz [6] introduced the concept of generalized decision reduct into the incomplete decision system. First, let us review the concept of the generalized decision.

Definition 2.5. [6] Let I be an incomplete decision system, define a function

$$\partial^{\mathrm{T}}_{AT} : U \rightarrow P(V_d) \tag{2.10}$$

as follows:

$$\partial^{\mathrm{T}}_{AT}(x) = \{f(y, d) : y \in TOL_A(x)\} \tag{2.11}$$

$\partial^{\mathrm{T}}_{AT}$ is referred to as the tolerance relation generalized decision in I, $P(V_d)$ is the power set of the domain of the decision attribute d.

Obviously, if $|\partial_{AT}(x)| = 1$ for each $x \in U$, then I is consistent, otherwise, I is inconsistent. The word "consistent" we used here is to represent the inclusion relation between binary relations $TOL(AT)$ and $IND(\{d\})$, i.e.

$$TOL(AT) \subseteq IND(\{d\}) \Leftrightarrow I \text{ is consistent.}$$

Example 2.3. Let us consider an incomplete decision system showed in Table 2.3, which is an expansion of Table 2.1 by adding the decision attribute d.

Similar to Example 2.1, P, M, S and MX will stand for *Price*, *Mileage*, *Size* and *Max-Speed*, respectively, $AT = \{P, M, S, MX\}$ is the set of condition attributes while d is the decision attribute.

Table 2.3 Incomplete decision system

Car	*Price*	*Mileage*	*Size*	*Max-Speed*	*d*
x_1	high	low	full	low	Good
x_2	low	*	full	low	Good
x_3	*	*	compact	low	Poor
x_4	high	*	full	high	Good
x_5	*	*	full	high	Excel
x_6	low	high	full	*	Good

By Definition 2.5, we obtain the result of the tolerance relation generalized decision in Table 2.3 as Table 2.4 shows.

Obviously, in Table 2.4, we have $|\partial_{AT}(x_4)|, |\partial_{AT}(x_5)|, |\partial_{AT}(x_6)| > 1$, then Table 2.3 is an inconsistent incomplete decision system in terms of the tolerance relation $TOL(AT)$.

Table 2.4 Tolerance relation generalized decision in Table 2.3

Car	∂_{AT}
x_1	{Good}
x_2	{Good}
x_3	{Poor}
x_4	{Good,Excel}
x_5	{Good,Excel}
x_6	{Good,Excel}

Definition 2.6. [6] Let I be an incomplete decision system in which $A \subseteq AT$:

1. if $\partial_A = \partial_{AT}$, then A is referred to as a tolerance relation generalized decision consistent attributes set in I;
2. if A is a tolerance relation generalized decision consistent attributes set in I and $\forall B \subset A$, B is not the tolerance relation generalized decision consistent attributes set in I, then A is referred to as a tolerance relation generalized decision reduct in I;
3. if $\partial_A(x) = \partial_{AT}(x)$, then A is referred to as a tolerance relation generalized decision consistent attributes set for x in I;
4. if A is a tolerance relation generalized decision consistent attributes set for x in I and $\forall B \subset A$, B is not the tolerance relation generalized decision consistent attributes set for x in I, then A is referred to as a tolerance relation generalized decision reduct for x in I.

By Definition 2.6, we can see that

- The tolerance relation generalized decision consistent attributes set in I is a subset of the condition attributes, which preserves the tolerance relation generalized decision function ∂_{AT} in I; the tolerance relation generalized decision reduct in I is a minimal subset of the condition attributes, which preserves the tolerance relation generalized decision function ∂_{AT} in I.
- The tolerance relation generalized decision consistent attributes set for x in I is a subset of the condition attributes, which preserves the tolerance relation generalized decision value $\partial_{AT}(x)$ in I; the tolerance relation generalized decision reduct for x in I is a minimal subset of the condition attributes, which preserves the tolerance relation generalized decision value $\partial_{AT}(x)$ in I .

Similar to the computing of tolerance relation reduct in incomplete information system, we first present the discernibility matrixes, which are corresponding to the above two tolerance relation generalized decision reducts. Given an incomplete decision system I,

$$M_G^{\mathrm{T}} = \{DIS^{\mathrm{T}}(x,y) : x \in U, f(y,d) \notin \partial_{AT}(x)\} \tag{2.12}$$

is referred to as the tolerance relation generalized decision discernibility matrix in I.

$\forall x \in U$,

$$M_G^{\mathrm{T}}(x) = \{DIS^{\mathrm{T}}(x,y) : f(y,d) \notin \partial_{AT}(x)\} \tag{2.13}$$

is referred to as the tolerance relation generalized decision discernibility matrix for x in I.

Theorem 2.3. *Let I be an incomplete decision system in which $A \subseteq AT$, then*

1. *A is a tolerance relation generalized decision consistent attributes set in $I \Leftrightarrow A \cap DIS^{\mathrm{T}}(x,y) \neq \emptyset$ for each $DIS^{\mathrm{T}}(x,y) \in \boldsymbol{M}_G^{\mathrm{T}}$;*
2. *A is a tolerance relation generalized decision consistent attributes set for x in $I \Leftrightarrow A \cap DIS^{\mathrm{T}}(x,y) \neq \emptyset$ for each $DIS^{\mathrm{T}}(x,y) \in \boldsymbol{M}_G^{\mathrm{T}}(x)$.*

Proof. We only prove 1, the proof of 2 is similar to the proof of 1.

$\Rightarrow$: Suppose $\exists DIS^{\mathrm{T}}(x,y) \in \boldsymbol{M}_G^{\mathrm{T}}$ such that $A \cap DIS^{\mathrm{T}}(x,y) = \emptyset$, then $(x,y) \in TOL(A)$, i.e. $y \in TOL_A(x)$, it follows that $f(y,d) \in \partial_A(x)$ by Definition 2.5. Moreover, since $\partial_A = \partial_{AT}$, then we have $\partial_A(x) = \partial_{AT}(x)$, thus $f(y,d) \in \partial_{AT}(x)$, such result is contradictive to the condition because $DIS^{\mathrm{T}}(x,y) \in \boldsymbol{M}_G^{\mathrm{T}} \Rightarrow f(y,d) \notin \partial_{AT}(x)$.

$\Leftarrow$: Since $A \subseteq AT$, then $TOL_A(x) \supseteq TOL_{AT}(x)$, it follows that $\partial_A(x) \supseteq \partial_{AT}(x)$ by Definition 2.5. Therefore, it must be proved that $\partial_A(x) \subseteq \partial_{AT}(x)$ for each $x \in U$. $\forall DIS^{\mathrm{T}}(x,y) \in \boldsymbol{M}_G^{\mathrm{T}}$, we have $f(y,d) \notin \partial_{AT}(x)$. Since $A \cap DIS^{\mathrm{T}}(x,y) \neq \emptyset$, then $(x,y) \notin TOL(A)$, i.e. $f(y,d) \notin \partial_A(x)$, from which we can conclude that $f(y,d) \notin \partial_{AT}(x) \Rightarrow f(y,d) \notin \partial_A(x)$, i.e. $\partial_A(x) \subseteq \partial_{AT}(x)$. □

Definition 2.7. Let I be an incomplete decision system, define

$$\Delta_G^{\mathrm{T}} = \bigwedge_{DIS^{\mathrm{T}}(x,y) \in \boldsymbol{M}_G^{\mathrm{T}}} \vee DIS^{\mathrm{T}}(x,y) \tag{2.14}$$

$$\Delta_G^{\mathrm{T}}(x) = \bigwedge_{DIS^{\mathrm{T}}(x,y) \in \boldsymbol{M}_G^{\mathrm{T}}(x)} \vee DIS^{\mathrm{T}}(x,y) \tag{2.15}$$

Δ_G^{T} is referred to as the tolerance relation generalized decision discernibility function in I, $\Delta_G^{\mathrm{T}}(x)$ is referred to as the tolerance relation generalized decision discernibility function for x in I.

By Boolean Reasoning technique, it is not difficult to obtain the following theorem by Theorem 2.3.

Theorem 2.4. *Let I be an incomplete decision system in which $A \subseteq AT$, then*

1. *A is a tolerance relation generalized decision reduct in I if and only if $\wedge A$ is the prime implicant of the tolerance relation generalized decision discernibility function in I, i.e. Δ_G^{T};*
2. *A is a tolerance relation generalized decision reduct for x in I if and only if $\wedge A$ is the prime implicant of the tolerance relation generalized decision discernibility function for x in I, i.e. $\Delta_G^{\mathrm{T}}(x)$.*

Example 2.4. Take for instance Table 2.3, following Example 2.3, we then obtain the tolerance relation generalized decision discernibility matrix $\boldsymbol{M}_G^{\mathrm{T}}$ in Table 2.3 as Table 2.5 shows.

Table 2.5 Tolerance relation generalized decision discernibility matrix $\boldsymbol{M}_G^{\mathrm{T}}$ in Table 2.3

	x_1	x_2	x_3	x_4	x_5	x_6
x_1			S, MX		MX	
x_2			S, MX		MX	
x_3	S, MX	S, MX		S	S	S
x_4			S			
x_5			S			
x_6			S			

By Definition 2.7, the tolerance relation generalized decision discernibility function in Table 2.3 is

$$\begin{aligned}\Delta_G^{\mathrm{T}} &= \bigwedge_{DIS^T(x,y)\in \boldsymbol{M}_G^{\mathrm{T}}} \vee DIS^{\mathrm{T}}(x,y)\\ &= (S \vee MX) \wedge MX \wedge S\\ &= S \wedge MX\end{aligned}$$

Then by Theorem 2.4, we know that set of attributes $\{S, MX\}$ is the tolerance relation generalized decision reduct in Table 2.3.

Similarly, we obtain

$$\Delta_G^{\mathrm{T}}(x_1) = MX$$

$$\Delta_G^{\mathrm{T}}(x_2) = MX$$

$$\Delta_G^{\mathrm{T}}(x_3) = S$$

$$\Delta_G^{\mathrm{T}}(x_4) = S$$

$$\Delta_G^{\mathrm{T}}(x_5) = S$$

$$\Delta_G^{\mathrm{T}}(x_6) = S$$

By the above results and Theorem 2.4, we can conclude that $\{MX\}$ is the tolerance relation generalized decision reduct for objects x_1 and x_2 in Table 2.3, while $\{S\}$ is the tolerance relation generalized decision reduct for objects x_3, x_4, x_5 and x_6 in Table 2.3.

Tolerance relation based rough set provides a direct approach to deal with the incomplete information systems. In such model, the unknown value is considered as to be equivalent to any other values in the domain of the corresponding attribute. From the viewpoint of such idea, many more expanded rough set models have been proposed to overcome the various limitations of the tolerance relation.

2.2 Valued Tolerance Relation Based Rough Set Approach

Following the tolerance relation, Stefanowski et al. [26, 27] further proposed the concept of valued tolerance relation for target approximation in incomplete information system. The valued tolerance approach may desire to express the intuitive idea that x is "more similar" or "less similar" to y than z where $x, y, z \in U$.

For example, in Table 2.1, we have $(x_2, x_6) \in TOL(AT)$ and $(x_5, x_6) \in TOL(AT)$. However, we have an intuitive idea that x_2 is more similar to x_6 than x_5. This is due to the fact that in the case of x_6 only one attribute value is unknown and the rest all are equal to the values in x_2 on the corresponding attributes, while in the case x_5, only one attribute value is equal to that in x_6 and the rest are unknown.

In Stefanowski's valued tolerance approach, assuming that the possible values on each attribute are discrete, then we may make the hypothesis that there exists a uniform probability distribution among such values, i.e. all values have the same probability to be associated to an object $x \in U$. For instance, given an incomplete information system I in which $a \in AT$, suppose that $V_a = \{a_1, a_2, a_3, a_4\}$ is the domain of the attribute a, if $f(x, a) = *$ where $*$ denotes the "do not care" unknown value, then $f(x, a)$ can be written such that

$$f(x, a) = \{a_1/0.25, a_2/0.25, a_3/0.25, a_4/0.25\}$$

It tells us that if the value that x holds on a is unknown, then we assume that x may hold any one of the values in V_a. Moreover, the probabilistic degrees that x holds each value are the same.

2.2.1 Valued Tolerance Relation

Definition 2.8. [26, 27] Let I be an incomplete information system in which $A \subseteq AT$, the valued tolerance relation in terms of A is defined as

$$VTOL(A) = \{VTOL_A(x, y) : x, y \in U\} \tag{2.16}$$

$VTOL_A(x, y)$ is the probability that x is tolerant with y,

$$VTOL_A(x, y) = \prod_{a \in A} VTOL_{(\{a\})}(x, y)$$

where $\forall a \in AT$,

$$VTOL_{(\{a\})}(x, y) = \begin{cases} \frac{1}{|V_a|}, & f(x, a) = * \wedge f(y, a) \neq * \\ \frac{1}{|V_a|}, & f(x, a) \neq * \wedge f(y, a) = * \\ \frac{1}{|V_a|^2}, & f(x, a) = * \wedge f(y, a) = * \wedge x \neq y \\ 0, & f(x, a) \neq * \wedge f(y, a) \neq * \wedge f(x, a) \neq f(y, a) \\ 1, & f(x, a) \neq * \wedge f(y, a) \neq * \wedge f(x, a) = f(y, a) \\ 1, & x = y \end{cases}$$

Obviously, the valued tolerance relation is reflexive and symmetric.

Example 2.5. Take for instance Table 2.1, by Definition 2.8, we then obtain the valued tolerance relation in Table 2.1 as Table 2.6 shows.

Table 2.6 Valued tolerance relation in Table 2.1

	x_1	x_2	x_3	x_4	x_5	x_6
x_1	1	0	0	0	0	0
x_2	0	1	0	0	0	1/4
x_3	0	0	1	0	0	0
x_4	0	0	0	1	1/8	1/4
x_5	0	0	0	1/8	1	1/8
x_6	0	1/4	0	1/4	1/8	1

2.2.2 Valued Tolerance Relation Based Fuzzy Rough Set

The open problem is how to define the concepts of lower and upper approximations by using the valued tolerance relation. The approach adopted by Stefanowski et al. is the well-known fuzzy rough set technique.

Definition 2.9. [26, 27] Let I be an incomplete information system in which $A \subseteq AT$, $\forall X \subseteq U$, the lower approximation and upper approximation of X in terms of the valued tolerance relation are denoted by $\underline{VTOL_A}(X)$ and $\overline{VTOL_A}(X)$, respectively, whose membership functions for each $x \in U$ are defined as

$$\mu_{\underline{VTOL_A}(X)}(x) = \wedge_{y\in U}\big(\mu_X(y) \vee (1 - VTOL_A(x,y))\big)$$
$$= \wedge\big\{1 - VTOL_A(x,y) : y \notin X\big\} \tag{2.17}$$

$$\mu_{\overline{VTOL_A}(X)}(x) = \vee_{y\in U}\big(\mu_X(y) \wedge VTOL_A(x,y)\big)$$
$$= \vee\big\{VTOL_A(x,y) : y \in X\big\} \tag{2.18}$$

The pair $\left[\underline{VTOL_A}(X), \overline{VTOL_A}(X)\right]$ is referred to as the fuzzy rough set of X in terms of the valued tolerance relation $VTOL(A)$.

Example 2.6. In Table 2.3, by the decision attribute d, the universe is partitioned into disjoint subsets such that

$$U/IND(\{d\}) = \big\{\{\text{Poor}\}, \{\text{Good}\}, \{\text{Excel}\}\big\} = \big\{\{x_3\}, \{x_1, x_2, x_4, x_6\}, \{x_5\}\big\}$$

Then by Definition 2.9, we obtain the lower and upper approximations of these decision classes as Table 2.7 shows.

Table 2.7 Valued tolerance relation based lower and upper approximation in Table 2.3

	$\mu_{\underline{VTOL_{AT}}(\text{Poor})}(x)$	$\mu_{\overline{VTOL_{AT}}(\text{Poor})}(x)$	$\mu_{\underline{VTOL_{AT}}(\text{Good})}(x)$
x_1	0	0	1
x_2	0	0	1
x_3	1	1	0
x_4	0	0	7/8
x_5	0	0	0
x_6	0	0	7/8

	$\mu_{\overline{VTOL_{AT}}(\text{Good})}(x)$	$\mu_{\underline{VTOL_{AT}}(\text{Excel})}(x)$	$\mu_{\overline{VTOL_{AT}}(\text{Excel})}(x)$
x_1	1	0	0
x_2	1	0	0
x_3	0	0	0
x_4	1	0	1/8
x_5	1/8	7/8	1
x_6	1	0	1/8

2.3 Maximal Consistent Block Based Rough Set Approach

2.3.1 Maximal Consistent Block and Its Reducts

Though the tolerance relation has been widely used to deal with incomplete information system, it has the following limitations [28]:

1. Firstly, different two tolerance classes may have inclusion relation, for instance, in Table 2.1, $TOL_{AT}(x_2) \subset TOL_{AT}(x_6)$ and $TOL_{AT}(x_4) \subset TOL_{AT}(x_5)$ hold, such case sometimes is unreasonable when defining approximate sets.
2. Secondly, for all objects in $TOL_{AT}(x)$, they may have no common values, for example, in Table 2.1, $TOL_{AT}(x_5) = \{x_4, x_5, x_6\}$, $f(x_4, P) = \{high\}$ while $f(x_6, P) = \{low\}$. Obviously, objects x_4 and x_6 should be distinguished with each other.

From discussions above, a more reasonable classification analysis in incomplete information system has become a necessity. To solve such problem, Leung et al. [29] proposed the maximal consistent block technique, which is adopted from discrete mathematics.

The maximal consistent block describes the maximal collection of objects, in which all objects are tolerant with each other, i.e., they are possibly indiscernible in terms of available information, which are provided by A. Mathematically, the form of maximal consistent block is showed in Definition 2.10.

Definition 2.10. Let I be an incomplete information system in which $A \subseteq AT$, the set of all maximal consistent blocks in terms of A is denoted by $\mathscr{M}(A)$ where

$$\mathscr{M}(A) = \{Y \subseteq U : Y^2 \subseteq TOL(A) \wedge (\forall x \notin Y \rightarrow (Y \cup \{x\})^2 \nsubseteq TOL(A))\} \quad (2.19)$$

Example 2.7. [29] Following the incomplete information system showed in Table 2.1, we obtain the set of all maximal consistent blocks such that

$$\mathcal{M}(AT) = \{Y_1 = \{x_1\}, Y_2 = \{x_2, x_6\}, Y_3 = \{x_3\}, Y_4 = \{x_4, x_5\}, Y_5 = \{x_5, x_6\}\}.$$

Remark 2.1. Generally speaking, all maximal consistent blocks form a covering in stead of a partition on the universe of discourse.

Theorem 2.5. [29] *Let I be an incomplete information system in which $A \subseteq AT$, then $\forall X \in \mathcal{M}(AT)$, there must be $Y \in \mathcal{M}(A)$ such that $X \subseteq Y$.*

Theorem 2.6. [29] Let I be an incomplete information system in which $A \subseteq AT$, then $\forall x \in U$, we have

$$TOL_A(x) = \cup\{Y : Y \in \mathcal{M}(A), x \in Y\} \tag{2.20}$$

Theorem 2.7. [29] *Let I be an incomplete information system in which $A \subseteq AT$, then*

$$Y \in \mathcal{M}(A) \Leftrightarrow Y = \bigcap_{x \in Y} TOL_A(x) \tag{2.21}$$

Theorem 2.5 shows the monotonicity property of maximal consistent block in terms of the variety of the attributes, while Theorem 2.6 and Theorem 2.7 show the relationships between maximal consistent block and tolerance class.

Definition 2.11. [29] Let I be an incomplete information system in which $A \subseteq AT$,

1. if $\mathcal{M}(A) = \mathcal{M}(AT)$, then A is referred to as a maximal consistent block consistent attributes set in I;
2. if A is a maximal consistent block consistent attributes set in I and $\forall B \subset A$, B is not the maximal consistent block consistent attributes set in I, then A is referred to as a maximal consistent block reduct in I.

By Definition 2.11, we can see that the maximal consistent block consistent attributes set in I is a subset of the attributes, which preserves the covering constituted by all maximal consistent blocks; the maximal consistent block reduct in I is a minimal subset of the attributes, which preserves the covering constituted by all maximal consistent blocks.

We can define the following discernibility matrix to compute the maximal consistent block reduct.

Firstly, $\forall X, Y \in \mathcal{M}(AT)$, let us denote

$$DIS(X, Y) = \begin{cases} \{B \subseteq AT : x \in X, y \in Y \text{ s.t. } B \cap DIS^{\mathrm{T}}(x, y) \neq \emptyset\}, & X \neq Y \\ \emptyset, & \text{otherwise} \end{cases}$$

then

$$\boldsymbol{M}^{M} = \{DIS(X, Y) : X, Y \in \mathcal{M}(AT) \wedge X \neq Y\} \tag{2.22}$$

is referred to as the maximal consistent block discernibility matrix in I.

Theorem 2.8. [29] *Let I be an incomplete information system in which $A \subseteq AT$, then we have*

A is a maximal consistent block consistent attributes set in $I \Leftrightarrow A \cap DIS(X,Y) \neq \emptyset$ for each $DIS(X,Y) \in \boldsymbol{M}^M$.

Proof. $\Rightarrow$: Suppose $\exists DIS(X,Y) \in \boldsymbol{M}^M$ such that $A \cap DIS(X,Y) = \emptyset$, then $\forall x \in X, \forall y \in Y, A \cap DIS^{\mathrm{T}}(x,y) = \emptyset$, i.e. x and y are tolerant in terms of A.

By condition, since $DIS(X,Y) \in \boldsymbol{M}^M$, then there must be $B \subseteq AT$ such that $B \cap DIS^{\mathrm{T}}(x,y) \neq \emptyset$, it follows that $DIS^{\mathrm{T}}(x,y) \neq \emptyset$, from which we can conclude that x and y are not tolerant in terms of AT. Moreover, since A is a maximal consistent block consistent attributes set in I, then $\mathcal{M}(AT) = \mathcal{M}(A)$, i.e. x and y are also not tolerant in terms of A, which is contradictive to what have been derived above.

$\Leftarrow$: $\forall X \in \mathcal{M}(AT)$, objects in X are tolerant with each other in terms of the attributes set A because $A \subseteq AT$. If $X \notin \mathcal{M}(A)$, then by Theorem 2.5, we know that there exists $X' \in \mathcal{M}(A)$ satisfying $X \neq X'$ and $y \in X'$ and $y \notin X$, i.e. $X' \supset X$. $y \notin X$ implies that that there exists $x \in X$ and $y \in Y$ ($Y \in \mathcal{M}(AT), X \neq Y$) such that $DIS^{\mathrm{T}}(x,y) \neq \emptyset$. Moreover, since $A \cap DIS(X,Y) \neq \emptyset$ for each $DIS(X,Y) \in \boldsymbol{M}^M$, then we have $A \cap DIS^{\mathrm{T}}(x,y) \neq \emptyset$, which is contradictive to that x and y are in the same block in terms of A because $x,y \in X'$ and $X' \in \mathcal{M}(A)$.

Inversely, $\forall X \in \mathcal{M}(A)$, if $X \notin \mathcal{M}(AT)$, then there must be $x,y \in X$ such that $DIS^{\mathrm{T}}(x,y) \neq \emptyset$. Moreover, it should be noticed that x and y belong to different maximal consistent blocks in terms of AT since $X \notin \mathcal{M}(AT)$. By condition we have $A \cap DIS(X,Y) \neq \emptyset$, i.e. $A \cap DIS^{\mathrm{T}}(x,y) \neq \emptyset$, it follows that $(x,y) \notin T_A$, which is contradictive to the assumption that $x,y \in X \in \mathcal{M}(A)$. □

Definition 2.12. Let I be an incomplete information system, define

$$\Delta^M = \bigwedge_{DIS(X,Y) \in \mathbf{M}} \Delta^M(X,Y) \tag{2.23}$$

where

$$\Delta^M(X,Y) = \bigwedge_{(x,y) X \times Y} \vee DIS^{\mathrm{T}}(x,y) \tag{2.24}$$

Δ^M is referred to as the maximal consistent block discernibility function in I.

By Boolean Reasoning technique, it is not difficult to obtain the following theorem by Theorem 2.8.

Theorem 2.9. *Let I be an incomplete information system in which $A \subseteq AT$, then*

A is a maximal consistent block reduct in I if and only if $\wedge A$ is the prime implicant of the maximal consistent block discernibility function Δ^M.

Example 2.8. Take for instance Table 2.1, following Example 2.7, we obtain the the maximal consistent block discernibility matrix $\boldsymbol{M}^M$ in Table 2.1 as Table 2.8 shows.

Table 2.8 Maximal consistent block discernibility matrix $\boldsymbol{M}^M$ in Table 2.1

	Y_1	Y_2	Y_3	Y_4	Y_5
Y_1		$\{P\},\{P,M\}$	$\{S\}$	$\{MX\}$	$\{MX\},\{P,M\}$
Y_2	$\{P\},\{P,M\}$		$\{S\}$	$\{P\},\{MX\}\{P,MX\}$	$\{MX\}$
Y_3	$\{S\}$	$\{S\}$		$\{S,MX\}$	$\{S\},\{S,MX\}$
Y_4	$\{MX\}$	$\{P\},\{MX\}\{P,MX\}$	$\{S,MX\}$		$\{P\}$
Y_5	$\{MX\},\{P,M\}$	$\{MX\}$	$\{S\},\{S,MX\}$	$\{P\}$	

By Definition 2.12, the maximal consistent block discernibility function is

$$\Delta^M = \bigwedge_{DIS(X,Y)\in \boldsymbol{M}} \Delta^M(X,Y)$$
$$= P \wedge S \wedge MX$$

By Theorem 2.9, we know that the set of attributes $\{P, S, MX\}$ is the maximal consistent block reduct in Table 2.1.

2.3.2 Maximal Consistent Block Based Rough Set and Approximate Distribution Reducts

By using the maximal consistent block technique, Leung et al. defined the following rough approximation.

Definition 2.13. [29] Let I be an incomplete information system in which $A \subseteq AT$, $\forall X \subseteq U$, the lower and upper approximations of X in terms of $\mathcal{M}(A)$ are denoted by $\underline{\mathcal{M}_A}(X)$ and $\overline{\mathcal{M}_A}(X)$, respectively, where

$$\underline{\mathcal{M}_A}(X) = \cup\{Y \in \mathcal{M}(A) : Y \subseteq X\} \tag{2.25}$$

$$\overline{\mathcal{M}_A}(X) = \cup\{Y \in \mathcal{M}(A) : Y \cap X \neq \emptyset\} \tag{2.26}$$

The pair $[\underline{\mathcal{M}_A}(X), \overline{\mathcal{M}_A}(X)]$ is referred to as the rough set of X in terms of the maximal consistent blocks in $\mathcal{M}(A)$.

Remark 2.2. It should be noticed that in Definition 2.13, Leung et al. used the "set-based" approach (see Remark 1.1) to define rough approximation. This is a difference between tolerance relation based rough set and maximal consistent block based rough set.

Theorem 2.10. [29] *Let I be an incomplete information system, if $A \subseteq AT$, then $\forall X \subseteq U$, we have*

$$\underline{TOL_A}(X) \subseteq \underline{\mathcal{M}_A}(X) \subseteq X \subseteq \overline{TOL_A}(X) = \overline{\mathcal{M}_A}(X) \tag{2.27}$$

Theorem 2.10 tells us that by comparing with the tolerance relation based rough set model, the maximal consistent block based rough set model can generate greater lower approximation and then the accuracy of approximation will be improved.

Example 2.9. Take for instance Table 2.3, by the decision attribute d, the universe is partitioned into disjoint subsets such that

$$U/IND(\{d\}) = \Big\{\{\text{Poor}\}, \{\text{Good}\}, \{\text{Excel}\}\Big\} = \Big\{\{x_3\}, \{x_1, x_2, x_4, x_6\}, \{x_5\}\Big\}$$

Then by Definition 2.13, the maximal consistent block based lower and upper approximations are

$$\underline{\mathscr{M}_{AT}}(\text{Poor}) = \{x_3\}$$
$$\overline{\mathscr{M}_{AT}}(\text{Poor}) = \{x_3\}$$
$$\underline{\mathscr{M}_{AT}}(\text{Good}) = \{x_1, x_2, x_6\}$$
$$\overline{\mathscr{M}_{AT}}(\text{Good}) = \{x_1, x_2, x_4, x_5, x_6\}$$
$$\underline{\mathscr{M}_{AT}}(\text{Excel}) = \varnothing$$
$$\overline{\mathscr{M}_{AT}}(\text{Excel}) = \{x_4, x_5, x_6\}$$

Following Leung's work, Qian et al. [30] introduced the concept of approximate distribution reduct into maximal consistent block based rough set and then proposed the following definition.

Definition 2.14. [30]Let I be an incomplete decision system in which $A \subseteq AT$, the decision attribute d partitioned the universe into disjoint subsets such that $U/IND\{d\} = \{X_1, X_2, \cdots, X_l\}$, then denote

$$\underline{\mathscr{M}_A}(d) = (\underline{\mathscr{M}_A}(X_1), \underline{\mathscr{M}_A}(X_2), \cdots, \underline{\mathscr{M}_A}(X_l))$$
$$\overline{\mathscr{M}_A}(d) = (\overline{\mathscr{M}_A}(X_1), \overline{\mathscr{M}_A}(X_2), \cdots, \overline{\mathscr{M}_A}(X_l))$$

1. If $\underline{\mathscr{M}_A}(d) = \underline{\mathscr{M}_{AT}}(d)$, then A is referred to as a maximal consistent block lower approximate distribution consistent set in I; if A is a maximal consistent block lower approximate distribution consistent set in I and $\forall B \subset A$, B is not the maximal consistent block lower approximate distribution consistent set in I, then A is referred to as a maximal consistent block lower approximate distribution reduct in I.
2. If $\overline{\mathscr{M}_A}(d) = \overline{\mathscr{M}_{AT}}(d)$, then A is referred to as a maximal consistent block upper approximate distribution consistent set in I; if A is a maximal consistent block upper approximate distribution consistent set in I and $\forall B \subset A$, B is not the maximal consistent block upper approximate distribution consistent set in I, then A is referred to as a maximal consistent block upper approximate distribution reduct in I.

By Definition 2.14, we can see that

- The maximal consistent block lower approximate distribution consistent set in I is a subset of the attributes, which preserves the lower approximations of all decision classes in terms of the maximal consistent block technique; the maximal consistent block lower approximate distribution reduct in I is a *minimal* subset of the attributes, which preserves the lower approximations of all decision classes in terms of the maximal consistent block technique.

- The maximal consistent block upper approximate distribution consistent set in I is a subset of the attributes, which preserves the upper approximations of all decision classes in terms of the maximal consistent block technique; the maximal consistent block upper approximate distribution reduct in I is a *minimal* subset of the attributes, which preserves the upper approximations of all decision classes in terms of the maximal consistent block technique.

Similar to the computing of the maximal consistent block reduct, $\forall X \in \mathcal{M}(AT)$ and $\forall x \in U$, denote

$$L_{AT}^{M}(X) = \{X_j \in U/IND(\{d\}) : X \subseteq X_j\} \quad (2.28)$$

$$U_{AT}^{M}(X) = \{X_j \in U/IND(\{d\}) : X \cap X_j \neq \emptyset\} \quad (2.29)$$

Then $\forall X, Y \in \mathcal{M}(AT)$, define

$$DIS_L^M(X,Y) = \begin{cases} \{a \in AT : (x,y) \notin TOL(\{a\}), x \in X, y \in Y\}, & L_{AT}^M(X) \neq L_{AT}^M(Y) \\ \emptyset, & \text{otherwise} \end{cases}$$

$$DIS_U^M(X,Y) = \begin{cases} \{a \in AT : (x,y) \notin TOL(\{a\}), x \in X, y \in Y\}, & U_{AT}^M(X) \neq U_{AT}^M(Y) \\ \emptyset, & \text{otherwise} \end{cases}$$

then

$$\boldsymbol{M}_L^M = \{DIS_M^L(X,Y) : L_{AT}^M(X) \neq L_{AT}^M(Y) \quad (2.30)$$

$$\boldsymbol{M}_U^M = \{DIS_M^U(X,Y) : U_{AT}^M(X) \neq U_{AT}^M(Y)\} \quad (2.31)$$

are referred to as the maximal consistent block lower approximate and upper approximate distribution discernibility matrixes in I, respectively.

Theorem 2.11. [30] *Let I be an incomplete information system in which $A \subseteq AT$, then*

1. *A is a maximal consistent block lower approximate distribution consistent set in I $\Leftrightarrow A \cap DIS_M^L(X,Y) \neq \emptyset$ for each $DIS_M^L(X,Y) \in \boldsymbol{M}_L^M$;*
2. *A is a maximal consistent block upper approximate distribution consistent set in I $\Leftrightarrow A \cap DIS_M^U(X,Y) \neq \emptyset$ for each $DIS_M^U(X,Y) \in \boldsymbol{M}_U^M$.*

Definition 2.15. Let I be an incomplete decision system, then define

$$\Delta_L^M = \bigwedge_{DIS_M^L(X,Y) \in \boldsymbol{M}_L^M} \vee DIS_M^L(X,Y) \quad (2.32)$$

$$\Delta_U^M = \bigwedge_{DIS_M^U(X,Y) \in \boldsymbol{M}_U^M} \vee DIS_M^U(X,Y) \quad (2.33)$$

Δ_L^M and Δ_U^M are referred to as the maximal consistent block lower approximate and upper approximate distribution discernibility functions, respectively.

By Boolean Reasoning technique, it is not difficult to obtain the following theorem by Theorem 2.11.

Theorem 2.12. *Let I be an incomplete decision system in which $A \subseteq AT$, then*

1. *A is a maximal consistent block lower approximate distribution reduct if and only if $\wedge A$ is the prime implicant of the maximal consistent block lower approximate distribution discernibility function Δ_L^M;*
2. *A is a maximal consistent block upper approximate distribution reduct if and only if $\wedge A$ is the prime implicant of the maximal consistent block upper approximate distribution discernibility function Δ_U^M.*

Example 2.10. Take for instance Table 2.3, by the computation results of Example 2.7 and Example 2.9, we have

$$L_{AT}^M(Y_1) = \{\text{Good}\}$$
$$L_{AT}^M(Y_2) = \{\text{Good}\}$$
$$L_{AT}^M(Y_3) = \{\text{Poor}\}$$
$$L_{AT}^M(Y_4) = \varnothing$$
$$L_{AT}^M(Y_5) = \varnothing$$

We then obtain the maximal consistent block lower approximate discernibility matrix $\boldsymbol{M}_L^M$ in Table 2.3 as Table 2.9 shows.

Table 2.9 Maximal consistent block lower approximate discernibility matrix $\boldsymbol{M}_L^M$ in Table 2.3

	Y_1	Y_2	Y_3	Y_4	Y_5
Y_1			S	MX	P, M, MX
Y_2			S	P, MX	MX
Y_3	S	S		S, MX	S, MX
Y_4	MX	P, MX	S, MX		
Y_5	P, M, MX	MX	S, MX		

By Definition 2.15, the maximal consistent block lower approximate distribution discernibility function is

$$\begin{aligned}\Delta_L^M &= \bigwedge_{DIS_M^L(X,Y)\in \boldsymbol{M}_L^M} \vee DIS_L^M(X,Y)\\ &= S \wedge MX \wedge (P \vee MX) \wedge (S \vee MX) \wedge (P \vee M \vee MX)\\ &= S \wedge MX\end{aligned}$$

Then by Theorem 2.12, we know that $\{S, MX\}$ is the maximal consistent block lower approximate distribution reduct in Table 2.3.

Similarly, it is not difficult to obtain that the set of the attributes $\{S, MX\}$ is also the maximal consistent block upper approximate distribution reduct in Table 2.3.

2.4 Descriptor Based Rough Set

2.4.1 Descriptor and Reduct Descriptor

Besides the maximal consistent block based rough set, Leung et al. [31] also proposed the descriptor based rough set in incomplete information system.

Given an incomplete information system I, the symbols $\wedge$ and $\vee$ can be used to denote the logical connectives "and" (conjunction) and "or" (disjunction), respectively. If $A \subseteq AT$, then any attribute-value pair (a, v) is called an A-atomic property where $a \in A$ and $v \in V_a$. Any A-atomic property or conjunction of different A-atomic properties are called the A-descriptors. An A-descriptor can be denoted by t [31].

Let t be an A-descriptor, the set of attributes occurring in t is denoted by $A(t)$. If t is an A-descriptor and $A(t) = A$, then t is called a full A-descriptor [31].

If (a, v) is the atomic property occurring in t, we simply say that $(a, v) \in t$. Let t_1 and t_2 are two A-descriptors, if for each $(a, v) \in t_1$, we have $(a, v) \in t_2$, that is, t_1 is constructed from a subset of atomic properties occurring in t_2, then we say t_1 is coarser than t_2 or t_2 is finer than t_1 and is denoted by $t_1 \succeq t_2$ or $t_2 \preceq t_1$. If t_1 is constructed from a proper subset of atomic properties occurring in t_2, then we say t_1 is proper coarser than t_2 and is denoted by $t_1 \succ t_2$ or $t_2 \prec t_1$ [31].

Here, suppose that $t = \wedge_{a_i \in A}(a_i, v_i)$ is a full A-descriptor, we denote

$$||t|| = || \wedge_{a_i \in A} (a_i, v_i)|| = \{x \in U : \forall a_i \in A, f(x, a_i) = * \vee f(x, a_i) = v_i\} \tag{2.34}$$

then $||t||$ is referred to as the support set of t [31].

Moreover, let us denote

$$DES(A) = \{t : t \textit{ is an } A\textit{-descriptor}, \ ||t|| \neq \emptyset\} \tag{2.35}$$

$$FDES(A) = \{t : t \in DES(A), A(t) = A\} \tag{2.36}$$

Proposition 2.3. [31] *Let I be an incomplete information system in which $t \in FDES(AT)$, $t' \in DES(AT)$, then we have*

$$||t|| = \bigcap_{(a_i, v_i) \in t} ||(a_i, v_i)|| \tag{2.37}$$

$$t' \succeq t \Rightarrow ||t'|| \supseteq ||t|| \tag{2.38}$$

Proof. It can be derived directly from Eq. (2.34). □

Proposition 2.3 tells us that the support sets of the descriptors are monotonous with respect to the coarser or finer relations between descriptors.

From the viewpoint of descriptors, the universe U could be partitioned into several subsets that may overlap, and the result is denoted by $\mathscr{D}(A)$ such that

$$\mathscr{D}(A) = \{||t|| : t \in FDES(A)\} \tag{2.39}$$

Definition 2.16. [31] Let I be an incomplete information system in which $t \in FDES(AT)$, $t' \in DES(AT)$, then define

1. if $t' \geq t$ and $||t'|| = ||t||$, then t' is referred to as a consistent descriptor of t;
2. if t' is a consistent descriptor of t and $\forall t'' > t'$, t' is not the consistent descriptor of t, then t' is referred to as a reduct descriptor of t.

By Definition 2.16, we can see that a consistent descriptor of t is a conjunction of the atomic properties, which preserves the support set of t; a reduct descriptor of t is a minimal conjunction of the atomic properties, which preserves the support set of t. The reduct descriptor allows us to classify objects with a smallest number of required atomic properties.

To compute the reduct descriptor of t, let us define

$$DIS(t,y) = \begin{cases} \{a \in AT : y \notin ||(a,v)||, (a,v) \in t\}, & y \notin ||t|| \\ \varnothing, & \text{otherwise} \end{cases}$$

then

$$\boldsymbol{M}_t = \{DIS(t,y) : y \notin ||t||\} \tag{2.40}$$

is referred to as the t discernibility matrix in I.

Theorem 2.13. *Let I be an incomplete information system in which $t \in FDES(AT)$, $t' \in DES(AT)$ and $t' \geq t$, then*

t' is a consistent descriptor of $t \Leftrightarrow AT(t') \cap DIS(t,y) \neq \varnothing$ for each $DIS(t,y) \in \boldsymbol{M}_t$.

Proof. $\Rightarrow$: Suppose $\exists DIS(t,y) \in \boldsymbol{M}_t$ such that $AT(t') \cap DIS(t,y) = \varnothing$, then by the definition of $DIS(t,y)$, we have $y \in ||t'||$ because $t' \geq t$. Moreover, since $||t'|| = ||t||$, then $y \in ||t||$, such result is contradictive to the condition $DIS(t,y) \in \boldsymbol{M}_t \Rightarrow y \notin ||t||$.

$\Leftarrow$: Since $t' \geq t$, then $||t'|| \supseteq ||t||$ holds obviously. Therefore, it must be proved that $||t'|| \subseteq ||t||$. For each $y \notin ||t||$, $DIS(t,y) \in \boldsymbol{M}_t$, since $AT(t') \cap DIS(t,y) \neq \varnothing$, then there must be $a \in AT(t')$ such that $y \notin ||(a,v)||$, it follows that $y \notin ||t'||$, from which we can conclude that $y \notin ||t|| \Rightarrow y \notin ||t'||$, i.e. $||t'|| \subseteq ||t||$. That completes the proof. □

Definition 2.17. Let I be an incomplete information system in which $t \in FDES(AT)$, then the t discernibility function is defined as

$$\Delta_t = \bigwedge_{DIS(t,y) \in \boldsymbol{M}_t} \vee DIS(t,y) \tag{2.41}$$

Theorem 2.14. *Let I be an incomplete information system in which $t \in FDES(AT)$, $t' \in DES(AT)$ and $t' \geq t$, then*

t' is a reduct descriptor of t if and only if $AT(t')$ is the prime implicant of the t discernibility function Δ_t.

Proof. $\Rightarrow$: By Theorem 2.13, we have $AT(t') \cap DIS(t,y) \neq \varnothing$ for each $DIS(t,y) \in \boldsymbol{M}_t$. We claim that for each $a \in AT(t')$, there must be $DIS(t,y) \in \boldsymbol{M}_t$ such that $AT(t') \cap DIS(t,y) = \{a\}$. If fact, if $\forall DIS(t,y) \in \boldsymbol{M}_t$, there exist $a \in DIS(t,y)$ such that $|AT(t') \cap DIS(t,y)| > 2$ where $a \in AT(t') \cap DIS(t,y)$, let $t'' > t'$ where $AT(t'') = AT(t') - \{a\}$, then by Theorem 2.13 we know $||t''|| = ||t||$, which contradicts that t' is a reduct descriptor of t, it follows that $AT(t')$ is the prime implicant of the Δ_t.

$\Leftarrow$: If $AT(t')$ is the prime implicant of the Δ_t, then by Definition 2.17, we have $AT(t') \cap DIS(t,y) \neq \emptyset$ for each $DIS(t,y) \in \boldsymbol{M}_t$. Moreover, for each $a \in AT(t')$, there exists $DIS(t,y) \in \boldsymbol{M}_t$ such that $AT(t') \cap DIS(t,y) = \{a\}$. Consequently, $\forall t'' \succ t'$ and $AT(t'') = AT(t') - \{a\}$, $||t''|| \neq ||t||$, we then conclude that t' is a reduct descriptor of t. □

Example 2.11. In Table 2.1, the covering $\mathscr{D}(AT)$ induced by descriptors is:

$\mathscr{D}(AT) = \{||t_1|| = \{x_4, x_5\}, ||t_2|| = \{x_3\}, ||t_3|| = \{x_4, x_5\}, ||t_4|| = \{x_1\}, ||t_5|| = \{x_3\}, ||t_6|| = \{x_5, x_6\}, ||t_7|| = \{x_2, x_6\}, ||t_8|| = \{x_3\}, ||t_9|| = \{x_5\}, ||t_{10}|| = \{x_2\}, ||t_{11}|| = \{x_3\}\}$

where

$$t_1 = (P, \text{high}) \wedge (M, \text{high}) \wedge (S, \text{full}) \wedge (MX, \text{high})$$

$$t_2 = (P, \text{high}) \wedge (M, \text{high}) \wedge (S, \text{compact}) \wedge (MX, \text{low})$$

$$t_3 = (P, \text{high}) \wedge (M, \text{low}) \wedge (S, \text{full}) \wedge (MX, \text{high})$$

$$t_4 = (P, \text{high}) \wedge (M, \text{low}) \wedge (S, \text{full}) \wedge (MX, \text{low})$$

$$t_5 = (P, \text{high}) \wedge (M, \text{low}) \wedge (S, \text{compact}) \wedge (MX, \text{low})$$

$$t_6 = (P, \text{low}) \wedge (M, \text{high}) \wedge (S, \text{full}) \wedge (MX, \text{high})$$

$$t_7 = (P, \text{low}) \wedge (M, \text{high}) \wedge (S, \text{full}) \wedge (MX, \text{low})$$

$$t_8 = (P, \text{low}) \wedge (M, \text{high}) \wedge (S, \text{compact}) \wedge (MX, \text{low})$$

$$t_9 = (P, \text{low}) \wedge (M, \text{low}) \wedge (S, \text{full}) \wedge (MX, \text{high})$$

$$t_{10} = (P, \text{low}) \wedge (M, \text{low}) \wedge (S, \text{full}) \wedge (MX, \text{low})$$

$$t_{11} = (P, \text{low}) \wedge (M, \text{low}) \wedge (S, \text{compact}) \wedge (MX, \text{low})$$

Then we can compute the reducts of these descriptors by the discernibility matrix approach.

Take for instance t_1, since $||t_1|| = \{x_4, x_5\}$, then $\forall y \notin ||t_1||$, we have

$$DIS(t_1, x_1) = \{M, MX\}$$

$$DIS(t_1, x_2) = \{P, MX\}$$

$$DIS(t_1, x_3) = \{S, MX\}$$

$$DIS(t_1, x_6) = \{P\}$$

By Definition 2.17, the t_1 discernibility function is

$$\begin{aligned} \Delta_{t_1} &= \bigwedge_{DIS(t_1,y)\in \boldsymbol{M}_{t_1}} \vee DIS(t_1, y) \\ &= P \wedge (M \vee MX) \wedge (P \vee MX) \wedge (S \vee MX) \\ &= (P \wedge MX) \vee (P \wedge S \wedge M) \end{aligned}$$

Then by Theorem 2.14, we know $(P, \text{high}) \wedge (MX, \text{high})$ and $(P, \text{high}) \wedge (M, \text{high}) \wedge (S, \text{full})$ are two reduct descriptors of t_1.

Similarly, we obtain the following results:

1. $(P, \text{high}) \wedge (M, \text{high}) \wedge (MX, \text{low})$ and $(S, \text{compact})$ are two reduct descriptors of t_2;
2. $(P, \text{high}) \wedge (MX, \text{high})$ and $(M, \text{low}) \wedge (MX, \text{high})$ are two reduct descriptors of t_3;
3. $(P, \text{high}) \wedge (S, \text{full}) \wedge (MX, \text{low})$ is the reduct descriptor of t_4;
4. $(S, \text{compact})$ is the reduct descriptor of t_5;
5. $(P, \text{low}) \wedge (MX, \text{high})$ is the reduct descriptor of t_6;
6. $(P, \text{low}) \wedge (S, \text{full}) \wedge (MX, \text{low})$ and $(M, \text{high}) \wedge (S, \text{full}) \wedge (MX, \text{low})$ are two reduct descriptors of t_7;
7. $(S, \text{compact})$ is the reduct descriptor of t_8;
8. $(P, \text{low}) \wedge (M, \text{low}) \wedge (MX, \text{high})$ is the reduct descriptor of t_9;
9. t_{10} is the reduct descriptor of itself;
10. $(S, \text{compact})$ is the reduct descriptor of t_{11}.

2.4.2 Descriptor Based Rough Set and Generalized Decision Reduct of Descriptor

Following the concept of support set of the descriptor, Leung et al. [31] further presented the descriptor based rough approximation as Definition 2.18 shows.

Definition 2.18. [31] Let I be an incomplete information system in which $A \subseteq AT$, then $\forall X \subseteq U$, the lower and upper approximations of X in terms of the descriptors are denoted by $\underline{DES_A}(X)$ and $\overline{DES_A}(X)$, respectively, where

$$\underline{DES_A}(X) = \{||t|| : ||t|| \subseteq X, t \in FDES(A)\} \tag{2.42}$$

$$\overline{DES_A}(X) = \{||t|| : ||t|| \cap X \neq \emptyset, t \in FDES(A)\} \tag{2.43}$$

The pair $\left[\underline{DES_A}(X), \overline{DES_A}(X)\right]$ is referred to as the rough set of X in terms of the descriptors in $FDES(A)$.

Remark 2.3. It should be noticed that in Definition 2.18, the lower and upper approximations are not subsets of the universe, but the families of the subsets of the universe of discourse.

Theorem 2.15. [31] *Let I be an incomplete information system in which $A \subseteq AT$, $\forall t \in FDES(A)$, there must be $x \in U$ such that $||t|| \in TOL_A(x)$.*

Theorem 2.16. [32] *Let I be an incomplete information system in which $A \subseteq AT$, then $\forall X \subseteq U$, we have*

$$\underline{TOL_A}(X) \subseteq \cup\underline{DES_A}(X) \subseteq X \subseteq \overline{TOL_A}(X) = \cup\overline{DES_A}(X) \tag{2.44}$$

Theorem 2.16 tells us that by comparing with the tolerance relation based rough set model, the descriptor based rough set model can generate greater lower approximations and then the accuracy of approximation will be improved.

Example 2.12. Take for instance Table 2.3, since

$$U/IND(\{d\}) = \Big\{\{\text{Poor}\}, \{\text{Good}\}, \{\text{Excel}\}\Big\} = \Big\{\{x_3\}, \{x_1, x_2, x_4, x_6\}, \{x_5\}\Big\}$$

then by the tolerance relation based rough set showed in Definition 2.4, we have

$$\begin{aligned}
\underline{TOL_{AT}}(\text{Poor}) &= \{x_3\}\\
\overline{TOL_{AT}}(\text{Poor}) &= \{x_3\}\\
\underline{TOL_{AT}}(\text{Good}) &= \{x_1, x_2\}\\
\overline{TOL_{AT}}(\text{Good}) &= \{x_1, x_2, x_4, x_5, x_6\}\\
\underline{TOL_{AT}}(\text{Excel}) &= \emptyset\\
\overline{TOL_{AT}}(\text{Excel}) &= \{x_4, x_5, x_6\}
\end{aligned}$$

On the other hand, if the descriptor approach is employed, then we obtain the following descriptor based lower and upper approximations:

$$\begin{aligned}
\underline{DES_{AT}}(\text{Poor}) &= \{||t_2||, ||t_5||, ||t_{11}||\} = \Big\{\{x_3\}\Big\}\\
\overline{DES_{AT}}(\text{Poor}) &= \{||t_2||, ||t_5||, ||t_{11}||\} = \Big\{\{x_3\}\Big\}\\
\underline{DES_{AT}}(\text{Good}) &= \{||t_4||, ||t_7||, ||t_{10}||\} = \Big\{\{x_1\}, \{x_2, x_6\}, \{x_2\}\Big\}\\
\overline{DES_{AT}}(\text{Good}) &= \{||t_1||, ||t_3||, ||t_4||, ||t_6||, ||t_7||, ||t_{10}||\}\\
&= \Big\{\{x_4, x_5\}, \{x_1\}, \{x_5, x_6\}, \{x_2, x_6\}, \{x_2\}\Big\}\\
\underline{DES_{AT}}(\text{Excel}) &= \{||t_9||\} = \Big\{\{x_5\}\Big\}\\
\overline{DES_{AT}}(\text{Excel}) &= \{||t_1||, ||t_3||, ||t_6||, ||t_9||\} = \Big\{\{x_4, x_5\}, \{x_5, x_6\}, \{x_5\}\Big\}
\end{aligned}$$

Obviously, the above results demonstrate the correctness of Theorem 2.16.

Definition 2.19. Let I be an incomplete decision system in which $t \in FDES(AT)$, $t' \in DES(AT)$, the partition determined by the decision attribute d is $U/IND(\{d\}) = \{X_1, X_2, \cdots, X_l\}$, define the generalized decision function of t such that

$$t^G = \{X_i : X_i \in U/IND(\{d\}), ||t|| \cap X_i \neq \emptyset\} \tag{2.45}$$

define

1. if $t^G = t'^G$, then t' is referred to as a generalized decision consistent descriptor of t;
2. if t' is a generalized decision consistent descriptor of t and $t'' \succ t'$, t'' is not the generalized decision consistent descriptor of t, then t' is referred to as a generalized decision reduct descriptor of t.

By Definition 2.19, we can see that a generalized decision consistent descriptor of t is a conjunction of the atomic properties, which preserves the generalized decision function of t; a generalized decision reduct descriptor of t is a minimal conjunction of the atomic properties, which preserves the generalized decision function of t.

To compute the generalized decision reduct descriptor of t, let us define

$$DIS^G(t,y) = \begin{cases} \{a \in AT : y \notin \|(a,v)\|, (a,v) \in t\}, & X_i \notin t^G, y \in X_i \\ \varnothing, & \text{otherwise} \end{cases}$$

then

$$\boldsymbol{M}_t^G = \{DIS^G(t,y) : X_i \notin t^G, y \in X_i\} \tag{2.46}$$

are referred to as the t generalized decision discernibility matrix in I.

Theorem 2.17. *Let I be an incomplete decision system in which $t \in FDES(AT)$, $t' \in DES(AT)$ and $t' \geq t$, then*

t' is a generalized decision consistent descriptor of $t \Leftrightarrow AT(t') \cap DIS^G(t,y) \neq \varnothing$ for each $DIS^G(t,y) \in \boldsymbol{M}_t^G$.

Proof. $\Rightarrow$: Suppose $\exists DIS^G(t,y) \in \boldsymbol{M}_t^G$ such that $AT(t') \cap DIS^G(t,y) = \varnothing$, then we have $y \in \|t'\|$ because $t' \succ t$. Moreover, by condition we have $t^G = t'^G$, i.e. $\|t\| \cap X_i = \varnothing \Leftrightarrow \|t'\| \cap X_i = \varnothing$ $(X_i \notin t^G)$, we know $y \notin X_i$, such result is contradictive to the assumption because $DIS^G(t,y) \in \boldsymbol{M}_t^G \Rightarrow y \in X_i$.

$\Leftarrow$: Since $t' \geq t$, then $\|t'\| \supseteq \|t\|$, from which we obtain $t^G \subseteq t'^G$. Thus, it must be proved that $t'^G \subseteq t^G$. $\forall X_i \notin t^G$, we have $\|t\| \cap X_i = \varnothing$. By condition we have $AT(t') \cap DIS^G(t,y) \neq \varnothing$ where $y \in X_i$, then there must be $a \in AT(t')$ such that $y \notin \|a,v\|$, it follows that $y \notin \|t'\|$. Since y is arbitrary in X_i, then we obtain $\|t'\| \cap X_i = \varnothing$, i.e. $X_i \notin t'^G$, it follows that $t'^G \subseteq t^G$. □

Definition 2.20. Let I be an incomplete decision system in which $t \in FDES(AT)$, then the t generalized decision discernibility function is defined as

$$\Delta_t^G = \bigwedge_{DIS^G(t,y) \in \boldsymbol{M}_t^G} \vee DIS^G(t,y) \tag{2.47}$$

Theorem 2.18. *Let I be an decision information system in which $t \in FDES(AT)$, $t' \in DES(AT)$ and $t' \geq t$, then*

t' is a generalized decision reduct descriptor of t if and only if $AT(t')$ is the prime implicant of the t generalized decision discernibility function Δ_t^G.

Proof. The proof of Theorem 2.18 is similar to the proof of Theorem 2.14. □

Example 2.13. Following Example 2.11 and Example 2.12, we have

$$
\begin{aligned}
t_1^G &= \{\text{Good,Excel}\}\\
t_2^G &= \{\text{Poor}\}\\
t_3^G &= \{\text{Good,Excel}\}\\
t_4^G &= \{\text{Good}\}\\
t_5^G &= \{\text{Poor}\}\\
t_6^G &= \{\text{Good,Excel}\}\\
t_7^G &= \{\text{Good}\}\\
t_8^G &= \{\text{Poor}\}\\
t_9^G &= \{\text{Excel}\}\\
t_{10}^G &= \{\text{Good}\}\\
t_{11}^G &= \{\text{Poor}\}
\end{aligned}
$$

Then by Eq. (2.46), we can obtain the t generalized decision discernibility matrix M_t^G in Table 2.3 as Table 2.10 shows.

By Definition 2.20, the t_1 generalized decision discernibility function is

$$
\begin{aligned}
\Delta_{t_1}^G &= \bigwedge_{DIS^G(t_1,y)\in M_{t_1}^G} \vee DIS^G(t_1,y)\\
&= S \vee MX
\end{aligned}
$$

Table 2.10 t generalized decision discernibility matrix M_t^G in Table 2.3

	x_1	x_2	x_3	x_4	x_5	x_6
t_1			S, MX			
t_2	M, S	P, S		S, MX	S, MX	P, S
t_3			S, MX			
t_4			S		MX	
t_5	S	P, S		S, MX	S, MX	P, M, S
t_6			S, MX			
t_7			S		MX	
t_8	P, M, S	S		P, S, MX	S, MX	S
t_9	P, MX	MX	S, MX	P		M
t_{10}			S		MX	
t_{11}	P, S	S		P, S, MX	S, MX	M, S

By Theorem 2.18, we know that (S, full) and (MX, high) are two generalized decision reduct descriptor of t_1.

Similarly, we obtain the following results:

1. $(S, \text{compact})$ and $(P, \text{high}) \wedge (M, \text{high}) \wedge (MX, \text{low})$ are two generalized decision reduct descriptor of t_2;
2. (S, full) and (MX, high) are two generalized decision reduct descriptor of t_3;

3. $(S, \text{full}) \wedge (MX, \text{low})$ is the generalized decision reduct descriptor of t_4;
4. $(S, \text{compact})$ is the generalized decision reduct descriptor of t_5;
5. (S, full) and (MX, high) are two generalized decision reduct descriptor of t_6;
6. $(S, \text{full}) \wedge (MX, \text{low})$ is the generalized decision reduct descriptor of t_7;
7. $(S, \text{compact})$ is the generalized decision reduct descriptor of t_8;
8. $(P, \text{low}) \wedge (M, \text{low}) \wedge (MX, \text{high})$ is the generalized decision reduct descriptor of t_9;
9. $(S, \text{full}) \wedge (MX, \text{low})$ is the generalized decision reduct descriptor of t_{10};
10. $(S, \text{compact})$ is the generalized decision reduct descriptor of t_{11}.

2.5 Similarity Relation Based Rough Set Approach

2.5.1 Similarity Relation and Similarity Based Rough Set

Obviously, tolerance relation, valued tolerance relation, maximal consistent block and descriptor techniques are all proposed to deal with the incomplete information systems, in which all unknown values are "do not care" conditions. However, it should be noticed that "lost" is also a semantic explanation of the unknown values in incomplete information system. If an unknown value is regarded as lost (denoted by "?"), then it can not be compared with any other values in the domain of the corresponding attribute. From this point of view, Stefanowski et al. [26, 27] proposed the similarity relation as Definition 2.21 shows.

Definition 2.21. [26, 27] Let I be an incomplete information system in which $A \subseteq AT$, the similarity relation in terms of A is denoted by $SIM(A)$ where

$$SIM(A) = \{(x, y) \in U^2 : f(x, a) = ? \vee f(x, a) = f(y, a)\} \tag{2.48}$$

$SIM(A)$ is reflexive and transitive, but not necessarily symmetric. By a similarity relation $SIM(A)$, we then can define for any object $x \in U$ the following two sets [26, 27]:

- $SIM_A(x) = \{y \in U : (y, x) \in SIM(A)\}$ is the set of objects similar to x;
- $SIM_A^{-1}(x) = \{y \in U : (x, y) \in SIM(A)\}$ is the set of objects to which x is similar.

Proposition 2.4. *Let I be an incomplete information system in which $A \subseteq AT$, we have*

$$SIM(AT) \subseteq SIM(A) \tag{2.49}$$

$$y \in SIM_A^{-1}(x) \Leftrightarrow SIM_A^{-1}(y) \subseteq SIM_A^{-1}(x) \tag{2.50}$$

$$SIM_A^{-1}(x) = \cup\{SIM_{AT}^{-1}(y) : y \in SIM_A^{-1}(x)\} \tag{2.51}$$

Proof.
1. Since $A \subseteq AT$, then $SIM(AT) \subseteq SIM(A)$ can be derived directly from Definition 2.21.

2. $\Rightarrow$: $\forall z \in SIM_A^{-1}(y)$, we have $(y,z) \in SIM(A)$. Since $y \in SIM_A^{-1}(x)$, then $(x,y) \in SIM(A)$. Moreover, the similarity relation $SIM(A)$ is transitive, then we can conclude that $(x,z) \in SIM(A)$, i.e. $z \in SIM_A^{-1}(x)$.
   $\Leftarrow$: Since $SIM_A^{-1}(y) \subseteq SIM_A^{-1}(x)$, then $y \in SIM_A^{-1}(x)$ holds obviously because similarity relation is reflexive.
3. $\forall y \in SIM_A^{-1}(x)$, we have $y \in SIM_{AT}^{-1}(y)$ because similarity relation is reflexive. Thus, $SIM_A^{-1}(x) \subseteq \cup\{SIM_{AT}^{-1}(y) : y \in SIM_A^{-1}(x)\}$ holds obviously.
   $\forall SIM_{AT}^{-1}(y) \subseteq \{SIM_{AT}^{-1}(y) : y \in SIM_A^{-1}(x)\}$, since $y \in SIM_A^{-1}(x)$, then by Eq. (2.49) and Eq. (2.50), we have $SIM_{AT}^{-1}(y) \subseteq SIM_A^{-1}(y) \subseteq SIM_A^{-1}(x)$, it follows that $SIM_{AT}^{-1}(y) \subseteq SIM_A^{-1}(x)$, i.e. $\cup\{SIM_{AT}^{-1}(y) : y \in SIM_A^{-1}(x)\} \subseteq SIM_A^{-1}(x)$. □

Definition 2.22. [26, 27] Let I be an incomplete information system in which $A \subseteq AT$, $\forall X \subseteq U$, the lower and upper approximations of X in terms of the similarity relation $SIM(A)$ are denoted by $\underline{SIM_A}(X)$ and $\overline{SIM_A}(X)$ respectively, where

$$\underline{SIM_A}(X) = \{x \in U : SIM_A^{-1}(x) \subseteq X\} \tag{2.52}$$

$$\overline{SIM_A}(X) = \bigcup_{x \in X} SIM_A(x) \tag{2.53}$$

The pair $[\underline{SIM_A}(X), \overline{SIM_A}(X)]$ is referred to as the rough set of X in terms of the similarity relation $SIM(A)$.

Theorem 2.19. *Let I be an incomplete information system in which $A \subseteq AT$, $\forall X \subseteq U$, we have*

$$\overline{SIM_A}(X) = \{x \in U : SIM_A^{-1}(x) \cap X \neq \varnothing\} \tag{2.54}$$

Proof. $\forall x \in U$, if $SIM_A^{-1}(x) \cap X \neq \varnothing$, then there must be $y \in U$ such that $y \in X$ and $y \in SIM_A^{-1}(x)$. Hence $x \in SIM_A(y)$, from which we can conclude that

$$\{x \in U : SIM_A^{-1}(x) \cap X \neq \varnothing\} \subseteq \bigcup_{y \in X} SIM_A(y)$$

Similarly, it is not difficult to prove that

$$\bigcup_{y \in X} SIM_A(y) \subseteq \{x \in U : SIM_A^{-1}(x) \cap X \neq \varnothing\}$$ □

Proposition 2.5. *Let I be an incomplete information system, if $A \subseteq AT$, then $\forall X \subseteq U$, we have*

$$\underline{SIM_A}(X) \subseteq \underline{SIM_{AT}}(X) \tag{2.55}$$

$$\overline{SIM_A}(X) \supseteq \overline{SIM_{AT}}(X) \tag{2.56}$$

Proposition 2.5 shows that the similarity relation based lower and upper approximations are monotonous with respect to the variety of the attributes.

Example 2.14. Table 2.11 shows an incomplete decision system, which was analyzed by Stefanowski et al. in Ref. [26, 27].

In Table 2.11,

- $U = \{x_1, x_2, \cdots, x_{12}\}$ is the universe of discourse;
- $AT = \{a_1, a_2, a_3, a_4\}$ is the set of condition attributes, d is the decision attribute.

Table 2.11 An incomplete decision system with lost unknown values

	a_1	a_2	a_3	a_4	d
x_1	3	2	1	0	Φ
x_2	2	3	2	0	Φ
x_3	2	3	2	0	Ψ
x_4	?	2	?	1	Φ
x_5	?	2	?	1	Ψ
x_6	2	3	2	1	Ψ
x_7	3	?	?	3	Φ
x_8	?	0	0	?	Ψ
x_9	3	2	1	3	Ψ
x_{10}	1	?	?	?	Φ
x_{11}	?	2	?	?	Ψ
x_{12}	3	2	1	?	Φ

By Definition 2.21, we have

$$SIM_{AT}^{-1}(x_1) = \{x_1\}$$

$$SIM_{AT}^{-1}(x_2) = \{x_2, x_3\}$$

$$SIM_{AT}^{-1}(x_3) = \{x_2, x_3\}$$

$$SIM_{AT}^{-1}(x_4) = \{x_4, x_5\}$$

$$SIM_{AT}^{-1}(x_5) = \{x_4, x_5\}$$

$$SIM_{AT}^{-1}(x_6) = \{x_6\}$$

$$SIM_{AT}^{-1}(x_7) = \{x_7, x_9\}$$

$$SIM_{AT}^{-1}(x_8) = \{x_8\}$$

$$SIM_{AT}^{-1}(x_9) = \{x_9\}$$

$$SIM_{AT}^{-1}(x_{10}) = \{x_{10}\}$$

$$SIM_{AT}^{-1}(x_{11}) = \{x_1, x_4, x_5, x_9, x_{11}, x_{12}\}$$

$$SIM_{AT}^{-1}(x_{12}) = \{x_1, x_9, x_{12}\}$$

Moreover, since the decision attribute determine the universe into partition such that

$$U/IND(\{d\}) = \{\Phi, \Psi\} = \Big\{\{x_1, x_2, x_4, x_7, x_{10}, x_{12}\}, \{x_3, x_5, x_6, x_8, x_9, x_{11}\}\Big\}$$

then by Definition 2.22, we obtain the following similarity relation based lower and upper approximations:

$$\underline{SIM_{AT}}(\Phi) = \{x_1, x_{10}\}$$
$$\overline{SIM_{AT}}(\Phi) = \{x_1, x_2, x_3, x_4, x_5, x_7, x_{10}, x_{11}, x_{12}\}$$
$$\underline{SIM_{AT}}(\Psi) = \{x_6, x_8, x_9\}$$
$$\overline{SIM_{AT}}(\Psi) = \{x_2, x_3, x_4, x_5, x_6, x_7, x_8, x_9, x_{11}, x_{12}\}$$

2.5.2 Approximate Distribution Reducts in Similarity Relation Based Rough Set

Definition 2.23. Let I be an incomplete decision system in which $A \subseteq AT$, the decision attribute d partitioned the universe into disjoint subsets such that $U/IND(\{d\}) = \{X_1, X_2, \cdots, X_l\}$, then denote

$$\underline{SIM_A}(d) = \{\underline{SIM_A}(X_1), \underline{SIM_A}(X_2), \cdots, \underline{SIM_A}(X_l)\} \tag{2.57}$$
$$\overline{SIM_A}(d) = \{\overline{SIM_A}(X_1), \overline{SIM_A}(X_2), \cdots, \overline{SIM_A}(X_l)\} \tag{2.58}$$

1. If $\underline{SIM_A}(d) = \underline{SIM_{AT}}(d)$, then A is referred to as a similarity relation lower approximate distribution consistent attributes set in I; if A is a similarity relation lower approximate distribution consistent attributes set and $\forall B \subset A$, B is not the similarity relation lower approximate distribution consistent attributes set, then A is referred to as a similarity relation lower approximate distribution reduct in I.
2. If $\overline{SIM_A}(d) = \overline{SIM_{AT}}(d)$, then A is referred to as a similarity relation upper approximate distribution consistent attributes set in I; if A is a similarity relation upper approximate distribution consistent attributes set and $\forall B \subset A$, B is not the similarity relation upper approximate distribution consistent attributes set, then A is referred to as a similarity relation upper approximate distribution reduct in I.

By Definition 2.23, we can see that

- The similarity relation lower approximate distribution consistent attributes set in I is a subset of the attributes, which preserves the lower approximations of all decision classes in terms of the similarity relation; the similarity relation lower approximate distribution reduct in I is a minimal subset of the attributes, which preserves the lower approximations of all decision classes in terms of the similarity relation.
- The similarity relation upper approximate distribution consistent attributes set in I is a subset of the attributes, which preserves the upper approximations of all decision classes in terms of the similarity relation; the similarity relation upper approximate distribution reduct in I is a minimal subset of the attributes, which preserves the upper approximations of all decision classes in terms of the similarity relation.

Theorem 2.20. *Let I be an incomplete decision system in which $A \subseteq AT$, $\forall x \in U$, denote*

$$L^S_{AT}(x) = \{X_i : X_i \in U/IND(\{d\}) : x \in \underline{SIM_{AT}}(X_i)\} \tag{2.59}$$

$$U^S_{AT}(x) = \{X_i : X_i \in U/IND(\{d\}) : x \in \overline{SIM_{AT}}(X_i)\} \tag{2.60}$$

then

1. *A is a similarity relation lower approximate distribution consistent attributes set $\Leftrightarrow L^S_A(x) = L^S_{AT}(x)$ for each $x \in U$;*
2. *A is a similarity relation upper approximate distribution consistent attributes set $\Leftrightarrow U^S_A(x) = U^S_{AT}(x)$ for each $x \in U$.*

Proof. It can be derived directly from Definition 2.23. □

Theorem 2.21. *Let I be an incomplete decision system in which $A \subseteq AT$, then*

1. *A is a similarity relation lower approximate distribution consistent attributes set $\Leftrightarrow \forall x, y \in U$, if $L^S_{AT}(x) \nsubseteq L^S_{AT}(y)$, then $SIM(A)^{-1}(y) \nsubseteq SIM(A)^{-1}(x)$;*
2. *A is a similarity relation upper approximate distribution consistent attributes set $\Leftrightarrow \forall x, y \in U$, if $U^S_{AT}(y) \nsubseteq U^S_{AT}(x)$, then $SIM(A)^{-1}(y) \nsubseteq SIM(A)^{-1}(x)$.*

Proof.

1. $\Rightarrow$: If $SIM^{-1}_A(y) \subseteq SIM^{-1}_A(x)$, then by the definition of lower approximation, we know that $L^S_A(x) \subseteq L^S_A(y)$. Since A is a similarity relation lower approximate distribution consistent attributes set, then $L^S_A(x) = L^S_{AT}(x)$ and $L^S_A(y) = L^S_{AT}(y)$, it follows that $L^S_{AT}(x) \subseteq L^S_{AT}(y)$.

   $\Leftarrow$: By condition we have $SIM^{-1}_A(y) \subseteq SIM^{-1}_A(x) \Rightarrow L^S_{AT}(x) \subseteq L^S_{AT}(y)$. Since $A \subseteq AT$, then $L^S_A(x) \subseteq L^S_{AT}(x)$ holds obviously. It must be proved that $L^S_{AT}(x) \subseteq L^S_A(x)$. If $L^S_{AT}(x) = \varnothing$, then $L^S_{AT}(x) \subseteq L^S_A(x)$ holds obviously. Otherwise, $\forall X_i \in L^S_{AT}(x)$, we have $SIM^{-1}_{AT}(x) \subseteq X_i$. $\forall y \in U$, if $SIM^{-1}_A(y) \subseteq SIM^{-1}_A(x)$, then by Eq. (2.51), we have $SIM^{-1}_A(x) = \cup\{SIM^{-1}_{AT}(y) : y \in SIM^{-1}_A(x)\}$. Moreover, by $SIM^{-1}_A(y) \subseteq SIM^{-1}_A(x) \Rightarrow L^S_{AT}(x) \subseteq L^S_{AT}(y)$ we know $X_i \in L^S_{AT}(y)$, i.e. $SIM^{-1}_{AT}(y) \subseteq X_i$. From discussion above, we have $SIM^{-1}_A(x) = \cup\{SIM^{-1}_{AT}(y) : y \in SIM^{-1}_A(x)\} \subseteq X_i$, i.e. $L^S_{AT}(x) \subseteq L^S_A(x)$.
2. $\Rightarrow$: If $SIM^{-1}_A(y) \subseteq SIM^{-1}_A(x)$, then by Theorem 2.20, we know that $U^S_A(y) \subseteq U^S_A(x)$. Since A is a similarity relation upper approximate distribution consistent attributes set, then $U^S_A(x) = U^S_{AT}(x)$ and $U^S_A(y) = U^S_{AT}(y)$, it follows that $U^S_{AT}(y) \subseteq U^S_{AT}(x)$.

   $\Leftarrow$: By condition we have $SIM^{-1}_A(y) \subseteq SIM^{-1}_A(x) \Rightarrow U^S_{AT}(y) \subseteq U^S_{AT}(x)$. Since $A \subseteq AT$, then $U^S_A(x) \supseteq U^S_{AT}(x)$ obviously. It must be proved that $U^S_A(x) \subseteq U^S_{AT}(x)$. $\forall X_i \in U^S_A(x)$, we have $SIM^{-1}_A(x) \cap X_i \neq \varnothing$. $\forall y \in SIM^{-1}_A(x) \cap X_i$, $y \in SIM^{-1}_A(x)$ and $y \in X_i$, then we have $SIM^{-1}_A(y) \subseteq SIM^{-1}_A(x)$ by Eq. (2.50). Since $y \in SIM^{-1}_{AT}(y) \cap X_i$, then $X_i \in U^S_{AT}(y)$, by $U^S_{AT}(y) \subseteq U^S_{AT}(x)$ we have $X_i \in U^S_{AT}(x)$, it follows that $U^S_A(x) \subseteq U^S_{AT}(x)$. □

Let us denote

$$DIS_L^S(x,y)=\begin{cases}\{a\in AT:(x,y)\notin SIM(\{a\})\}, & L_{AT}^S(x)\not\subseteq L_{AT}^S(y)\\ \varnothing, & \text{otherwise}\end{cases}$$

$$DIS_U^S(x,y)=\begin{cases}\{a\in AT:(x,y)\notin SIM(\{a\})\}, & U_{AT}^S(y)\not\subseteq U_{AT}^S(x)\\ \varnothing, & \text{otherwise}\end{cases}$$

then

$$\boldsymbol{M}_L^S=\{DIS_L^S(x,y):L_{AT}^S(x)\not\subseteq L_{AT}^S(y)\} \tag{2.61}$$
$$\boldsymbol{M}_U^S=\{DIS_U^S(x,y):U_{AT}^S(y)\not\subseteq U_{AT}^S(x)\} \tag{2.62}$$

are referred to as the similarity relation lower approximate and upper approximate distribution discernibility matrixes in I, respectively.

Theorem 2.22. *Let I be an incomplete decision system in which $A\subseteq AT$, then*

1. *A is a similarity relation lower approximate distribution consistent attributes set $\Leftrightarrow A\cap DIS_L^S(x,y)\neq\varnothing$ for each $DIS_L^S(x,y)\in\boldsymbol{M}_L^S$;*
2. *A is a similarity relation upper approximate distribution consistent attributes set $\Leftrightarrow A\cap DIS_U^S(x,y)\neq\varnothing$ for each $DIS_U^S(x,y)\in\boldsymbol{M}_U^S$.*

Proof. We only prove 1, the proof of 2 is similar to the proof of 1.

$\Rightarrow$: Since A is a similarity relation lower approximate distribution consistent attributes set, by Theorem 2.21, $\forall x,y\in U$, if $L_{AT}^S(x)\not\subseteq L_{AT}^S(y)$, then $SIM_A^{-1}(y)\not\subseteq SIM_A^{-1}(x)$, that is, there must be $z\in U$ such that $z\in SIM_A^{-1}(y)$ and $z\notin SIM_A^{-1}(x)$. Suppose that $(x,y)\in SIM(A)$, then we have $(x,z)\in SIM(A)$ because $z\in SIM_A^{-1}(y)$ and similarity relation is transitive. Such result is contradictive to $z\notin SIM_A^{-1}(x)$, from which we can conclude that there must be $a\in A$ such that $(x,y)\notin SIM(\{a\})$, i.e. $A\cap DIS_L^S(x,y)\neq\varnothing$.

$\Leftarrow$: Since $DIS_L^S(x,y)\in\boldsymbol{M}_L^S$, then $L_{AT}^S(x)\not\subseteq L_{AT}^S(y)$. By condition we have $A\cap DIS_L^S(x,y)\neq\varnothing$, then there must be $a\in A$ such that $(x,y)\notin SIM(\{a\})$, $y\notin SIM_A^{-1}(x)$. Due to the reflexive of similarity relation, $y\in SIM_A^{-1}(y)$, thus $SIM_A^{-1}(y)\not\subseteq SIM_A^{-1}(x)$, i.e. $L_{AT}^S(x)\not\subseteq L_{AT}^S(y)\Rightarrow SIM_A^{-1}(y)\not\subseteq SIM_A^{-1}(x)$, by Theorem 2.21, we know that A is a similarity lower approximate distribution consistent attributes set. □

Definition 2.24. Let I be an incomplete decision system in which $A\subseteq AT$, define

$$\varDelta_L^S=\bigwedge_{DIS_L^S(x,y)\in\boldsymbol{M}_L^S}\vee DIS_L^S(x,y) \tag{2.63}$$

$$\varDelta_U^S=\bigwedge_{DIS_U^S(x,y)\in\boldsymbol{M}_U^S}\vee DIS_U^S(x,y) \tag{2.64}$$

$\varDelta_L^S$ is referred to as the similarity relation lower approximate distribution discernibility function in I, $\varDelta_U^S$ is referred to as the similarity relation upper approximate distribution discernibility function in I.

By Boolean Reasoning technique, it is not difficult to obtain the following theorem by Theorem 2.22.

Theorem 2.23. *Let I be an incomplete decision system in which $A \subseteq AT$, then*

1. *A is a similarity relation lower approximate distribution reduct in I if and only if $\wedge A$ is the prime implicant of the similarity relation lower approximate distribution discernibility function in I, i.e. Δ_L^S;*
2. *A is a similarity relation upper approximate distribution reduct in I if and only if $\wedge A$ is the prime implicant of the similarity relation upper approximate distribution discernibility function in I, i.e. Δ_U^S.*

Example 2.15. Take for instance Table 2.11, following Example 2.14, we obtain $L_{AT}^S(x)$ and $U_{AT}^S(x)$ for each object in Table 2.11 as Table 2.12 shows.

Table 2.12 $L_{AT}^S(x)$ and $U_{AT}^S(x)$ for each object in Table 2.11

	$L_{AT}^S(x)$	$U_{AT}^S(x)$
x_1	Φ	Φ
x_2	$\varnothing$	$\{\Phi, \Psi\}$
x_3	$\varnothing$	$\{\Phi, \Psi\}$
x_4	$\varnothing$	$\{\Phi, \Psi\}$
x_5	$\varnothing$	$\{\Phi, \Psi\}$
x_6	Ψ	Ψ
x_7	$\varnothing$	$\{\Phi, \Psi\}$
x_8	Ψ	Ψ
x_9	Ψ	Ψ
x_{10}	Φ	Φ
x_{11}	$\varnothing$	$\{\Phi, \Psi\}$
x_{12}	$\varnothing$	$\{\Phi, \Psi\}$

Then by Eqs. (2.61) and (2.62) , we obtain the similarity relation lower approximate and upper approximate distribution discernibility matrixes in Table 2.11 as Table 2.13 and Table 2.14 show, respectively.

Table 2.13 Similarity relation lower approximate distribution discernibility matrix $\boldsymbol{M}_L^S$ in Table 2.11

	x_1	x_2	x_3	x_4	x_5	x_6	x_7	x_8	x_9	x_{10}	x_{11}	x_{12}
x_1		a_1, a_2, a_3	a_1, a_2, a_3	a_1, a_3, a_4	a_1, a_3, a_4	AT	a_2, a_3, a_4	AT	a_4		a_1, a_3, a_4	a_4
x_2												
x_3												
x_4												
x_5												
x_6	AT	a_4	a_4	a_1, a_2, a_3	a_1, a_2, a_3		AT			AT	AT	AT
x_7												
x_8	a_2, a_3	a_2, a_3	a_2, a_3	a_2, a_3	a_2, a_3		a_2, a_3			a_2, a_3	a_2, a_3	a_2, a_3
x_9	a_4	AT	AT	a_1, a_3, a_4	a_1, a_3, a_4		a_2, a_3			AT	a_1, a_3, a_4	a_4
x_{10}		a_1	a_1	a_1	a_1	a_1	a_1	a_1	a_1		a_1	a_1
x_{11}												
x_{12}												

Table 2.14 Similarity relation upper approximate distribution discernibility matrix $\boldsymbol{M}_U^S$ in Table 2.11

	x_1	x_2	x_3	x_4	x_5	x_6	x_7	x_8	x_9	x_{10}	x_{11}	x_{12}
x_1						AT		a_2, a_3	a_4			
x_2	a_1, a_2, a_3					a_4		a_2, a_3	AT	a_1		
x_3	a_1, a_2, a_3					a_4		a_2, a_3	AT	a_1		
x_4	a_1, a_3, a_4					a_1, a_2, a_3		a_2, a_3	a_1, a_3, a_4	a_1		
x_5	a_1, a_3, a_4					a_1, a_2, a_3		a_2, a_3	a_1, a_3, a_4	a_1		
x_6	AT									a_1		
x_7	a_2, a_3, a_4					AT		a_2, a_3	a_2, a_3	a_1		
x_8	AT									a_1		
x_9	a_4									a_1		
x_{10}						AT		a_2, a_3	AT			
x_{11}	a_1, a_3, a_4					AT		a_2, a_3	a_1, a_3, a_4	a_1		
x_{12}	a_4					AT		a_2, a_3	a_4	a_1		

By Definition 2.24,

$$\begin{aligned}
\Delta_L^S = & \bigwedge_{DIS_L^S(x,y)\in \boldsymbol{M}_L^S} \vee DIS_L^S(x,y) \\
& = a_1 \wedge a_4 \wedge (a_1 \vee a_2 \vee a_3) \wedge (a_1 \vee a_3 \vee a_4) \wedge \\
& \quad (a_2 \vee a_3 \vee a_4) \wedge (a_2 \vee a_3) \wedge AT \\
& = (a_1 \wedge a_2 \wedge a_4) \vee (a_1 \wedge a_3 \wedge a_4)
\end{aligned}$$

$$\begin{aligned}
\Delta_U^S = & \bigwedge_{DIS_U^S(x,y)\in \boldsymbol{M}_U^S} \vee DIS_U^S(x,y) \\
& = a_1 \wedge a_4 \wedge (a_1 \vee a_2 \vee a_3) \wedge (a_1 \vee a_3 \vee a_4) \wedge \\
& \quad (a_2 \vee a_3 \vee a_4) \wedge (a_2 \vee a_3) \wedge AT \\
& = (a_1 \wedge a_2 \wedge a_4) \vee (a_1 \wedge a_3 \wedge a_4)
\end{aligned}$$

Then by Theorem 2.23, we know that the sets of the attributes $\{a_1, a_2, a_4\}$ and $\{a_1, a_3, a_4\}$ are two similarity relation lower approximate distribution reducts in Table 2.11, they are also the similarity upper approximate distribution reducts in Table 2.11.

2.6 Difference Relation Based Rough Set Approach

2.6.1 Difference Relation and Its Reducts

By the similarity relation based rough set approach, the knowledge hidden in data contained in an incomplete decision system may be discovered and expressed in the form of decision rules from each training example $x \in U$ such that

$$r : u \to v \tag{2.65}$$

where $u = \wedge(a, f(x,a))$, $a \in A \subseteq AT \wedge a(x) \neq ?$, and $v = \vee(d, i)$, $i \in V_d$. In such rule, u and v represent the condition and decision part of rule $r : u \to v$ respectively.

In this section, we may desire to induce the negative rules [33, 34] from the incomplete decision system by using the rough set technique. A negative rule in terms of a training example $x \in U$ may be expressed in the form such that

$$r_x : u \rightarrow v \tag{2.66}$$

where $u = \wedge(\sim (a, f(x,a)))$, $a \in A \subseteq AT \wedge a(x) \neq ?$, and $v = \sim (d, i)$, $i \in V_d$.

For each rule r_x, we can define $C_A(r_x)$ the certainty factor of such rule, i.e.

$$C_A(r_x) = \frac{\left| ||u|| \cap ||v|| \right|}{\left| ||u|| \right|} \tag{2.67}$$

where $||u||$ and $||v||$ are support sets of the condition part and decision part of the rule r_x respectively, i.e.

$$||u|| = \{y \in U : \forall a \in A \wedge f(x,a) \neq ?, f(y,a) \neq f(x,a)\}$$
$$||v|| = \{y \in U : f(y,d) \neq i\}$$

The negative rule is different from the decision rules generated from the similarity relation based rough set model because it has a different meaning: if an object does not satisfy with the attribute-value pairs in the condition part of such rule, then we can exclude the decision part from such object in some degree.

An negative rule $r_x : u \rightarrow v$ with $C_A(r_x) = 1$ is referred to as a certain negative rule; while an negative rule $r_x : u \rightarrow v$ with $0 < C_A(r_x) < 1$ is referred to as an possible negative rule.

To generate definite and indefinite negative rules from the incomplete decision table, Yang et al. [35] have proposed the difference relation as Definition 2.25 shows.

Definition 2.25. [35] Let I be an incomplete information system in which $A \subseteq AT$, the difference relation in terms of A is defined as

$$DIF(A) = \{(x,y) \in U^2 : \forall a \in A, f(x,a) = ? \vee f(x,a) \neq f(y,a)\}$$

Obviously, the difference relation $DIF(A)$ is not necessarily reflexive, symmetric and transitive. Thus, it is a generalized binary relation in the incomplete information system without any additional constraints.

Proposition 2.6. *Let I be an incomplete information system in which $A \subseteq AT$, we have*

$$DIF(AT) \subseteq DIF(A) \tag{2.68}$$
$$DIF(A) = \bigcap_{a \in A} DIF(\{a\}) \tag{2.69}$$

$\forall(x, y) \in DIF(A)$, we say that y is $DIF(A)$ related to x, x is a predecessor of y, and y is a successor of x. Therefore, $\forall x \in U$, the difference class of x in terms of $DIF(A)$ is denoted by $DIF_A(x)$ such that

$$DIF_A(x) = \{y \in U : DIF(A)\} \tag{2.70}$$

Remark 2.4. $DIF_A(x)$ may be interpreted as a successor neighborhood of x. It must be noticed that since difference relation $DIF(A)$ is not necessarily reflexive, $DIF_A(x)$ may be an empty set in the incomplete information system.

Example 2.16. Let us consider Table 2.15, which is an incomplete information system with "?" unknown values.

In Table 2.15,

- $U = \{x_1, x_2, \cdots, x_{12}\}$ is the universe of discourse;
- $AT = \{a_1, a_2, a_3, a_4, a_5\}$ is the set of the attributes where $V_{a_1} = V_{a_2} = V_{a_3} = V_{a_4} = V_{a_5} = \{0, 1, 2\}$.

Table 2.15 An example of incomplete information system

U	a_1	a_2	a_3	a_4	a_5
x_1	1	0	?	2	0
x_2	2	2	0	0	?
x_3	0	1	2	?	1
x_4	2	2	0	1	1
x_5	1	0	1	2	0
x_6	2	2	0	?	1
x_7	?	?	1	0	1
x_8	0	1	2	1	?
x_9	1	0	0	2	0
x_{10}	?	0	?	2	0
x_{11}	0	1	1	1	0
x_{12}	0	1	1	1	0

By Definition 2.25, we obtain the following difference classes:

$$DIF_{AT}(x_1) = \{x_2, x_3, x_4, x_6, x_7, x_8\}$$
$$DIF_{AT}(x_2) = \{x_1, x_3, x_5, x_8, x_9, x_{10}, x_{11}, x_{12}\}$$
$$DIF_{AT}(x_3) = \{x_1, x_2, x_5, x_9, x_{10}\}$$
$$DIF_{AT}(x_4) = \{x_1, x_5, x_{10}\}$$
$$DIF_{AT}(x_5) = \{x_2, x_3, x_4, x_6, x_8\}$$
$$DIF_{AT}(x_6) = \{x_1, x_5, x_8, x_{10}, x_{11}, x_{12}\}$$
$$DIF_{AT}(x_7) = \{x_1, x_8, x_9, x_{10}\}$$
$$DIF_{AT}(x_8) = \{x_1, x_2, x_5, x_6, x_7, x_9, x_{10}\}$$
$$DIF_{AT}(x_9) = \{x_3, x_7, x_8\}$$

$$DIF_{AT}(x_{10}) = \{x_2, x_3, x_4, x_6, x_7, x_8\}$$
$$DIF_{AT}(x_{11}) = \{x_2, x_6\}$$
$$DIF_{AT}(x_{12}) = \{x_2, x_6\}$$

Definition 2.26. Let I be an incomplete information system in which $A \subseteq AT$,

1. if $DIF(A) = DIF(AT)$, then A is referred to as a difference relation consistent set in I;
2. if A is a difference relation consistent attributes set in I and $\forall B \subset A$, B is not the difference relation consistent attributes set in I, then B is referred to as a difference relation reduct in I.

By Definition 2.26, we can see that a difference relation consistent attributes set in I is a subset of the attributes, which preserves the original difference relation in I; a difference relation reduct in I is a minimal subset of the attributes, which preserves the original difference relation in I. To present the practical approach to compute such difference relation reduct, we need the following notions.

$$DIS^D(x,y) = \begin{cases} \{a \in AT : (x,y) \notin DIF(\{a\})\}, & (x,y) \notin DIF(AT) \\ \varnothing, & \text{otherwise} \end{cases}$$

$DIS^D(x,y)$ is referred to as the difference discernibility attributes set in I w.r.p. to the pair of objects (x,y). Moreover, the matrix

$$\boldsymbol{M}^D = \{DIS^D(x,y) : (x,y) \notin DIF(AT)\} \tag{2.71}$$

is referred to as the difference relation discernibility matrix in I.

Theorem 2.24. *Let I be an incomplete information system in which $A \subseteq AT$, we have*
A is a difference relation consistent attributes set in $I \Leftrightarrow A \cap DIS^D(x,y) \neq \varnothing$ for each $DIS^D(x,y) \in \boldsymbol{M}^D$.

Proof. $\Rightarrow$: Suppose $\exists DIS^D(x,y) \in \boldsymbol{M}^D$ such that $A \cap DIS^D(x,y) = \varnothing$. Then, by the definition of $DIS^D(x,y)$, we have $(x,y) \in DIF(A)$. Moreover, since A is a difference relation consistent attributes set in I, then $DIF(A) = DIF(AT)$, from which we can conclude that $(x,y) \in DIF(AT)$, which is contradictive to the assumption because $DIS^D(x,y) \in \boldsymbol{M}^D \Rightarrow (x,y) \notin DIF(AT)$.

$\Leftarrow$: By Proposition 2.6, we have $A \subseteq AT \Rightarrow DIF(AT) \subseteq DIF(A)$. Thus, it must be proved that $DIF(A) \subseteq DIF(AT)$. Since for each $DIS^D(x,y) \in \boldsymbol{M}^D(DIS^D(x,y) \in \boldsymbol{M}^D \Rightarrow (x,y) \notin DIF(AT))$, we have $A \cap DIS^D(x,y) \neq \varnothing$, then $(x,y) \notin DIF(A)$, from which we can conclude that $(x,y) \notin DIF(AT) \Rightarrow (x,y) \notin DIF(A)$, i.e. $DIF(A) \subseteq DIF(AT)$. That completes the proof. □

Definition 2.27. Let I be an incomplete information system, define

$$\Delta^D = \bigwedge_{DIS^D(x,y) \in \boldsymbol{M}^D} \vee DIS^D(x,y) \tag{2.72}$$

Δ^D is referred to as the difference relation discernibility function in I.

By Boolean Reasoning technique, it is not difficult to obtain the following theorem by Theorem 2.24.

Theorem 2.25. *Let I be an incomplete information system in which $A \subseteq AT$, then*

A is a difference relation reduct in I if and only if $\wedge A$ is the prime implicant of the difference relation discernibility function in I, i.e. Δ^D.

Example 2.17. Take for instance Table 2.15, by Eq. (2.71), we obtain the difference relation discernibility matrix $\boldsymbol{M}^D$ in Table 2.15 as Table 2.16 shows.

By Definition 2.27, the difference relation discernibility function is

$$\begin{aligned}\Delta^D = & \bigwedge_{DIS^D(x,y)\in \boldsymbol{M}^D} \vee DIS^D(x,y) \\ = & (a_1 \vee a_2 \vee a_4 \vee a_5) \wedge (a_2 \vee a_4 \vee a_5) \wedge (a_1 \vee a_2 \vee a_3 \vee a_4) \wedge \\ & (a_1 \vee a_2 \vee a_3) \wedge (a_1 \vee a_2 \vee a_3 \vee a_5) \wedge (a_1 \vee a_2) \wedge (a_2 \vee a_4 \vee a_5) \wedge \\ & (a_3 \vee a_5) \wedge (a_3 \vee a_4 \vee a_5) \wedge \wedge (a_1 \vee a_2 \vee a_4) \wedge (a_3 \vee a_5) \wedge \\ & (a_1 \vee a_2 \vee a_4) \wedge a_3 \wedge a_4 \wedge a_5 \wedge AT \\ = & (a_1 \wedge a_3 \wedge a_4 \wedge a_5) \vee (a_2 \wedge a_3 \wedge a_4 \wedge a_5)\end{aligned}$$

Then by Theorem 2.25, we know that the sets of the attributes $A_1 = \{a_1, a_3, a_4, a_5\}$ and $A_2 = \{a_2, a_3, a_4, a_5\}$ are two difference relation reducts in Table 2.15.

2.6.2 Rough Set Based on Difference Relation

Given an incomplete decision system, to generate the negative rules by the difference relation $DIF(A)$, the set to be approximated is not the decision class induced by the decision attribute d, but the complement of such decision class. Here, all complements of decision classes do not constitute a partition, but a covering on the universe of discourse. Such covering can be expressed such that

$$\sim \big(U/IND(\{d\})\big) = \{\sim X_1, \sim X_2, \cdots, \sim X_l\} = \{U - X_1, U - X_2, \cdots, U - X_l\}$$

Definition 2.28. Let I be an incomplete decision system, $\forall A \subseteq AT$, then $\forall \sim X_i \in \sim \big(U/IND(\{d\})\big)$, the lower and upper approximations of $\sim X_i$ in terms of the difference relation $DIF(A)$ are denoted by $\underline{DIF_A}(\sim X_i)$ and $\overline{DIF_A}(\sim X_i)$, respectively, where

$$\underline{DIF_A}(\sim X_i) = \{x \in U : DIF_A(x) \neq \varnothing \wedge DIF_A(x) \subseteq (\sim X_i)\} \tag{2.73}$$

$$\overline{DIF_A}(\sim X_i) = \{x \in U : DIF_A(x) \cap (\sim X_i) \neq \varnothing\} \tag{2.74}$$

The pair $[\underline{DIF_A}(\sim X_i), \overline{DIF_A}(\sim X_i)]$ is referred to as the rough set of $\sim X_i$ in terms of the difference relation $DIF(A)$. The boundary region of $\sim X_i$ is then defined by

$$BN_{DIF_A}(\sim X_i) = \overline{DIF_A}(\sim X_i) - \underline{DIF_A}(\sim X_i) \tag{2.75}$$

Table 2.16 Difference relation discernibility matrix $\boldsymbol{M}^D$ in Table 2.15

	x_1	x_2	x_3	x_4	x_5	x_6	x_7	x_8	x_9	x_{10}	x_{11}	x_{12}
x_1	a_1,a_2,a_4,a_5				a_1,a_2,a_4,a_5				a_1,a_2,a_4,a_5	a_2,a_4,a_5	a_5	a_5
x_2		a_1,a_2,a_3,a_4		a_1,a_2,a_3		a_1,a_2,a_3	a_4					
x_3			a_1,a_2,a_3,a_5	a_5		a_5	a_5	a_1,a_2,a_3			a_1,a_2	a_1,a_2
x_4		a_1,a_2,a_3	a_5	AT		a_1,a_2,a_3,a_5	a_5	a_4	a_3		a_4	a_4
x_5	a_1,a_2,a_4,a_5				AT		a_3		a_1,a_2,a_4,a_5	a_2,a_4,a_5	a_3,a_5	a_3,a_5
x_6		a_1,a_2,a_3	a_5	a_1,a_2,a_3,a_5		a_1,a_2,a_3,a_5	a_5		a_3			
x_7		a_4	a_5	a_5	a_3	a_5	a_3,a_4,a_5				a_3	a_3
x_8			a_1,a_2,a_3	a_4				a_1,a_2,a_3,a_4			a_1,a_2,a_4	a_1,a_2,a_4
x_9	a_1,a_2,a_4,a_5	a_3			a_1,a_2,a_4,a_5	a_3			AT	a_2,a_4,a_5	a_5	a_5
x_{10}	a_2,a_4,a_5				a_2,a_4,a_5				a_2,a_4,a_5	a_2,a_4,a_5	a_5	a_5
x_{11}	a_5		a_1,a_2	a_4	a_3,a_5		a_3	a_1,a_2,a_4	a_5	a_5	AT	AT
x_{12}	a_5		a_1,a_2	a_4	a_3,a_5		a_3	a_1,a_2,a_4	a_5	a_5	AT	AT

Remark 2.5. It must be noticed that $\underline{DIF_A}(U) = \overline{DIF_A}(U) = U$ does not hold in general because $DIF_A(x)$ may be an empty set in the incomplete decision system. For example, if $DIF_A(x) = \emptyset$, then $x \notin \underline{DIF_A}(U)$ and $x \notin \overline{DIF_A}(U)$ by Definition 2.28.

Theorem 2.26. *Let I be an incomplete decision system in which* $A \subseteq AT$, $\forall x \in U$, *we have*

1. $x \in \underline{DIF_A}(\sim X_i) \Rightarrow r_x : u \rightarrow v$ *is a certain negative rule;*
2. $x \in \overline{BN_{DIF_A}}(\sim X_i) \Rightarrow r_x : u \rightarrow v$ *is a possible negative rule.*

Proof. We only prove 1, the proof of 2 is similar to the proof of 1.

Suppose that $u = \wedge(\sim (a, f(x,a)))$, $a \in A \subseteq AT \wedge a(x) \neq ?$ and $v = \sim (d,i)$, $i \in V_d$, then we have

$$\begin{aligned} x \in \underline{DIF_A}(\sim X_i) &\Rightarrow DIF_A(x) \neq \emptyset, DIF_A(x) \subseteq \sim X_i \\ &\Rightarrow \forall y \in DIF_A(x), d(y) \neq i \\ &\Rightarrow \forall y \in \|u\|, d(y) \neq i \\ &\Rightarrow C_A(r_x) = 1 \end{aligned}$$

□

By Theorem 2.26, one can induce all certain and possible negative rules from an incomplete decision system by using the lower approximations and boundary regions.

Table 2.17 An example of incomplete decision system

U	a_1	a_2	a_3	a_4	a_5	d
x_1	1	0	?	2	0	1
x_2	2	2	0	0	?	0
x_3	0	1	2	?	1	0
x_4	2	2	0	1	1	0
x_5	1	0	1	2	0	1
x_6	2	2	0	?	1	0
x_7	?	?	1	0	1	2
x_8	0	1	2	1	?	2
x_9	1	0	0	2	0	2
x_{10}	?	0	?	2	0	1
x_{11}	0	1	1	1	0	0
x_{12}	0	1	1	1	0	1

Example 2.18. Let us consider Table 2.17, which is an expansion of Table 2.15 by adding the decision attribute d.

By decision d, the universe can be partitioned into disjoint subsets such that

$$U/(\{d\}) = \{X_0, X_1, X_2\} = \Big\{\{x_2, x_3, x_4, x_6, x_{11}\}, \{x_1, x_5, x_{10}, x_{12}\}, \{x_7, x_8, x_9\}\Big\}$$

Thus, the sets to be approximated are:

$$\begin{aligned} \sim X_0 &= \{x_1, x_5, x_7, x_8, x_9, x_{10}, x_{12}\} \\ \sim X_1 &= \{x_2, x_3, x_4, x_6, x_7, x_8, x_9, x_{11}\} \\ \sim X_2 &= \{x_1, x_2, x_3, x_4, x_5, x_6, x_{10}, x_{11}, x_{12}\} \end{aligned}$$

By Definition 2.28, we obtain the following difference relation based lower and upper approximations:

$$\underline{DIF_{AT}}(\sim X_0) = \{x_4, x_7\}$$
$$\overline{DIF_{AT}}(\sim X_0) = \{x_1, x_2, x_3, x_4, x_5, x_6, x_7, x_8, x_9, x_{10}\}$$
$$\underline{DIF_{AT}}(\sim X_1) = \{x_1, x_5, x_9, x_{10}, x_{11}, x_{12}\}$$
$$\overline{DIF_{AT}}(\sim X_1) = \{x_1, x_2, x_3, x_5, x_6, x_7, x_8, x_9, x_{10}, x_{11}, x_{12}\}$$
$$\underline{DIF_{AT}}(\sim X_2) = \{x_4, x_{11}, x_{12}\}$$
$$\overline{DIF_{AT}}(\sim X_2) = U$$

By Theorem 2.26, we present here the certain negative rules in Table 2.17, which can be generated from the lower approximations $\underline{DIF_{AT}}(\sim X_0)$, $\underline{DIF_{AT}}(\sim X_1)$ and $\underline{DIF_{AT}}(\sim X_2)$. These rules are:

1. $f(x,a) \neq 2 \wedge f(x,b) \neq 2 \wedge f(x,c) \neq 0 \wedge f(x,e) \neq 1 \wedge f(x,f) \neq 1 \rightarrow f(x,d) \neq 0$ // supported by $x_4 \in \underline{DIF_{AT}}(\sim X_0)$;
2. $f(x,c) \neq 1 \wedge f(x,e) \neq 0 \wedge f(x,f) \neq 1 \rightarrow f(x,d) \neq 0$ // supported by $x_7 \in \underline{DIF_{AT}}(\sim X_0)$;
3. $f(x,a) \neq 1 \wedge f(x,b) \neq 0 \wedge f(x,e) \neq 2 \wedge f(x,f) \neq 0 \rightarrow f(x,d) \neq 1$ // supported by $x_1 \in \underline{DIF_{AT}}(\sim X_1)$;
4. $f(x,a) \neq 1 \wedge f(x,b) \neq 0 \wedge f(x,c) \neq 1 \wedge f(x,e) \neq 2 \wedge f(x,f) \neq 0 \rightarrow f(x,d) \neq 1$ // supported by $x_5 \in \underline{DIF_{AT}}(\sim X_1)$;
5. $f(x,a) \neq 1 \wedge f(x,b) \neq 0 \wedge f(x,c) \neq 0 \wedge f(x,e) \neq 2 \wedge f(x,f) \neq 0 \rightarrow f(x,d) \neq 1$ // supported by $x_9 \in \underline{DIF_{AT}}(\sim X_1)$;
6. $f(x,b) \neq 0 \wedge f(x,e) \neq 2 \wedge f(x,f) \neq 0 \rightarrow f(x,d) \neq 1$ // supported by $x_{10} \in \underline{DIF_{AT}}(\sim X_1)$;
7. $f(x,a) \neq 0 \wedge f(x,b) \neq 1 \wedge f(x,c) \neq 1 \wedge f(x,e) \neq 1 \wedge f(x,f) \neq 0 \rightarrow f(x,d) \neq 1$ // supported by $x_{11} \in \underline{DIF_{AT}}(\sim X_1)$;
8. $f(x,a) \neq 0 \wedge f(x,b) \neq 1 \wedge f(x,c) \neq 1 \wedge f(x,e) \neq 1 \wedge f(x,f) \neq 0 \rightarrow f(x,d) \neq 1$ // supported by $x_{12} \in \underline{DIF_{AT}}(\sim X_1)$;
9. $f(x,a) \neq 2 \wedge f(x,b) \neq 2 \wedge f(x,c) \neq 0 \wedge f(x,e) \neq 1 \wedge f(x,f) \neq 1 \rightarrow f(x,d) \neq 2$ // supported by $x_4 \in \underline{DIF_{AT}}(\sim X_2)$;
10. $f(x,a) \neq 0 \wedge f(x,b) \neq 1 \wedge f(x,c) \neq 1 \wedge f(x,e) \neq 1 \wedge f(x,f) \neq 0 \rightarrow f(x,d) \neq 2$ // supported by $x_{11} \in \underline{DIF_{AT}}(\sim X_2)$;
11. $f(x,a) \neq 0 \wedge f(x,b) \neq 1 \wedge f(x,c) \neq 1 \wedge f(x,e) \neq 1 \wedge f(x,f) \neq 0 \rightarrow f(x,d) \neq 2$ // supported by $x_{12} \in \underline{DIF_{AT}}(\sim X_2)$.

2.6.3 Approximate Distribution Reducts in Difference Relation Based Rough Set

Definition 2.29. Let I be an incomplete decision system in which $A \subseteq AT$, $\forall x \in U$, if $DIF_A(x) \neq \emptyset$, then define

$$L^D_{AT}(x) = \left\{ \sim X_i \in \left(\sim \left(U/IND(\{d\}) \right) \right) : x \in \underline{DIF_{AT}}(\sim X_i) \right\}$$

$$U^D_{AT}(x) = \left\{ \sim X_i \in \left(\sim \left(U/IND(\{d\}) \right) \right) : x \in \overline{DIF_{AT}}(\sim X_i) \right\}$$

$$BN^D_{AT}(x) = \left\{ \sim X_i \in \left(\sim \left(U/IND(\{d\}) \right) \right) : x \in BN_{DIF_{AT}}(\sim X_i) \right\}$$

1. If $L^D_A(x) = L^D_{AT}(x)$, then A is referred to as a difference relation lower approximate distribution consistent attributes set for x in I; if A is a difference relation lower approximate distribution consistent attributes set for x in I and $\forall B \subset A$, B is not the difference relation lower approximate distribution consistent attributes set of x in I, then A is referred to as the difference relation lower approximate distribution reduct for x in I.
2. If $U^D_A(x) = U^D_{AT}(x)$, then A is referred to as a difference relation upper approximate distribution consistent attributes set for x in I; if A is a difference relation upper approximate distribution consistent attributes set for x in I and $\forall B \subset A$, B is not the difference relation upper approximate distribution consistent attributes set for x in I, then A is referred to as the difference relation upper approximate distribution reduct for x in I.
3. If $BN^D_A(x) = BN^D_{AT}(x)$, then A is referred to as a difference relation rough distribution consistent attributes set for x in I; if A is a difference relation rough distribution consistent attributes set for x in I and $\forall B \subset A$, B is not the difference relation rough distribution consistent attributes set for x in I, then A is referred to as the difference relation rough distribution reduct for x in I.

By Definition 2.29, we can see that

- The difference relation lower approximate distribution consistent attributes set for x in I is a subset of the attributes, which preserves belongingness of difference class of x to the difference relation based lower approximations of the decision classes; the difference relation lower approximate distribution reduct for x in I is a minimal subset of the attributes, which preserves belongingness of difference class of x to the difference relation based lower approximations of the decision classes; by using the difference relation lower approximate distribution reduct of x, one can generate simplified certain negative rules, which are compatible with the certain negative rules generated from the original decision system.
- The difference relation upper approximate distribution consistent attributes set for x in I is a subset of the attributes, which preserves belongingness of difference class of x to the difference relation based upper approximations of the decision classes; the difference relation upper approximate distribution reduct for x in I is a minimal subset of the attributes, which preserves belongingness of difference class of x to the difference relation based upper approximations of the decision classes.

- The difference relation rough distribution consistent attributes set for x in I is a subset of the attributes, which preserves belongingness of difference class of x to the difference relation based boundary regions of the decision classes; the difference relation rough distribution reduct for x in I is a minimal subset of the attributes, which preserves belongingness of difference class of x to the difference relation based boundary regions of the decision classes; by using the difference relation rough distribution reduct of x, one can generate simplified possible negative rules, which are compatible with the possible negative rules generated from the original decision system.

Remark 2.6. It must be noticed that if $DIF_{AT}(x) = \emptyset$, then $DIF_{AT}(x)$ does not belong to any difference relation based lower (upper) approximations or boundary regions of the decision classes, it follows that no negative rules (supported by x) can be generated from the incomplete decision system. From this point of view, it is useless to compute reducts for x if $DIF_{AT}(x) = \emptyset$.

Remark 2.7. It must be noticed that if $L^D_{AT}(x) = \emptyset(U^D_{AT}(x) = \emptyset, BN^D_{AT}(x) = \emptyset)$, then $DIF_{AT}(x)$ does not belong to any difference relation based lower approximations (upper approximations, boundary regions) of the decision classes, it follows that no negative rules supported by x can be generated from the incomplete decision system, i.e. it is useless to compute reducts for x if $L^D_{AT}(x) = \emptyset(U^D_{AT}(x) = \emptyset, BN^D_{AT}(x) = \emptyset)$. From this point of view, the reducts will be computed in the following are all based on the assumption that $L^D_{AT}(x) \neq \emptyset(U^D_{AT}(x) \neq \emptyset, BN^D_{AT}(x) \neq \emptyset)$.

Theorem 2.27. *Let I be an incomplete decision system in which $A \subseteq AT$, if $DIF_{AT}(x) \neq \emptyset$, then we have*

A is a difference relation rough consistent attributes set for x in $I \Leftrightarrow A$ is both the difference relation lower and upper approximate distribution consistent attributes sets of x in I.

Proof. $\Rightarrow$: Suppose that A is not the difference relation lower or upper approximate distribution consistent attributes sets for x in I, then there must be $L^D_A(x) \neq L^D_{AT}(x)$ or $U^D_A(x) \neq U^D_{AT}(x)$. By Eq. (2.75), it is not difficult to conclude that $BN^D_A(x) \neq BN^D_{AT}(x)$, which contradicts that A is a difference relation rough consistent attributes set for x in I.

$\Leftarrow$: Since A is both the difference relation lower and upper approximate distribution consistent attributes sets for x in I, we have $L^D_A(x) = L^D_{AT}(x)$ and $U^D_A(x) = U^D_{AT}(x)$. Thus, $BN^D_A(x) = BN^D_{AT}(x)$ holds. □

Given an incomplete decision system I in which $A \subseteq AT$, $\forall x \in U$, if $N_{AT}(x) \neq \emptyset$, then denote

$$P_{AT}(x) = \{y \in X_i : \sim X_i \in L_{AT}(x)\}$$
$$Q_{AT}(x) = \{y \in \sim X_i : DIF_{AT}(x) \cap (\sim X_i) = \emptyset\}$$
$$R_{AT}(x) = P_{AT}(x) \cup Q_{AT}(x)$$

where

$$DIS_L^D(x,y) = \begin{cases} \{a \in AT : (x,y) \notin DIF(\{a\})\}, & y \in P_{AT}(x) \\ \emptyset, & \text{otherwise} \end{cases}$$

$$DIS_U^D(x,y) = \begin{cases} \{a \in AT : (x,y) \notin DIF(\{a\})\}, & y \in Q_{AT}(x) \\ \emptyset, & \text{otherwise} \end{cases}$$

$$DIS_{BN}^D(x,y) = \begin{cases} \{a \in AT : (x,y) \notin DIF(\{a\})\}, & y \in R_{AT}(x) \\ \emptyset, & \text{otherwise} \end{cases}$$

then

$$\boldsymbol{M}_L^D(x) = \{DIS_L^D(x,y) : y \in P_{AT}(x)\} \tag{2.76}$$
$$\boldsymbol{M}_U^D(x) = \{DIS_U^D(x,y) : y \in Q_{AT}(x)\} \tag{2.77}$$
$$\boldsymbol{M}_{BN}^D(x) = \{DIS_{BN}^D(x,y) : y \in Q_{AT}(x)\} \tag{2.78}$$

are referred to as the difference relation lower approximate, upper approximate and rough distribution discernibility matrixes of x in I, respectively.

Theorem 2.28. *Let I be an incomplete decision system in which $A \subseteq AT$, if $DIF_{AT}(x) \neq \emptyset$, then*

1. *A is a difference relation lower approximate distribution consistent attributes set for x in I $\Leftrightarrow$ $A \cap DIS_L^D(x,y) \neq \emptyset$ for each $y \in P_{AT}(x)$;*
2. *A is a difference relation upper approximate distribution consistent attributes set for x in I $\Leftrightarrow$ $A \cap DIS_U^D(x,y) \neq \emptyset$ for each $y \in Q_{AT}(x)$;*
3. *A is a difference relation rough distribution consistent attributes set for x in I $\Leftrightarrow$ $A \cap DIS_{BN}^D(x,y) \neq \emptyset$ for each $y \in R_{AT}(x)$.*

Proof.
1. $\Rightarrow$: Suppose $\exists y \in P_{AT}(x)$ such that $A \cap DIS_L^D(x,y) = \emptyset$, then $y \in DIF_A(x)$. Since A is a difference relation lower approximate distribution consistent attributes set for x in I, we have $L_A(x) = L_{AT}(x)$, thus, there must be $\sim X_i \in L_{AT}(x)$ such that $DIF_A(x) \in (\sim X_i)$, it follows that $y \in \sim X_i$, which is contradictive to $y \in P_{AT}(x)$ because $y \in P_{AT}(x) \Rightarrow y \in X_i$.

   $\Leftarrow$: Since $A \subseteq AT$ and $DIF_{AT}(x) \neq \emptyset$, we have $DIF_A(x) \subseteq DIF_{AT}(x)$. Suppose that A is not a difference relation lower approximate distribution consistent attributes set for x in I, then $L_A(x) \neq L_{AT}(x)$, by $DIF_A(x) \subseteq DIF_{AT}(x)$ we can conclude that there must be $\sim X_i \in \left(\sim \left(U/IND(\{d\}\right)\right)$ such that $\sim X_i \in L_{AT}(x)$ and $\sim X_i \notin L_A(x)$, thus, $\exists y \in X_i$ and $y \in DIF_A(x)$, here $A \cap DIS_L^D(x,y) = \emptyset$. From discussions above, we conclude that if $A \cap DIS_L^D(x,y) \neq \emptyset$ for each $y \in P_{AT}(x)$, then A is a difference relation lower approximate distribution consistent attributes set for x in I.
2. The proof of 2 is similar to the proof of 1.

3. $\Rightarrow$: By Theorem 2.27, A is a difference relation rough consistent attributes set for x in $I \Rightarrow L_A(x) = L_{AT}(x)$ and $U_A(x) = U_{AT}(x)$. By 1 and 2, $A \cap R_{AT}(x,y) \neq \emptyset$ holds obviously for each $y \in R_{AT}(x)$ because $R_{AT}(x) = P_{AT}(x) \cup Q_{AT}(x)$.

   $\Leftarrow$: Since $R_{AT}(x) = P_{AT}(x) \cup Q_{AT}(x)$, by 1 and 2, we conclude that: if $A \cap DIS^D_{BN}(x,y) \neq \emptyset$, $\forall y \in R_{AT}(x)$, then A is both the difference relation lower and upper approximate distribution consistent attributes set for x in I, from which we conclude that A is a difference relation rough consistent attributes set for x in I. □

Definition 2.30. Let I be an incomplete decision system, define

$$\Delta^D_L(x) = \bigwedge_{y \in P_{AT}(x)} \vee DIS^D_L(x,y) \tag{2.79}$$

$$\Delta^D_U(x) = \bigwedge_{y \in Q_{AT}(x)} \vee DIS^D_U(x,y) \tag{2.80}$$

$$\Delta^D_{BN}(x) = \bigwedge_{y \in R_{AT}(x)} \vee DIS^D_{BN}(x,y) \tag{2.81}$$

$\Delta^L_D(x)$ is referred to as the difference relation lower approximate distribution discernibility function for x in I, $\Delta^U_D(x)$ is referred to as the difference relation relation upper approximate distribution discernibility function for x in I, $\Delta^{BN}_D(x)$ is referred to as the difference relation rough distribution discernibility function for x in I.

By Boolean Reasoning technique, it is not difficult to obtain the following theorem by Theorem 2.28.

Theorem 2.29. *Let I be an incomplete decision system in which $A \subseteq AT$, then*

1. *A is a difference relation lower approximate distribution reduct for x in I if and only if $\wedge A$ is the prime implicant of the difference relation lower approximate distribution discernibility function for x in I, i.e. $\Delta^D_L(x)$;*
2. *A is a difference relation upper approximate distribution reduct for x in I if and only if $\wedge A$ is the prime implicant of the difference relation upper approximate distribution discernibility function for x in I, i.e. $\Delta^D_U(x)$;*
3. *A is a difference relation rough distribution reduct for x in I if and only if $\wedge A$ is the prime implicant of the difference relation rough distribution discernibility function for x in I, i.e. $\Delta^D_{BN}(x)$.*

Example 2.19. Take for instance Table 2.17, since $L_{AT}(x_1) = \emptyset$, then by Remark 2.7, it is meaningless to compute the difference relation lower approximate distribution reduct of x_1.

Moreover, take for instance x_4, $L_{AT}(x_4) = \sim X_2$, then we have $P_{AT}(x_4) = \{x_7, x_8, x_9\}$, by Definition 2.30 we obtain that

$$\begin{aligned}\Delta^D_L(x_4) &= \big(\vee DIS^D_L(x_4, x_7)) \wedge (\vee DIS^D_L(x_4, x_8)) \wedge (\vee DIS^D_L(x_4, x_9)\big) \\ &= a_3 \wedge a_4 \wedge a_5\end{aligned}$$

By Theorem 2.29, we know that $\{a_3, a_4, a_5\}$ is the difference relation lower approximate distribution reduct for x_4 in Table 2.17. Thus, rule 1 in Example 2.18 can be simplified as:

$$c(x) \neq 0 \wedge e(x) \neq 1 \wedge f(x) \neq 1 \rightarrow d(x) \neq 2$$

Similarly, we can obtain the following results:

- $\{a_3, a_4, a_5\}$ is the difference relation lower approximate distribution reduct for x_7;
- $\{a_5\}$ is the difference relation lower approximate distribution reduct for x_9;
- $\{a_1, a_3, a_5\}$, $\{a_2, a_3, a_5\}$ and $\{a_3, a_4, a_5\}$ are the difference relation lower approximate distribution reducts for x_{11} and x_{12};
- $\{a_1\}$, $\{a_2\}$, $\{a_3\}$, $\{a_4\}$ and $\{a_5\}$ are the difference relation upper approximate distribution reducts for $x_1, x_2, x_3, x_5, x_6, x_7, x_8, x_9$ and x_{10};
- $\{a_3, a_4, a_5\}$ is the difference relation upper approximate distribution reduct for x_4;
- $\{a_1, a_3, a_5\}$, $\{a_2, a_3, a_5\}$ and $\{a_3, a_4, a_5\}$ are the difference relation upper approximate distribution reducts for x_{11} and x_{12};
- $\{a_1\}$, $\{a_2\}$, $\{a_3\}$, $\{a_4\}$ and $\{a_5\}$ are the difference relation rough distribution reducts for $x_1, x_2, x_3, x_5, x_6, x_8$ and x_{10};
- $\{a_3, a_4, a_5\}$ is the difference relation rough distribution reduct for x_4 and x_7;
- $\{a_5\}$ is the difference relation rough distribution reduct for x_9;
- $\{a_1, a_3, a_5\}$, $\{a_2, a_3, a_5\}$ and $\{a_3, a_4, a_5\}$ are the difference relation rough distribution reducts for x_{11} and x_{12}.

2.7 Limited Tolerance Relation Based Rough Set Approach

2.7.1 Limited Tolerance Relation

By further investigations of tolerance relation, valued tolerance relation and similarity relation, Wang [20, 21] presented the following limitations of these relations:

1. In tolerance relation, two objects may be considered as indiscernible while they have no equal attribute value and then there is just a little probability for their attribute values to be equal to each other.
2. In valued tolerance relation, we must know the probability distribution of the information system in advance, unfortunately, this is difficult when dealing with a new system; for a new system, we do not know even the total information about it, how can we know its accurate probability distribution?
3. In similarity relation, two similar objects may be considered as distinct because of a little missing information.

To overcome the above limitations, Wang proposed his limited tolerance relation as Definition 2.31 shows.

Definition 2.31. [20, 21] Let I be an incomplete decision system in which $A \subseteq AT$, the limited tolerance relation in terms of A is denoted by L_A where

$$\begin{aligned} LTOL(A) = \{(x,y) \in U^2 : &(\forall a \in A, f(x,a) \text{ and } f(y,a) \text{ are unknown}) \\ &\vee((P_A(x) \cap P_A(y) \neq \emptyset) \wedge_{\forall a \in A} (f(x,a) \text{ and } f(y,a) \text{ are known}) \Rightarrow \\ &(f(x,a) = f(y,a)))\} \end{aligned} \quad (2.82)$$

where $P_A(x) = \{a \in A : f(x) \text{ is known}\}$.

In the above limited tolerance relation, two cases have been considered:

1. firstly, two objects have no known values on each attribute in A;
2. secondly, two objects have attributes, on which the values are known for both of them, moreover, on each of these attributes, the values should be same for such two objects.

Obviously, the limited tolerance relation is reflexive and symmetric, while it is not necessarily transitive.

2.7.2 Limited Tolerance Relation Based Rough Set

Definition 2.32. *Let I be an incomplete decision system in which $A \subseteq AT$, $\forall x \in U$, the limited tolerance class of x in terms of A is denoted by $LTOL_A(x)$ where*

$$LTOL_A(x) = \{y \in U : (x,y) \in LTOL(A)\} \quad (2.83)$$

Definition 2.33. *Let I be an incomplete information system in which $A \subseteq AT$, $\forall X \subseteq U$, the lower approximation and upper approximation of X in terms of the limited tolerance relation $LTOL(A)$ are denoted by $\underline{LTOL_A}(X)$ and $\overline{LTOL_A}(X)$, respectively, where*

$$\underline{LTOL_A}(X) = \{x \in U : LTOL_A(x) \subseteq X\} \quad (2.84)$$

$$\overline{LTOL_A}(X) = \{x \in U : LTOL_A(x) \cap X \neq \emptyset\} \quad (2.85)$$

The pair $[\underline{LTOL_A}(X), \overline{LTOL_A}(X)]$ is referred to as the rough set of X in terms of the limited tolerance relation $LTOL(A)$.

Theorem 2.30. [20] *Let I be an incomplete information system in which $A \subseteq AT$, then $\forall X \subseteq U$, we have*

$$\underline{TOL_A}(X) \subseteq \underline{LTOL_A}(X) \quad (2.86)$$

$$\overline{LTOL_A}(X) \subseteq \overline{TOL_A}(X) \quad (2.87)$$

The above theorem tells us that by comparing with the tolerance relation, we can obtain the greater lower approximation and the smaller upper approximation if the limited tolerance relation is employed in the incomplete information system.

Theorem 2.31. [20] *Let I be an incomplete information system in which* $A \subseteq AT$, $\forall X \subseteq U$, *if* $P_A(x) \neq \emptyset (\forall x \in U)$, *then we have*

$$\underline{TOL_A}(X) \subseteq \underline{LTOL_A}(X) \subseteq \underline{SIM_A}(X) \tag{2.88}$$

$$\overline{SIM_A}(X) \subseteq \overline{LTOL_A}(X) \subseteq \overline{TOL_A}(X) \tag{2.89}$$

Theorem 2.30 still holds while Theorem 2.31 will not hold without the condition $P_A(x) \neq \emptyset$ ($\forall x \in U$). For objects without any known attribute value, we can deal with them separately. We do not need to deal with them together with the other objects. Thus, Theorem 2.31 is also valuable in practical applications.

Example 2.20. Take for instance the incomplete decision system showed in Table 2.11, by the limited tolerance relation, we obtain the following limited tolerance classes:

$$LTOL_{AT}(x_1) = \{x_1, x_{11}, x_{12}\}$$
$$LTOL_{AT}(x_2) = \{x_2, x_3\}$$
$$LTOL_{AT}(x_3) = \{x_2, x_3\}$$
$$LTOL_{AT}(x_4) = \{x_4, x_5, x_{11}, x_{12}\}$$
$$LTOL_{AT}(x_5) = \{x_4, x_5, x_{11}, x_{12}\}$$
$$LTOL_{AT}(x_6) = \{x_6\}$$
$$LTOL_{AT}(x_7) = \{x_7, x_9, x_{12}\}$$
$$LTOL_{AT}(x_8) = \{x_8\}$$
$$LTOL_{AT}(x_9) = \{x_7, x_9, x_{11}, x_{12}\}$$
$$LTOL_{AT}(x_{10}) = \{x_{10}\}$$
$$LTOL_{AT}(x_{11}) = \{x_1, x_4, x_5, x_9, x_{11}, x_{12}\}$$
$$LTOL_{AT}(x_{12}) = \{x_1, x_4, x_5, x_7, x_9, x_{11}, x_{12}\}$$

Then by Definition 2.33, since the decision attribute determine the universe into partition such that

$$U/IND(\{d\}) = \{\Phi, \Psi\} = \Big\{\{x_1, x_2, x_4, x_7, x_{10}, x_{12}\}, \{x_3, x_5, x_6, x_8, x_9, x_{11}\}\Big\}$$

then we obtain the following limited tolerance relation based lower and upper approximations:

$$\underline{LTOL_{AT}}(\Phi) = \{x_{10}\}$$
$$\overline{LTOL_{AT}}(\Phi) = \{x_1, x_2, x_3, x_4, x_5, x_7, x_9, x_{10}, x_{11}, x_{12}\}$$
$$\underline{LTOL_{AT}}(\Psi) = \{x_6, x_8\}$$
$$\overline{LTOL_{AT}}(\Psi) = \{x_1, x_2, x_3, x_4, x_5, x_6, x_7, x_8, x_9, x_{11}, x_{12}\}$$

2.8 Characteristic Relation Based Rough Set Approach

From discussions above, we can see that in the tolerance relation, valued tolerance relation, maximal consistent block, descriptor based rough sets, all of the unknown values can be compared with any other values in the domains of the corresponding attributes; in the similarity relation, difference relation based rough sets, all of the unknown values cannot be compared with any other values in the domains of the corresponding attributes; the limited tolerance relation based rough set is on the assumption that all unknown values can be compared with any other values or cannot be compared with any other values in the domains of the corresponding attributes. Therefore, all of these expansions of the rough sets can only be used to deal with the incomplete information systems, in which all of the unknown values have one and only one type of the semantic explanation.

Obviously, all of the above rough sets do not take a more general incomplete information system into account, i.e. lost and "do not care" unknown values are coexisting in one information system, such incomplete information system is referred to as a Generalized Incomplete Information System in the following. To deal with generalized information system by rough technique, Grzymala-Busse proposed his characteristic relation [23, 24]. Such binary relation is a generalized form of the tolerance and similarity relations, while tolerance and similarity relations are two special forms of the characteristic relation.

2.8.1 Characteristic Relation and Characteristic Relation Based Rough Set

Definition 2.34. [23, 24] Let I be a generalized incomplete information system in which $A \subseteq AT$, the characteristic relation in terms of A is denoted by $K(A)$ where

$$K(A) = \Big\{(x,y) \in U^2 : \forall a \in A, f(x,a) = ? \vee \Big(f(x,a) = f(y,a) \vee f(x,a) = * \vee f(y,a) = *\Big)\Big\} \tag{2.90}$$

Obviously, characteristic relation $K(A)$ is only reflexive, but it is not necessarily symmetric and transitive.

Given a generalized incomplete information system I, $\forall x \in U$, the characteristic class [23, 24] of x in terms of $K(A)$ is denoted by $K_A(x)$ such that

$$K_A(x) = \{y \in U : (x,y) \in K(A)\} \tag{2.91}$$

Proposition 2.7. Let I be a generalized incomplete information system in which $A \subseteq AT$, we have

$$K(A) = \bigcap_{a \in A} K(\{a\}) \tag{2.92}$$

Example 2.21. Table 2.18 shows a generalized incomplete information system.

In Table 2.18,

- $U = \{x_1, x_2, x_3, x_4, x_5, x_6\}$ is the universe of discourse;

- $AT = \{a_1, a_2, a_3, a_4\}$ is the set of the attributes, the domains of these attributes are $V_{\{a_1\}} = V_{\{a_2\}} = V_{\{a_3\}} = V_{\{a_4\}} = \{1, 2\}$.

Table 2.18 A generalized incomplete information system

U	a_1	a_2	a_3	a_4
x_1	1	1	1	1
x_2	2	*	1	1
x_3	*	?	2	2
x_4	1	2	1	2
x_5	?	*	1	2
x_6	2	1	1	*

By Definition 2.34, we obtain the following characteristic classes:

$$K_{AT}(x_1) = \{x_1\}$$
$$K_{AT}(x_2) = \{x_2, x_6\}$$
$$K_{AT}(x_3) = \{x_3\}$$
$$K_{AT}(x_4) = \{x_4\}$$
$$K_{AT}(x_5) = \{x_4, x_5, x_6\}$$
$$K_{AT}(x_6) = \{x_2, x_6\}$$

Definition 2.35. Let I be a generalized incomplete information system in which $A \subseteq AT$, $\forall X \subseteq U$, the lower approximation and upper approximation of X in terms of the characteristic relation $K(A)$ are denoted by $\underline{K_A}(X)$ and $\overline{K_A}(X)$, respectively, where

$$\underline{K_A}(X) = \{x \in U : K_A(x) \subseteq X\} \tag{2.93}$$
$$\overline{K_A}(X) = \{x \in U : K_A(x) \cap X \neq \emptyset\} \tag{2.94}$$

Proposition 2.8. *Let I be a generalized incomplete information system in which $A \subseteq AT$, $\forall X \subseteq U$, we have*

$$\underline{K_A}(X) \subseteq \underline{K_{AT}}(X) \tag{2.95}$$
$$\overline{K_A}(X) \supseteq \overline{K_{AT}}(X) \tag{2.96}$$

Proposition 2.8 shows that the tolerance relation based lower and upper approximations are monotonous with respect to the variety of the attributes.

Example 2.22. Let us consider a generalized incomplete decision system (Table 2.19), which is an expansion of Table 2.18 by adding the decision attribute d.

Table 2.19 A generalized incomplete decision system

U	a_1	a_2	a_3	a_4	d
x_1	1	1	1	1	1
x_2	2	*	1	1	1
x_3	*	?	2	2	2
x_4	1	2	1	2	2
x_5	?	*	1	2	2
x_6	2	1	1	*	2

In Table 2.19, by the decision attribute d, the universe is partitioned into disjoint subsets such that

$$U/IND(\{d\}) = \{X_1, X_2\} = \Big\{\{x_1, x_2\}, \{x_3, x_4, x_5, x_6\}\Big\}$$

Then by Definition 2.35, we obtain the characteristic relation based lower and upper approximations such that

$$\underline{K_{AT}}(X_1) = \{x_1\}$$
$$\overline{K_{AT}}(X_1) = \{x_1, x_2, x_6\}$$
$$\underline{K_{AT}}(X_2) = \{x_3, x_4, x_5\}$$
$$\overline{K_{AT}}(X_2) = \{x_2, x_3, x_4, x_5, x_6\}$$

2.8.2 Approximate Distribution Reducts in Characteristic Relation Based Rough Set

Definition 2.36. Let I be an incomplete decision system in which $A \subseteq AT$, the partition determined by decision attribute d is $U/IND(\{d\}) = \{X_1, X_2, \cdots, X_l\}$, then denote

$$\underline{K_A}(d) = \{\underline{K_A}(X_1), \underline{K_A}(X_2), \cdots, \underline{K_A}(X_l)\} \tag{2.97}$$
$$\overline{K_A}(d) = \{\overline{K_A}(X_1), \overline{K_A}(X_2), \cdots, \overline{K_A}(X_l)\} \tag{2.98}$$

1. If $\underline{K_A}(d) = \underline{K_{AT}}(d)$, then A is referred to as a characteristic relation lower approximate distribution consistent attributes set in I; if A is a characteristic relation lower approximate distribution consistent attributes set and $\forall B \subset A$, B is not the characteristic relation lower approximate distribution consistent attributes set, then A is referred to as a characteristic relation lower approximate distribution reduct in I.
2. If $\overline{K_A}(d) = \overline{K_{AT}}(d)$, then A is referred to as a characteristic relation upper approximate distribution consistent attributes set in I; if A is a characteristic relation upper approximate distribution consistent attributes set and $\forall B \subset A$, B is not the characteristic relation upper approximate distribution consistent attributes set, then A is referred to as a characteristic relation upper approximate distribution reduct in I.

By Definition 2.36, we can see that

- The characteristic relation lower approximate distribution consistent attributes set in I is a subset of the attributes, which preserves the lower approximations of all decision classes in terms of the characteristic relation; a characteristic relation lower approximate distribution reduct in I is a minimal subset of the attributes, which preserves the lower approximations of all decision classes in terms of the characteristic relation.
- The characteristic relation upper approximate distribution consistent attributes set in I is a subset of the attributes, which preserves the upper approximations of all decision classes in terms of the characteristic relation; a characteristic relation upper approximate distribution reduct in I is a minimal subset of the attributes, which preserves the upper approximations of all decision classes in terms of the characteristic relation.

Theorem 2.32. *Let I be an incomplete decision system in which $A \subseteq AT$, $\forall x \in U$, denote*

$$L_{AT}^{K}(x) = \{X_k : X_k \in U/IND(\{d\}) : x \in \underline{K_{AT}}(X_k)\} \tag{2.99}$$

$$U_{AT}^{K}(x) = \{X_k : X_k \in U/IND(\{d\}) : x \in \overline{K_{AT}}(X_k)\} \tag{2.100}$$

then

1. *A is a characteristic relation lower approximate distribution consistent attributes set $\Leftrightarrow L_{A}^{K}(x) = L_{AT}^{K}(x)$ for each $x \in U$;*
2. *A is a characteristic relation upper approximate distribution consistent attributes set $\Leftrightarrow U_{A}^{K}(x) = U_{AT}^{K}(x)$ for each $x \in U$.*

Proof. It can be derived directly from Definition 2.36. □

Let us denote

$$DIS_{L}^{K}(x,y) = \begin{cases} \{a \in AT : (x,y) \notin K(\{a\})\}, & X_k \in L_{AT}^{K}(x), y \notin X_k \\ \emptyset, & \text{otherwise} \end{cases}$$

$$DIS_{U}^{K}(x,y) = \begin{cases} \{a \in AT : (x,y) \notin K(\{a\})\}, & X_k \notin U_{AT}^{K}(y), x \in X_k \\ \emptyset, & \text{otherwise} \end{cases}$$

then

$$\boldsymbol{M}_{L}^{K} = \{DIS_{L}^{K}(x,y) : X_k \in L_{AT}^{K}(x), y \notin X_k\} \tag{2.101}$$

$$\boldsymbol{M}_{U}^{K} = \{DIS_{U}^{K}(x,y) : X_k \notin U_{AT}^{K}(y), x \in X_k\} \tag{2.102}$$

are referred to as the characteristic relation lower and upper approximate distribution discernibility matrixes in I, respectively.

Theorem 2.33. *Let I be an incomplete decision system in which $A \subseteq AT$, then*

1. *A is a characteristic relation lower approximate distribution consistent attributes set in $I \Leftrightarrow A \cap DIS_{L}^{K}(x,y) \neq \emptyset$ for each $DIS_{L}^{K}(x,y) \in \boldsymbol{M}_{L}^{K}$;*
2. *A is a characteristic relation upper approximate distribution consistent attributes set in $I \Leftrightarrow A \cap DIS_{U}^{K}(x,y) \neq \emptyset$ for each $DIS_{U}^{K}(x,y) \in \boldsymbol{M}_{U}^{K}$.*

Proof. We only prove 1, the proof of 2 is similar to the proof of 1.

$\Rightarrow$: Suppose $\exists DIS_{L}^{K}(x,y) \in \boldsymbol{M}_{L}^{K}$ such that $A \cap DIS_{L}^{K}(x,y) = \emptyset$, then we have $(x,y) \in K(A)$, i.e. $y \in K_A(x)$. Since $DIS_{L}^{K}(x,y) \in \boldsymbol{M}_{L}^{K}$, then there exists $X_k \in U/IND(\{d\})$ such that $X_k \in L_{AT}^{K}(x)$ and $y \notin X_k$. By condition we know that A is a characteristic relation lower approximate distribution consistent attributes set in I, then by Theorem 2.32 we also have $X_k \in L_{A}^{K}(x)$, it follows that $K_A(x) \subseteq X_k$, $y \in X_k$, which is contradictive to the assumption because $DIS_{L}^{K}(x,y) \in \boldsymbol{M}_{L}^{K} \Rightarrow y \notin X_k$.

$\Leftarrow$: Since $A \subseteq AT$, then by Proposition 2.8 we have $\underline{K_A}(X_k) \subseteq \underline{K_{AT}}(X_k)$ for each $X_k \in U/IND(\{d\})$. Suppose that A is not a characteristic relation lower approximate distribution consistent attributes set in I, then there must be $X_k \in U/IND(\{d\})$ such that $\underline{K_A}(X_k) \subset \underline{K_{AT}}(X_k)$, i.e. $\exists x \in U$ such that $X_k \in L^K_{AT}(x)$ and $X_k \notin L^K_A(x)$. Since $X_k \notin L^K_A(x)$, then there must be $y \notin X_k$ such that $y \in K_A(x)$, i.e. $(x,y) \in K(A)$, it follows that $A \cap DIS^K_L(x,y) = \emptyset$. From discussions above, if $\forall X_k \in L^K_{AT}(x)$ and $y \notin X_k$, i.e. $\forall DIS^K_L(x,y) \in \boldsymbol{M}^K_L$, $A \cap DIS^K_L(x,y) \neq \emptyset$, then A is a characteristic relation lower approximate distribution consistent attributes set in I. □

Definition 2.37. Let I be an incomplete decision system in which $A \subseteq AT$, then the characteristic relation lower and upper approximate distribution discernibility functions for computing the characteristic relation lower and upper approximate distribution reducts in I are defined as:

$$\Delta^K_L = \bigwedge_{DIS^K_L(x,y)\in \boldsymbol{M}^K_L} \vee DIS^K_L(x,y) \tag{2.103}$$

$$\Delta^K_U = \bigwedge_{DIS^K_U(x,y)\in \boldsymbol{M}^K_U} \vee DIS^K_U(x,y) \tag{2.104}$$

By Boolean Reasoning technique, it is not difficult to obtain the following theorem by Theorem 2.33.

Theorem 2.34. *Let I be an incomplete decision system in which $A \subseteq AT$, then*

1. *A is a characteristic relation lower approximate distribution reduct in I if and only if $\wedge A$ is the prime implicant of the characteristic relation lower approximate distribution discernibility function Δ^K_L;*
2. *A is a characteristic relation upper approximate distribution reduct in I if and only if $\wedge A$ is the prime implicant of the characteristic relation upper approximate distribution discernibility function Δ^K_U.*

Example 2.23. Take for instance Table 2.19, following Example 2.22, we then obtain the characteristic relation lower approximate distribution discernibility matrix $\boldsymbol{M}^K_L$ in Table 2.19 as Table 2.20 shows.

Table 2.20 Characteristic relation lower approximate distribution discernibility matrix $\boldsymbol{M}^K_L$ in Table 2.19

	x_1	x_2	x_3	x_4	x_5	x_6
x_1			a_2, a_3, a_4	a_2, a_4	a_1, a_4	a_1
x_2						
x_3	a_3, a_4	a_3, a_4				
x_4	a_2, a_4	a_1, a_4				
x_5	a_4	a_4				
x_6						

By Definition 2.37, the characteristic relation lower approximate distribution discernibility function is

$$\begin{aligned}\Delta_L^K &= \bigwedge_{DIS_L^K(x,y)\in \boldsymbol{M}_L^K} \vee DIS_L^K(x,y)\\ &= (a_2 \vee a_3 \vee a_4) \wedge (a_3 \vee a_4) \wedge (a_2 \vee a_4) \wedge (a_1 \vee a_4) \wedge a_1 \wedge a_4\\ &= a_1 \wedge a_4\end{aligned}$$

Then by Theorem 2.34, we know that $\{a_1, a_4\}$ is the characteristic relation lower approximate distribution reduct in Table 2.19. Similarly, it is not difficult to obtain that $\{a_1, a_4\}$ is also the characteristic relation upper approximate distribution reduct in Table 2.19.

2.9 Conclusions

In this chapter, several important expanded rough set models, which can be used to deal with the incomplete information system, have been introduced. In these models,

- tolerance relation based rough set, valued tolerance relation based rough set, maximal consistent block based rough set and descriptor based rough set can be used to deal with the incomplete information system with "*" unknown values;
- similarity relation based rough set, difference relation based rough set can be used to deal with the incomplete information system with "?" unknown values;
- limited tolerance relation based rough set can be used to deal with the incomplete information system with either "*" or "?" unknown values;
- characteristic relation based rough set can be used to deal with the incomplete information system with both "*" and "?" unknown values.

It can be seen that most of these expanded rough set models in the incomplete information system are constructed on the basis of the weaker binary relations. For example, the tolerance and valued tolerance relations are not to be transitive; the similarity relation is not to be symmetric; the difference relation is not to be reflexive, symmetric and transitive; the characteristic relation is a generalization of the tolerance and similarity relations, then, it is not to be symmetric and transitive. From this point of view, how to generalize the binary relations plays a crucial role in the expansions of the rough sets in incomplete information system.

References

[1] Gore, A.: Earth in the balance. New York: Plume Books (1992)

[2] Ebenbach, D. H., Moore, C. F.: Incomplete information, inferences, and individual differences: the case of environmental judgements. Organ. Behav. Hum. Dec. **81**, 1–27 (2000)

[3] Grzymala-Busse, J. W., Rząsa, W.: Local and global approximations for incomplete data. Transactions on Rough Sets VIII, LNCS, **5084**, 21–34 (2008)

[4] Hong, T. P., Tseng, L. H., Wang, S. L.: Learning rules from incomplete training examples by rough sets. Expert Syst. Appl. **22**, 285–293 (2002)

[5] Kryszkiewicz, M.: Rough set approach to incomplete information systems. In: Proceedings of the Second Annual Joint Conference on Information Sciences, pp. 194–197 (1995)

[6] Kryszkiewicz, M.: Rough set approach to incomplete information systems. Informa. Sci. **112**, 39–49 (1998)
[7] Kryszkiewicz, M.: Rules in incomplete information systems. Inform. Sci. **113**, 271–292 (1999)
[8] Latkowski, R.: On Decomposition for incomplete data. Fund. Informa. **54**, 1–16 (2003)
[9] Latkowski, R.: Flexible indiscernibility relations for missing attribute values. Fund. Inform. **67**, 131–147 (2005)
[10] Liang J. Y., Shi, Z. Z., Li, D. Y., Wierman, M. J.: Information entropy, rough entropy and knowledge granulation in incomplete information systems. Int. J. Gen. Syst. **35**, 641–654 (2006)
[11] Nakamura, A.: A rough logic aased on incomplete information and its application. Int. J. Approx. Reason. **15**, 367–378 (1995)
[12] Qian, Y. H., Liang, J. Y., Pedrycz, W., Dang, C. Y.: An efficient accelerator for attribute reduction from incomplete data in rough set framework. Pattern Recognit. **44**, 1658–1670 (2011)
[13] Salama, A. S.: Topological solution of missing attribute values problem in incomplete information tables. Inform. Sci. **180**, 631–639 (2010)
[14] Wu, W. Z.: Attribute reduction based on evidence theory in incomplete decision systems. Inform. Sci. **178**, 1355–1371 (2008)
[15] Wu, C., Wang, L. J.: An improved limited and variable precision rough set model for rule acquisition based on pansystems methodology. Kybernetes. 37, 1264–1271 (2008)
[16] Wu, W. Z., Zhang, W. X., Li, H. Z.: Knowledge acquisition in incomplete fuzzy information systems via the rough set approach. Expert Syst. **20**, 280–286 (2003)
[17] Yang, X. P.: Fuzziness in incomplete information systems. In: Proceedmgs of the Third International Conference on Machine Learning and Cybemetics, pp.1599–1603 (2004)
[18] Yang, X. B., Qu, F., Yang, J. Y., Xie J.: A novel extension of rough set model in incomplete information system. In: Third International Conference on Innovative Computing, Information and Control, pp. 306–306 (2008)
[19] Zhang, W. X., Mi, J. S.: Incomplet information system and its optimal selections. Comput. Math. Appl. **48**, 691–698 (2004)
[20] Wang, G. Y.: Extension of rough set under incomplete information systems. In: Proceeding of the 11th IEEE International Conference on Fuzzy Systems, pp. 1098–1103 (2002)
[21] Wang, G. Y., Guan L. H., Hu F.: Rough set extensions in incomplete information systems. Front. Electr. Electron. Eng. China, **3**, 399–405 (2008)
[22] Grzymala-Busse, J. W.: On the unknown attribute values in learning from examples. In: Proceeding of the Sixth International Symposium on Methodologies for Intelligent Systems, pp. 368–377 (1991)
[23] Grzymala-Busse, J. W.: Characteristic relations for incomplete data: A generalization of the indiscernibility relation. In: Proceeding of the Third International Conference on Rough Sets and Current Trends in Computing, pp. 244–253 (2004)
[24] Grzymala-Busse, J. W.: Data with missing attribute values: Generalization of indiscernibility relation and rule reduction. Transactions on Rough Sets I, LNCS, **3100**, 78–95 (2004)
[25] Grzymala-Busse, J. W., Wang, A. Y.: Modified algorithms LEM1 and LEM2 for rule induction from data with missing attribute values. In: Proceeding of the Fifth International Workshop on Rough Sets and Soft Computing at the Third Joint Conference on Information Sciences, pp. 69–72 (1997)

[26] Stefanowski, J., Tsoukiàs, A.: On the extension of rough sets under incomplete information. In: Proceeding of New directions in rough sets, data mining and granular-soft computing, pp. 73–82 (1999)

[27] Stefanowski, J., Tsoukiàs, A.: Incomplete information tables and rough classification. Comput. Intell. **17**, 545–566 (2001)

[28] Guan, Y. Y., Wang, H. K.: Set-valued information systems. Inform. Sci. **176**, 2507–2525 (2006)

[29] Leung, Y., Li, D. Y.: Maximal consistent block technique for rule acquisition in incomplete information systems. Inform. Sci. **115**, 85–106 (2003)

[30] Qian, Y. H., Liang, J. Y., Li, D. Y., Wang, F., Ma, N. N.: Approximation reduction in inconsistent incomplete decision tables. Knowl.-Based Syst. **23**, 427–433 (2010)

[31] Leung, Y., Wu, W. Z., Zhang, W. X.: Knowledge acquisition in incomplete information systems: a rough set approach. Eur. J. Oper. Res. **168**, 464–473 (2006)

[32] Wu, W. Z., Xu, Y. H.: On two types of generalized rough set approximations in incomplete information systems. In: 2005 IEEE International Conference on Granular Computing, pp. 303–306 (2005)

[33] Tsumoto, S.: Automated discovery of positive and negative knowledge in clinical databases, IEEE Eng. Med. Biol. **19**, 56–62 (2000)

[34] Tsumoto, S.: Automated extraction of medical expert system rules from clinical databases on rough set theory. Inform. Sci. **112**, 67–84 (1998)

[35] Yang, X. B., Yu, D. J., Yang J. Y., Song, X. N.: Difference relation-based rough set and negative rrules in incomplete information system. Int. J. Uncertain. Fuzz. **17**, 649–665 (2009)

Chapter 3

Neighborhood System and Rough Set in Incomplete Information System

Abstract As the first model for Granular Computing, neighborhood system has been widely investigated. In this chapter, the neighborhood system approach is introduced into the incomplete information system. By employing the coverings induced by maximal consistent blocks and support sets of descriptors, two different neighborhood systems can be obtained, respectively. By using the knowledge engineering view in Granular Computing, a new knowledge operation is defined on the neighborhood system, which can help us obtain more knowledge through the known knowledge. Furthermore, by using neighborhood system based rough set model, we can obtain the same lower approximations and smaller upper approximations than the maximal consistent block and descriptor based rough sets.

3.1 Neighborhood System

3.1.1 From Granular Computing to Neighborhood System

The concept of Granular Computing (GrC), was jointly proposed by Lin and Zadeh in 1990s. Although many excellent research contributions have been made in this theory, granular computing is still in its inception stage. To further push the granular computing theory into practical applications, Lin summarized eight formal granular computing models [1]:

1. local granular computing model;
2. global granular computing model;
3. binary granular computing model;
4. multi-binary granular computing model;
5. relational granular computing model;
6. function granular computing model;
7. Turing granular computing model;
8. category based granular computing model.

It should be noticed that the first granular computing model, i.e. local granular computing model, is defined on the neighborhood system [2–10]. Therefore, the first granular computing model is also referred to as the neighborhood system based granular computing model.

Mathematically, neighborhood system is a model that formalizes an ancient intuition, infinitesimal granules, which led to the invention of calculus, topology and non-standard analysis. Roughly speaking, a neighborhood system assigns each object a (possibly finite or infinite) family of subsets. Each subset is referred to as a neighborhood, which can be used to represent the semantics of "near" [3]. Obviously, the classification analysis in the rough set theory can be included into neighborhood system theory because for each object in the universe, one or more classes can be considered as the neighborhoods of such object. For instance, in Pawlak's rough set, each object is associated with one and only one equivalence class, such equivalence class is the neighborhood; in covering based rough approximations, each object is associated with one or more blocks in the covering, these blocks are also neighborhoods.

Since neighborhood system is the first granular computing model, the formalized definition of neighborhood system is then necessary. In the following, we first present the fifth granular computing model, i.e. relational granular computing model, to generate neighborhood system.

Definition 3.1. [9] The fifth granular computing model:

1. $\mathscr{U} = \{U_1, U_2, \cdots\}$ is a family of classical sets, which is referred to as the family of the universes;
2. $\underbrace{U_i \times U_j \times \cdots \times U_k}_{n}$ is a Cartesian product of n universes in which $U_i, U_j, \cdots, U_k \in \mathscr{U}$, it should be noticed that distinct indices do not imply the universes are distinct;
3. a n-ary relation is a subset such that $Rel^n \subseteq \underbrace{U_i \times U_j \times \cdots \times U_k}_{n}$;
4. $\mathfrak{R}^n = \{Rel_1^n, Rel_2^n, \cdots\}$ is a family of n-ary relations (n could be vary);

then the pair $(\mathscr{U}, \mathfrak{R}^n)$ is referred to as the relational granular computing model, is the formal definition of the fifth granular computing model.

Now, we shall consider only one universe, i.e. $\mathscr{U} = \{U\}$ and n=2. In this case the pair $(\mathscr{U}, \mathfrak{R}^n) = (U, \mathfrak{R}^2)$ is referred to as the fourth granular computing model (multi-binary granular computing model), in which $\mathfrak{R}^2$ is a collection of all *binary relations* on U. From this point of view, we can see that the fourth granular computing model is a special form of the fifth granular computing model, while fifth granular computing model is a generalization of the fourth granular computing model. In the following, to simplify our discussion, the collection of all binary relations in fourth granular computing model is denoted by $\mathfrak{R}^2 = \{Rel_1^2, Rel_2^2, \cdots\} = \{Rel_1, Rel_2, \cdots\}$.

Remark 3.1. Let $(U, \mathfrak{R}^2)$ be a fourth granular computing model, if $\mathfrak{R}^2$ is a family of indiscernibility relations (equivalence relations), then $(U, \mathfrak{R}^2)$ is the concept of knowledge base (See Section 1.1.1), which has been defined in Pawlak's rough set theory.

Definition 3.2. Let $(U, \Re^2)$ be a fourth granular computing model where $(U, \Re^2) = \{Rel_1, Rel_2, \cdots\}$ is a family of binary relations, then for each $x \in U$, we can obtain a family of subsets such that:

$$N_1(x) = \{y \in U : (x, y) \in Rel_1\}$$
$$N_2(x) = \{y \in U : (x, y) \in Rel_2\}$$
$$\vdots$$

Then define

1. $N_1(x), N_2(x), \cdots$ are referred to as the neighborhoods of x.
2. (a) collection such that $\{N_1(x), N_2(x), \cdots\}$ is referred to as the neighborhood system of x, it is denoted by $NS(x)$; in other words, the mapping
$$NS : U \to 2^{2^U}$$
associates with each object x a family of subsets, which may be empty, finite or infinite;
   (b) the neighborhoods in $NS(x)$ are referred to as the elementary neighborhoods of x;
   (c) if $NS(x)$ is an empty family, we simple say that x has no neighborhood.
3. The collection $\{NS(x) : x \in U\}$ is referred to as the neighborhood system of U, it is denoted by $NS(U)$, we may also refer $NS(U)$ as a pre-topology of U.
4. The universe U together with $NS(U)$ is referred to as a neighborhood system space, it is denoted by a pair $(U, NS(U))$, we may also refer it as a pre-topological space.
5. A subset X of U is open if for every object x in X, there is an elementary neighborhood of x, which is a subset of X, a subset X is closed if its complement is open.
6. An object x is a limit point of a set X, if every neighborhood of x contains a point of X other than x; the set of all limit points of X is called derived set of X.

Definition 3.2 is the formalized definition of neighborhood system. Such definition also tells us that we can use the fourth granular computing model to generate the neighborhood system based granular computing model, such model is also referred to as the first granular computing model, i.e. local granular computing model.

Example 3.1. Suppose that $X = \left\{\frac{1}{n} : n \in \mathbf{Z} - \{0\}\right\}$ where $\mathbf{Z}$ denotes the set of all integers, $NS(0) = \{[-\epsilon, +\epsilon] : \epsilon \in \mathbf{R}\}$ where $\mathbf{R}$ denotes the set of all real numbers.

$\forall \epsilon \in \mathbf{R}$, there must be $[-\epsilon, +\epsilon] \cap X \neq \emptyset$. Moreover, since $0 \notin [-\epsilon, +\epsilon] \cap X$, we can conclude that the number 0 is the limit point of set X.

We list some specific neighborhood systems:

1. $NS(U)$ is discrete $\Leftrightarrow NS(U) = 2^U$;
2. $NS(U)$ is indiscrete $\Leftrightarrow NS(U) = \{U\}$;
3. $NS(U)$ is serial $\Leftrightarrow \forall x \in U, \exists N(x) \in NS(x)$ such that $N(x) \neq \emptyset$;

4. $NS(U)$ is reflexive $\Leftrightarrow \forall x \in U, \exists N(x) \in NS(x)$ such that $x \in N(x)$;
5. $NS(U)$ is symmetric $\Leftrightarrow \forall x, y \in U, \exists N(x) \in NS(x), N(y) \in NS(y)$ such that $x \in N(y) \Rightarrow y \in N(x)$;
6. $NS(U)$ is transitive $\Leftrightarrow \forall x, y, z \in U, \exists N(y) \in NS(y), N(z) \in NS(z)$ such that $x \in N(y) \wedge y \in N(z) \Rightarrow x \in N(z)$.

As the first model of the granular computing, neighborhood system has been studied by many researchers. For example, Wu and Zhang [11] presented a k-step neighborhood systems to study the generalized rough set approach. Yao investigated the rough approximation problem from the viewpoint of the binary neighborhood system in Ref. [12]. Moreover, Yao [13–16] also defined some operations on the neighborhood system to study the partial ordering of the neighborhood systems. Qian et al. presented the partial ordering of binary and fuzzy binary neighborhood systems in Ref. [17] and Ref. [18], respectively. In this chapter, we are mainly focusing on the mathematical structure of neighborhood system and the application of neighborhood system approach to incomplete information system.

3.1.2 Binary Neighborhood System

Presently, a special neighborhood system has been widely used in many fields, it is referred to as the binary neighborhood system. Suppose that $NS(U)$ is the neighborhood system of U, $\forall x \in U$, if $NS(x)$ consists of singleton, we may induces a mapping such that

$$BNS : U \to 2^U \tag{3.1}$$

the collection $\{BNS(x) : x \in U\}$ is referred to as a binary neighborhood system of U, such binary neighborhood system is denoted by $BNS(U)$.

Theorem 3.1. *Binary relation induces and is induced by a binary neighborhood system.*

Proof. $\Rightarrow$: Given a binary relation Rel^2 on U, then $\forall x \in U$, we can induce a binary neighborhood system of x such that

$$BNS(x) = \{y \in U : (x, y) \in Rel^2\}$$

Therefore, $\{BNS(x) : x \in U\}$ is a binary neighborhood system of U.

$\Leftarrow$: Given a binary neighborhood system $BNS(U)$ in which $BNS(x)$ is the binary neighborhood system of x. By Eq. (3.1), we know that $BNS(x)$ is a subset of U. Therefore, the set of pairs $Rel^2 = \{(x, y) : x \in U, y \in BNS(x)\}$ is a binary relation on U. □

3.1.3 Covering and Neighborhood System

In set-based approach (see Remark 1.1), Pawlak's rough approximations are unions of some equivalence classes with respect to the inclusion and intersection conditions. In such case, different equivalence classes may be disjoint to each other. However, it should be noticed that covering is also a widely used concept in the applied mathematics. Generally speaking, given a covering, two different blocks in such covering may have overlap. Therefore, covering is a generalized concept of the partition.

Definition 3.3. Let U be a universe of discourse, $\mathbf{C}$ is a family of subsets of U, and none subsets in $\mathbf{C}$ be empty. If $\cup\mathbf{C} = U$, then $\mathbf{C}$ is referred to as a covering of U.

Informally speaking, covering is also a special form of neighborhood system. Let $\mathbf{C}$ be a given covering on U, $\forall x \in U$, the covering-based neighborhood system of x is defined as:

$$CNS(x) = \{C : C \in \mathbf{C}, x \in C\} \tag{3.2}$$

Therefore, by a covering $\mathbf{C}$, one can induce a neighborhood system of U such that

$$CNS(U) = \{CNS(x) : x \in U\}$$

Remark 3.2. Different from the general neighborhood system, neighborhood system $CNS(U)$ is a special form because neighborhoods in $CNS(U)$ are all open.

Example 3.2. Given an incomplete information system I in which $A \subseteq AT$,

1. since $\mathscr{M}(A)$ (see Eq. (2.19)) is a covering, then $\forall x \in U$, the maximal consistent block based neighborhood system of x in I is

$$MNS_A(x) = \{Y : Y \in \mathscr{M}(A), x \in Y\} \tag{3.3}$$

the maximal consistent block based neighborhood system of U in I is

$$MNS_A(U) = \{MNS_A(x) : x \in U\} \tag{3.4}$$

2. since $\mathscr{D}(A)$ (see Eq. (2.39)) is a covering, then $\forall x \in U$, the descriptor based neighborhood system of x in I is

$$DNS_A(x) = \{||t|| : ||t|| \in \mathscr{D}(A), x \in ||t||\} \tag{3.5}$$

the descriptor based neighborhood system of U in I is

$$DNS_A(U) = \{DNS_A(x) : x \in U\} \tag{3.6}$$

Let us consider Table 2.1, following Example 2.7, we have

$$\mathscr{M}(AT) = \left\{Y_1 = \{x_1\}, Y_2 = \{x_2, x_6\}, Y_3 = \{x_3\}, Y_4 = \{x_4, x_5\}, Y_5 = \{x_5, x_6\}\right\}$$

Then by Eq. (3.3), we can obtain the following maximal consistent block based neighborhood systems in Table 2.1:

$$MNS_{AT}(x_1) = \{Y_1\} = \left\{\{x_1\}\right\}$$
$$MNS_{AT}(x_2) = \{Y_2\} = \left\{\{x_2, x_6\}\right\}$$

$$MNS_{AT}(x_3) = \{Y_3\} = \{\{x_3\}\}$$
$$MNS_{AT}(x_4) = \{Y_4\} = \{\{x_4, x_5\}\}$$
$$MNS_{AT}(x_5) = \{Y_4, Y_5\} = \{\{x_4, x_5\}, \{x_5, x_6\}\}$$
$$MNS_{AT}(x_6) = \{Y_2, Y_5\} = \{\{x_2, x_6\}, \{x_5, x_6\}\}$$

On the other hand, following Example 2.11, the covering $\mathscr{D}(AT)$ induced by descriptors is:

$$\mathscr{D}(AT) = \Big\{||t_1|| = \{x_4, x_5\}, ||t_2|| = \{x_3\}, ||t_3|| = \{x_4, x_5\}, ||t_4|| = \{x_1\}, ||t_5|| = \{x_3\}, ||t_6|| = \{x_5, x_6\}, ||t_7|| = \{x_2, x_6\}, ||t_8|| = \{x_3\}, ||t_9|| = \{x_5\}, ||t_{10}|| = \{x_2\}, ||t_{11}|| = \{x_3\}\Big\}$$

in which

$$t_1 = (P, \text{high}) \wedge (M, \text{high}) \wedge (S, \text{full}) \wedge (MX, \text{high})$$
$$t_2 = (P, \text{high}) \wedge (M, \text{high}) \wedge (S, \text{compact}) \wedge (MX, \text{low})$$
$$t_3 = (P, \text{high}) \wedge (M, \text{low}) \wedge (S, \text{full}) \wedge (MX, \text{high})$$
$$t_4 = (P, \text{high}) \wedge (M, \text{low}) \wedge (S, \text{full}) \wedge (MX, \text{low})$$
$$t_5 = (P, \text{high}) \wedge (M, \text{low}) \wedge (S, \text{compact}) \wedge (MX, \text{low})$$
$$t_6 = (P, \text{low}) \wedge (M, \text{high}) \wedge (S, \text{full}) \wedge (MX, \text{high})$$
$$t_7 = (P, \text{low}) \wedge (M, \text{high}) \wedge (S, \text{full}) \wedge (MX, \text{low})$$
$$t_8 = (P, \text{low}) \wedge (M, \text{high}) \wedge (S, \text{compact}) \wedge (MX, \text{low})$$
$$t_9 = (P, \text{low}) \wedge (M, \text{low}) \wedge (S, \text{full}) \wedge (MX, \text{high})$$
$$t_{10} = (P, \text{low}) \wedge (M, \text{low}) \wedge (S, \text{full}) \wedge (MX, \text{low})$$
$$t_{11} = (P, \text{low}) \wedge (M, \text{low}) \wedge (S, \text{compact}) \wedge (MX, \text{low})$$

Then by Eq. (3.5), we can obtain the following descriptor based neighborhood systems in Table 2.1:

$$DNS_{AT}(x_1) = \{||t_4||\} = \{\{x_1\}\}$$
$$DNS_{AT}(x_2) = \{||t_7||, ||t_{10}||\} = \{\{x_2, x_6\}, \{x_2\}\}$$
$$DNS_{AT}(x_3) = \{||t_2||, ||t_5||, ||t_8||, ||t_{11}||\} = \{\{x_3\}\}$$
$$DNS_{AT}(x_4) = \{||t_1||, ||t_3||\} = \{\{x_4, x_5\}\}$$
$$DNS_{AT}(x_5) = \{||t_1||, ||t_3||, ||t_6||, ||t_9||\} = \{\{x_4, x_5\}, \{x_5, x_6\}, \{x_5\}\}$$
$$DNS_{AT}(x_6) = \{||t_6||, ||t_7||\} = \{\{x_5, x_6\}, \{x_2, x_6\}\}$$

3.1.4 Fuzzy Neighborhood System

So far implicitly all discussions are in the crisp world. Zadeh notes that human information processing is often fuzzy. So it is desirable these notions are fuzzified. From the viewpoint of fuzzy set theory, which was proposed by Zadeh, we should have the definition of fuzzy neighborhood system.

A fuzzy neighborhood is to each object $x \in U$, we associate a family of fuzzy subsets, denoted by $FNS(x)$. In other words, we have a mapping such that

$$FNS : U \rightarrow 2^{FZ(U)}$$

where $FZ(U)$ represents all fuzzy sets on U. $FNS(x)$ is referred to as a fuzzy neighborhood system of x.

$$FNS(U) = \{FNS(x) : x \in U\}$$

is then a fuzzy neighborhood system of U.

3.1.5 Neighborhood System and Topological Space

In this subsection, the relationship between neighborhood system and topological space will be studied. First, let us consider the following example.

Example 3.3. Suppose that $U = \{x_1, x_2, x_3\}$ is the universe of discourse:

1. If

$$NS(x_1) = \left\{\{x_1\}, \{x_1, x_3\}\right\}$$
$$NS(x_2) = \left\{\{x_2\}, \{x_1\}\right\}$$
$$NS(x_3) = \left\{\{x_1, x_3\}, \{x_2, x_3\}\right\}$$

then we obtain the open sets such that: $\varnothing$, $\{x_1\}$, $\{x_2\}$, $\{x_1, x_3\}$, $\{x_1, x_2\}$, $\{x_2, x_3\}$, $\{x_1, x_2, x_3\}$. It should be noticed that $\{x_3\}$ is not open. $NS(U)$ is serial but not reflexive, and it is not topological.

2. If

$$NS'(x_1) = \left\{\{x_1\}\right\}$$
$$NS'(x_2) = \left\{\{x_2\}\right\}$$
$$NS'(x_3) = \left\{\{x_3\}\right\}$$

then we obtain the open sets such that: $\varnothing$, $\{x_1\}$, $\{x_2\}$, $\{x_3\}$, $\{x_1, x_2\}$, $\{x_1, x_3\}$, $\{x_2, x_3\}$, $\{x_1, x_2, x_3\}$. Here, the neighborhood system $NS'(U)$ can induce a topological space.

By the above example, we can see not every neighborhood system can induce a topological space, and then what are the conditions for inducing a topological space? This is what will be discussed in this subsection.

Definition 3.4. Let U be the universe, x is an object in U, $NS(x)$ is a neighborhood system of x, $N(x)$ and $N'(x)$ are two arbitrary neighborhoods of x, if the intersection of $N(x)$ and $N'(x)$ is also the neighborhood of x, then we say the neighborhood system $NS(x)$ satisfies with the intersection condition.

Theorem 3.2. *If a neighborhood system satisfies with the intersection condition, then it can induce a topological space.*

Proof. Let $\tau = \{X \subseteq U : X \text{ is an open set}\}$.

1. It is not difficult to observe that $\varnothing, U \in \tau$.
2. $\forall X, Y \in \tau$, if the intersection of X and Y is not empty, it must be proved that the intersection of X and Y is also an open set.
   $\forall x \in X \cap Y$, since X and Y are open sets, there there are two neighborhoods of x, such that $N(x) \subseteq X$ and $N'(x) \subseteq Y$. By the condition of intersection, we know that the intersection of $N(x) \cap N'(x)$ is also a neighborhood of x, it follows that $N(x) \cap N'(x) \subseteq X \cap Y$, since x is arbitrary, then the intersection of X and Y is an open set.
3. For the arbitrary union of subsets of U, we denote it by X, it must be proved that X is an open set.

   $\forall x \in X$, there is a open set Y, such that $x \in Y$, and $Y \subseteq X$. Since Y is open, there exists a neighborhood of x, denoted by $N(x)$, such that $N(x) \subseteq Y$, it follows that $N(x) \subseteq X$. By the arbitrary of x, we can conclude that X is an open set.

From discussions above, we can see that (U, τ) is a topological space. □

Remark 3.3. We should inform the readers that there is another way to induce a topology. We can take the neighborhood system as a subbase of a topological space, then it can also induce a topological space.

Example 3.4. Suppose that $U = \{x_1, x_2, x_3\}$ is the universe,

$$NS(x_1) = \Big\{\{x_1\}, \{x_1, x_3\}\Big\}$$
$$NS(x_2) = \Big\{\{x_2\}, \{x_1\}\Big\}$$
$$NS(x_3) = \Big\{\{x_1, x_3\}, \{x_2, x_3\}\Big\}$$

If we take the neighborhood system as a subbase, then the open sets are: $\varnothing$, $\{x_1\}$, $\{x_2\}$, $\{x_3\}$, $\{x_1, x_3\}$, $\{x_1, x_2\}$, $\{x_2, x_3\}$, $\{x_1, x_2, x_3\}$, it is a topological space.

3.1.6 Knowledge Operation in Neighborhood System

In Lin's granular computing theory, formal granular computing model has three semantic views [6]:

1. Knowledge engineering;
2. Uncertainty theory;
3. How to solve (computing) it.

Though neighborhood system was motivated from uncertainty, we will view it from the prospective of knowledge engineering. Thus, each neighborhood (element of neighborhood system) is regarded as a set of data that carries a unit of basic knowledge.

Since we take the knowledge engineering view to consider neighborhood system, the open problem is can the known knowledge be propagated, or, can we obtain new knowledge from the known knowledge? To solve this problem, Yang and Lin [6, 19] have proposed a new knowledge operation as Definition 3.5 shows.

Definition 3.5. Let $NS(U)$ be a neighborhood system of U, $\forall x_1, x_2, \cdots, x_m \in U$, define

$$N(x_1) \odot \cdots \odot N(x_m) = \begin{cases} N(x_1) \cap \cdots \cap N(x_m), & C_N(x_1) \cap \cdots \cap C_N(x_m) \neq \varnothing \\ \varnothing, & C_N(x_1) \cap \cdots \cap C_N(x_m) = \varnothing \end{cases}$$

where $N(x_i) \in NS(x_i)$ $(1 \leqslant i \leqslant m)$, $C_N(x_i)$ is referred to as the center set of $N(x_i)$, it consists of all those points that have $N(x_i)$ as neighborhood.

In the above definition of knowledge operation, the following cases have been considered:

1. $x_1 = x_2 = \cdots = x_m$, i.e. the knowledge operation is defined on the neighborhoods of one object only, thus, the result of $N(x_1)\odot\cdots\odot N(x_m)$ is actually a new neighborhood by using the *finite intersection* operation on the elementary neighborhoods of such object. Since such new neighborhood is derived from the neighborhoods of an initial neighborhood system, then the new obtained neighborhood should also be added into the initial neighborhood system for expanding the neighborhood system.
2. The knowledge operation is defined on the neighborhoods, which belong to different objects.
   For example, if $x_i \neq x_j$, then $N(x_i) \odot N(x_j) = N(x_i) \cap N(x_j)$ if and only if $\exists x_k \in U$ such that $N(x_i), N(x_j) \in NS(x_k)$, i.e. both $N(x_i)$ and $N(x_j)$ are elementary neighborhoods of x_k. From this point of view, $N(x_i) \odot N(x_j)$ is the intersection of the neighborhoods of x_k, it is actually the first case we discussed above.
3. If $C_N(x_1) \cap \cdots \cap C_N(x_m) = \varnothing$, then $N(x_1) \odot \cdots \odot N(x_m) = \varnothing$. Such case tells us that $\forall x \in U$, x does not take $N(x_1), \cdots, N(x_m)$ as its neighborhoods. Therefore, $N(x_1)\cap\cdots\cap N(x_m)$ should not be considered as a new neighborhood of x, it follows that $N(x_1) \odot \cdots \odot N(x_m) = \varnothing$.

From discussions above, we can see that by knowledge operation showed in Definition 3.5, new knowledge (neighborhoods) have been obtained. If $N(x_1) \odot \cdots \odot N(x_m) \neq \varnothing$, then the result of $N(x_1) \odot \cdots \odot N(x_m)$ should be added into the initial neighborhood system $NS(x)$ where $N(x_1), \cdots, N(x_m) \in NS(x)$. The new obtained neighborhood system is referred to as the expanded neighborhood system of x and denoted by $NS^{\odot}(x)$.

Since $N(x_1) \odot \cdots \odot N(x_m) = N(x_1) \cap \cdots \cap N(x_m)$ is the finite intersection of some neighborhoods of x, then formally, we have

$$\begin{aligned} NS^{\odot}(x) &= NS(x) \bigcup \Big\{ \bigcap_{\text{finitely many } i} N_i(x) \Big\} \\ &= \Big\{ \bigcap_{\text{finitely many } i} N_i(x) : i = 1, 2, \cdots \Big\} \end{aligned} \tag{3.7}$$

where $N_i(x) \in NS(x)$. By considering all expanded neighborhood systems, we obtain the expanded neighborhood system on U such that

$$NS^{\odot}(U) = \{NS^{\odot}(x) : x \in U\}.$$

Example 3.5. Following Example 3.2, take for instance x_1 and x_2, since $MNS_{AT}(x_1) = \{Y_1\}$, $MNS_{AT}(x_2) = \{Y_2\}$, then Y_1 is the neighborhood of x_1 and Y_2 is the neighborhood of x_2.

By Definition 3.5, we obtain the center set of Y_1, i.e. $C_{Y_1}(x_1) = \{x_1\}$, $C_{Y_2}(x_2) = \{x_2, x_6\}$. Obviously, $C_{Y_1}(x_1) \cap C_{Y_2}(x_2) = \varnothing$, thus by Definition 3.5, $Y_1 \odot Y_2 = \varnothing$. It is neither the neighborhood of x_1 nor the neighborhood of x_2.

Take for instance x_4 and x_5, we have $MNS_{AT}(x_4) = \{Y_4\}$, $MNS_{AT}(x_5) = \{Y_4, Y_5\}$. By Definition 3.5, $C_{Y_4}(x_4) = \{x_4, x_5\}$ because Y_4 is both the neighborhood of objects x_4 and x_5. Similarly, $C_{Y_5}(x_5) = \{x_5, x_6\}$, then $C_{Y_4}(x_4) \cap C_{Y_5}(x_5) = \{x_5\}$. By Definition 3.5, we have $Y_4 \odot Y_5 = Y_4 \cap Y_5 = x_5$. By such result, we know that the set $\{x_5\}$ is an expanded neighborhood of object x_5.

Similarly, we can obtain the following expanded neighborhood systems which are induced from the initial neighborhood systems $MNS_{AT}(U)$ and $DNS_{AT}(U)$ in Table 2.1.

$$MNS^{\odot}_{AT}(x_1) = \big\{\{x_1\}\big\}$$

$$MNS^{\odot}_{AT}(x_2) = \big\{\{x_2, x_6\}\big\}$$

$$MNS^{\odot}_{AT}(x_3) = \big\{\{x_3\}\big\}$$

$$MNS^{\odot}_{AT}(x_4) = \big\{\{x_4, x_5\}\big\}$$

$$MNS^{\odot}_{AT}(x_5) = \big\{\{x_4, x_5\}, \{x_5, x_6\}, \{x_5\}\big\}$$

$$MNS^{\odot}_{AT}(x_6) = \big\{\{x_2, x_6\}, \{x_5, x_6\}, \{x_6\}\big\}$$

$$DNS^{\odot}_{AT}(x_1) = \big\{\{x_1\}\big\}$$

$$DNS^{\odot}_{AT}(x_2) = \big\{\{x_2, x_6\}, \{x_2\}\big\}$$

$$DNS^{\odot}_{AT}(x_3) = \big\{\{x_3\}\big\}$$

$$DNS^{\odot}_{AT}(x_4) = \big\{\{x_4, x_5\}\big\}$$

$$DNS^{\odot}_{AT}(x_5) = \big\{\{x_4, x_5\}, \{x_5, x_6\}, \{x_5\}\big\}$$

$$DNS^{\odot}_{AT}(x_6) = \big\{\{x_5, x_6\}, \{x_2, x_6\}, \{x_6\}\big\}$$

Theorem 3.3. *Let U be the universe of discourse, if $BNS(U)$ is a binary neighborhood system on U, then we have*

1. $BNS(x) = BNS^{\odot}(x)$ $(\forall x \in U)$;
2. $BNS(U) = BNS^{\odot}(U)$.

Proof. Since $BNS(U)$ is a binary neighborhood system, then $\forall x \in U$, there is one and only one neighborhood of x (empty set is an acceptable neighborhood). By Eq. (3.7), it is a trivial to prove $BNS(x) = BNS^{\odot}(x)$ and $BNS(U) = BNS^{\odot}(U)$. □

Theorem 3.3 tells us that though the knowledge operation is employed in the binary neighborhood system, no new knowledge are derived since each object is associative with one and only one neighborhood.

In the following theorems, for technical reason, we add the empty set into the expanded neighborhood system $NS^{\odot}(x)$, i.e. $\varnothing \in NS^{\odot}(x)$.

Theorem 3.4. *Let U be the universe of discourse, $\forall x \in U$, $NS(x)$ is the initial neighborhood system of x, the pair $< NS^{\odot}(x), \odot >$ is a semigroup.*

Proof. To prove that $< NS^{\odot}(x), \odot >$ is a semigroup, what should be done is to verify closure and associativity.

1. Closure.
   $\forall N_1(x), N_2(x) \in NS^{\odot}(x)$, since
   $$NS^{\odot}(x) = \left\{ \bigcap_{\text{finitely many } i} N_i(x) : i = 1, 2, \cdots \right\}$$
   we have
   $$N_1(x) = \bigcap_{\text{finitely many } i} N_i(x), (i = 1, 2, \cdots)$$
   $$N_2(x) = \bigcap_{\text{finitely many } j} N_j(x), (j = 1, 2, \cdots)$$
   If $i, j = 1$, then $N_1(x), N_2(x) \in NS(x)$, it follows that $N_1(x) \cap N_2(x) \in NS^{\odot}(x)$ by the definition of $NS^{\odot}(x)$; otherwise,
   $$\begin{aligned} N_1(x) \cap N_2(x) &= \left(\bigcap_{\text{finitely many } i} N_i(x) \right) \bigcap \left(\bigcap_{\text{finitely many } j} N_j(x) \right) \\ &= \bigcap_{\text{finitely many } k} N_k(x) \end{aligned}$$
   from which we can conclude that $N_1(x) \cap N_2(x)$ is also a result of finite intersection of neighborhoods in $NS(x)$, i.e. $N_1(x) \cap N_2(x) \in NS^{\odot}(x)$.
2. Associativity.
   Since each element in $NS^{\odot}(x)$ is a finite intersection of neighborhoods in $NS(x)$ and associative law holds for intersection of sets, it follows that $(N_1(x) \odot N_2(x)) \odot N_3(x) = N_1(x) \odot (N_2(x) \odot N_3(x))$ where $N_1(x), N_2(x), N_3(x) \in NS^{\odot}(x)$. □

Theorem 3.5. *Let U be the universe of discourse, $< \cup NS^{\odot}(U), \odot >$ is a semigroup.*

Proof.
1. Closure.
   $\forall N(x), N(y) \in \cup NS^{\odot}(U)$,
   a. if $C_N(x) \cap C_N(y) = \varnothing$, then by Definition 3.5, we have $N(x) \odot N(y) = \varnothing$, $\varnothing \in \cup NS^{\odot}(U)$ because $\varnothing \in \cup NS^{\odot}(x)$ ($\forall x \in U$);

b. if $C_N(x) \cap C_N(y) \neq \varnothing$, there must be $z \in C_N(x) \cap C_N(y)$ such that $N(x)$ and $N(y)$ are both neighborhoods of z. By Definition 3.5,

$$\begin{aligned} N(x) \odot N(y) &= N(x) \cap N(y) \\ &= \left(\bigcap_{\text{finitely many } i} N_i(z) \right) \bigcap \left(\bigcap_{\text{finitely many } j} N_j(z) \right) \\ &= \bigcap_{\text{finitely many } k} N_k(z) \end{aligned}$$

where $i, j, k = 1, 2, \cdots$, thus, the result of $N(x) \odot N(y)$ is also a neighborhood of z, i.e. $N(x) \cap N(y) \in NS^{\odot}(z)$, it follows that $N(x) \odot N(y) \in \cup NS^{\odot}(U)$.

2. Associativity.

$\forall N(x), N(y), N(z) \in \cup NS^{\odot}(U)$, we have

$$(C_N(x) \cap C_N(y)) \cap C_N(z) = C_N(x) \cap (C_N(y) \cap C_N(z))$$

a. If $(C_N(x) \cap C_N(y)) \cap C_N(z) \neq \varnothing$, then $(N(x) \odot N(y)) \odot N(z) = (N(x) \cap N(y)) \cap N(z) = N(x) \cap (N(y)) \cap N(z)) = N(x) \odot (N(y)) \odot N(z))$;
b. If $(C_N(x) \cap C_N(y)) \cap C_N(z) = C_N(x) \cap (C_N(y) \cap C_N(z)) = \varnothing$, then by Definition 3.5, $(N(x) \odot N(y)) \odot N(z) = N(x) \odot (N(y) \odot N(z)) = \varnothing$ holds.

From discussions above, we know that "$\odot$" operation satisfies with the associative law on $\cup NS^{\odot}(U)$. □

3.2 Neighborhood System and Rough Approximations

3.2.1 Neighborhood System Based Rough Sets

By using the concepts of closure and interior in topological space, Lin presented the neighborhood system based rough set as Definition 3.6 shows.

Definition 3.6. [1] Let U be the universe of discourse, $NS(U)$ be a given neighborhood system of U, $\forall X \subseteq U$, the lower and upper approximations of X in terms of the neighborhood system $NS(U)$ are denoted by $\underline{apr}(X)$ and $\overline{apr}(X)$, respectively, where

$$\underline{apr}(X) = \{x \in U : \exists N(x) \in NS(x) \text{ s.t. } N(x) \neq \varnothing \wedge N(x) \subseteq X\}$$
$$\overline{apr}(X) = \{x \in U : \forall N(x) \in NS(x) \text{ s.t. } N(x) \cap X \neq \varnothing\}$$

$[\underline{apr}(X), \overline{apr}(X)]$ is referred to as the rough set of X in terms of the neighborhood system $NS(U)$.

Proposition 3.1. *Let U be the universe of discourse, $NS(U)$ be a given neighborhood system of U, $\forall X, Y \subseteq U$, we have*

1. $\underline{apr}(\varnothing) = \varnothing, \overline{apr}(\varnothing) = \varnothing$;
2. $\underline{apr}(X \cap Y) \subseteq \underline{apr}(X) \cap \underline{apr}(Y), \overline{apr}(X \cap Y) \subseteq \overline{apr}(X) \cap \overline{apr}(Y)$;

3. $\underline{apr}(X \cup Y) \supseteq \underline{apr}(X) \cup \underline{apr}(Y), \overline{apr}(X \cup Y) \supseteq \overline{apr}(X) \cup \overline{apr}(Y)$;
4. $X \subseteq Y \Rightarrow \underline{apr}(X) \subseteq \underline{apr}(Y), \overline{apr}(X) \subseteq \overline{apr}(Y)$.

Proof.

1. It can be derived directly from Definition 3.6.
2. By Definition 3.6,

$$\begin{aligned} x \in \underline{apr}(X \cap Y) &\Rightarrow \exists N(x) \in NS(x) \text{ s.t. } N(x) \neq \varnothing \wedge N(x) \subseteq X \cap Y \\ &\Rightarrow N(x) \neq \varnothing, N(x) \subseteq X \wedge N(x) \subseteq Y \\ &\Rightarrow x \in \underline{apr}(X), p \in \underline{apr}(Y) \\ &\Rightarrow x \in \underline{apr}(X) \cap \underline{apr}(Y) \end{aligned}$$

   Similarly, it is not difficult to prove $\overline{apr}(X \cap Y) \subseteq \overline{apr}(X) \cap \overline{apr}(Y)$.
3. The proof of 3 is similar to the proof of 2.
4. $\forall x \in \underline{apr}(X)$, there must be $N(x) \in NS(x)$ such that $N(x) \neq \varnothing$ and $N(x) \subseteq X$. Since $X \subseteq Y$, we then obtain $N(x) \subseteq Y$, i.e. $x \in \underline{apr}(Y)$. Similarly, it is not difficult to prove $\overline{apr}(X) \subseteq \overline{apr}(Y)$. □

Remark 3.4. The following properties do not hold for neighborhood system based rough set:

1. $\underline{apr}(U) = U$;
2. $\overline{apr}(U) = U$;
3. $\underline{apr}(X) \subseteq X$;
4. $X \subseteq \overline{apr}(X)$;
5. $\underline{apr}(X \cap Y) \supseteq \underline{apr}(X) \cap \underline{apr}(Y)$;
6. $\overline{apr}(X \cup Y) \subseteq \overline{apr}(X) \cup \overline{apr}(Y)$;
7. $\underline{apr}(\underline{apr}(X)) = \underline{apr}(X)$;
8. $\overline{apr}(\overline{apr}(X)) = \overline{apr}(X)$;
9. $\underline{apr}(X) = \sim(\overline{apr}(\sim X))$;
10. $\overline{apr}(X) = \sim(\underline{apr}(\sim X))$;
11. $\underline{apr}(\sim X) = \sim \overline{apr}(X)$.

3.2.2 Relationship Between Neighborhood System Based Rough Set and VPRS

From the viewpoint of inclusion error in variable precision rough set (see Section 1.2), if the value of β is given, we can define a special neighborhood system as Definition 3.7 shows.

Definition 3.7. Let U be the universe, $U/R = \{X_1, X_2, \cdots, X_m\}$ denotes the partition on U by indiscernibility relation R, $0 \leqslant \beta \leqslant 0.5$, for each case $x \in U$, we can define the following neighborhoods for x:

- Let m_i be the maximal positive integer such that $m_i \leqslant \beta|X_i|$;
- T varies through subsets of X_i such that $|T| = m_i$;
- Let T varies through all subsets of X_i such that $|T| = m_i$;

we then define

1. $N_T^\beta(x) = X_i - T$, for every $x \in X_i$;
2. $NS^\beta(x) = \{N_T^\beta(x) : T \text{ varies in its domain}\}$

to be the neighborhood systems of x under the given threshold β.

Definition 3.8. Let X_i varies through all equivalence classes, T varies its domain within each X_i, and x varies through $X_i - T$, then $NS^\beta(x)$ defines the neighborhood system $NS^\beta(U)$ with given threshold β such that

$$NS^\beta(U) = \{NS^\beta(x) : x \in U\}$$

Following Definition 3.6 and Definition 3.7, it is not difficult to obtain the following neighborhood system based rough set:

$$\underline{apr}^\beta(X) = \{x \in U : \exists N_T^\beta(P) \in NS^\beta(p) \text{ s.t. } N_T^\beta(x) \neq \varnothing \wedge N_T^\beta(x) \subseteq X\} \tag{3.8}$$

$$\overline{apr}^\beta(X) = \{x \in U : \forall N_T^\beta(x) \in NS^\beta(x) \text{ s.t. } N_T^\beta(x) \cap X \neq \varnothing\} \tag{3.9}$$

Theorem 3.6. *Let U be the universe of discourse, $U/R = \{X_1, X_2, \cdots, Y_m\}$ denotes the partition on U by indiscernibility relation R, $0 \leqslant \beta \leqslant 0.5$, then $\forall X \subseteq U$, we have*

$$\underline{R}^\beta(X) = \underline{apr}^\beta(X) \tag{3.10}$$

$$\overline{R}^\beta(X) = \overline{apr}^\beta(X) \tag{3.11}$$

Proof.

1. $\forall x \in U$,

$$\begin{aligned} x \in \underline{R}^\beta(X) &\Leftrightarrow X_i \in U/R, x \in X_i, 1 - \frac{|X_i \cap X|}{|X_i|} \leqslant \beta \\ &\Leftrightarrow \frac{|X_i| - |X_i \cap X|}{|X_i|} \leqslant \beta \\ &\Leftrightarrow \frac{|X_i - X_i \cap X|}{|X_i|} \leqslant \beta \\ &\Leftrightarrow \frac{|T|}{|X_i|} \leqslant \beta \text{ where we choose} T = X_i - X_i \cap X \\ &\Leftrightarrow \exists (X_i - T) = N_T^\beta(x) \in NS^\beta(x) \text{ s.t. } N_T^\beta(x) \subseteq X \\ &\Leftrightarrow x \in \underline{apr}^\beta(X) \end{aligned}$$

2. $\forall x \in U$,

$$x \notin \overline{R}^\beta(X) \Leftrightarrow X_i \in U/R, x \in X_i, 1 - \frac{|X_i \cap X|}{|X_i|} \geqslant 1 - \beta$$

$$\Leftrightarrow \frac{|X_i - T|}{|X_i|} \geqslant 1 - \beta \text{ where we choose} T = X_i \cap X$$
$$\Leftrightarrow \frac{|T|}{|X_i|} \leqslant \beta, (X_i - T) \cap X = \varnothing$$
$$\Leftrightarrow \exists (X_i - T) = N_T^\beta(x) \in NS^\beta(x) \text{ s.t. } N_T^\beta(x) \cap X = \varnothing$$
$$\Leftrightarrow x \notin \overline{apr}^\beta(X).$$
□

Theorem 3.6 tells us that the variable precision rough set model is actually a special form of the neighborhood system based rough set model while the neighborhood system based rough set is more generalized than the variable precision rough set.

Example 3.6. Suppose that $U = \{x_1, x_2, x_3, x_4, x_5, x_6\}$ is the universe of discourse,

$$U/R = \{P_1, P_2\} = \Big\{\{x_1, x_2, x_3, x_4\}, \{x_5, x_6\}\Big\}$$

is a partition on U, if $X = \{x_2, x_3, x_4, x_5\}$, then $e(P_1, X) = 0.25$, $e(P_2, X) = 0.5$. Suppose that $\beta = 0.4$, by the variable precision rough set model, we have

$$\underline{R}^{0.4}(X) = \{x_1, x_2, x_3, x_4\}$$
$$\overline{R}^{0.4}(X) = \{x_1, x_2, x_3, x_4, x_5, x_6\}$$

On the other hand, by Definition 3.7, we have $0.4|P_1| = 1.6$, $0.4|P_2| = 0.8$, thus $m_1 = 1$, $m_2 = 0$, from which we obtain the following neighborhood systems:

$$NS^{0.4}(x_m) = \Big\{\{x_1, x_2, x_3\}, \{x_1, x_2, x_4\}, \{x_1, x_3, x_4\}, \{x_2, x_3, x_4\}\Big\}, \quad m = 1, 2, 3, 4$$

$$NS^{0.4}(x_n) = \Big\{\{x_5, x_6\}\Big\}, \quad n = 5, 6$$

Therefore, by the neighborhood system based rough set model, we have

$$\underline{apr}^{0.4}(X) = \{x_1, x_2, x_3, x_4\}$$
$$\underline{apr}^{0.4}(X) = \{x_1, x_2, x_3, x_4, x_5, x_6\}$$

Such results are same to the above variable precision rough set.

3.2.3 Neighborhood System Based Rough Approximations in Incomplete Information System

It has been shown that in Example 3.2, we can construct two different types of the neighborhood systems in incomplete information system through the coverings induced by maximal consistent blocks and support sets of the descriptors. Therefore, it is interesting to study the neighborhood system based rough set in incomplete information system. Following Example 3.2 and Definition 3.6, it is not difficult to obtain the neighborhood system based rough set with respect to $\mathscr{M}(A)$ as following:

$$\underline{apr_A}(X) = \{x \in U : \exists MN_A(x) \in MNS_A(x) \text{ s.t. } MN_A(x) \neq \varnothing \wedge MN_A(x) \subseteq X\} \tag{3.12}$$

$$\overline{apr_A}(X) = \{x \in U : \forall MN_A(x) \in MNS_A(x), MN_A(x) \cap X \neq \varnothing\} \tag{3.13}$$

where $MN_A(x)$ is the neighborhood of x in neighborhood system $MNS_A(x)$.

Similarly, we obtain the neighborhood system based rough set with respect to $\mathscr{D}(A)$ as following:

$$\begin{aligned}\underline{apr'_A}(X) &= \{x \in U : \exists DN_A(x) \in DNS_A(x) \text{ s.t. } DN_A(x) \neq \varnothing \wedge DN_A(x) \subseteq X\}\\ &= \{x \in U : \exists \|t\| \in DNS_A(x) \text{ s.t. } \|t\| \subseteq X\}\end{aligned} \tag{3.14}$$

$$\begin{aligned}\overline{apr'_A}(X) &= \{x \in U : \forall DN_A(x) \in DNS_A(x), DN_A(x) \cap X \neq \varnothing\}\\ &= \{x \in U : \forall \|t\| \in DNS_A(x) \text{ s.t. } \|t\| \cap X \neq \varnothing\}\end{aligned} \tag{3.15}$$

where $DN_A(x)$ is the neighborhood of x in neighborhood system $DNS_A(x)$.

The pairs $\left[\underline{apr_A}(X), \overline{apr_A}(X)\right]$ is referred to as the maximal consistent block based neighborhood systems rough set of X in I; the pair $\left[\underline{apr'_A}(X), \overline{apr'_A}(X)\right]$ is referred to as the descriptor based neighborhood systems rough set of X in I.

Theorem 3.7. *Let I be an incomplete information system in which $A \subseteq AT$, $\forall X \subseteq U$, we have*

$$\underline{\mathscr{M}_A}(X) = \underline{apr_A}(X) \tag{3.16}$$

$$\overline{\mathscr{M}_A}(X) \supseteq \overline{apr_A}(X) \tag{3.17}$$

$$\cup\underline{DES_A}(X) = \underline{apr'_A}(X) \tag{3.18}$$

$$\cup\overline{DES_A}(X) \supseteq \overline{apr'_A}(X) \tag{3.19}$$

Proof.

$$\begin{aligned}x \in \underline{\mathscr{M}_A}(X) &\Leftrightarrow x \in Y \wedge Y \subseteq X (Y \in \mathscr{M}(A))\\ &\Leftrightarrow Y = MN_A(x) \in MNS_A(x) \wedge Y \subseteq X (Y \neq \varnothing)\\ &\Leftrightarrow x \in \underline{apr_A}(X)\end{aligned}$$

$$\begin{aligned}x \in \overline{apr_A}(X) &\Rightarrow \forall MN_A(x) \in MNS_A(x), MN_A(x) \cap X \neq \varnothing\\ &\Rightarrow x \in Y \wedge Y \cap X \neq \varnothing (Y \in \mathscr{M}(A))\\ &\Rightarrow x \in \overline{\mathscr{M}_A}(X)\end{aligned}$$

$$\begin{aligned}x \in \cup\underline{DES_A}(X) &\Leftrightarrow x \in \|t\| \wedge \|t\| \subseteq X (t \in FDES(A))\\ &\Leftrightarrow \|t\| = DN_A(x) \in DNS_A(x) \wedge \|t\| \subseteq X (\|t\| \neq \varnothing)\\ &\Leftrightarrow x \in \underline{apr'_A}(X)\end{aligned}$$

$$\begin{aligned}x \in \overline{apr'_A}(X) &\Rightarrow \forall \|t\| \in DNS_A(x), \|t\| \cap X \neq \varnothing\\ &\Rightarrow x \in \|t\| \wedge \|t\| \cap X \neq \varnothing (t \in FDES(A))\\ &\Rightarrow \|t\| \in \overline{DES_A}(X)\\ &\Rightarrow x \in \cup\overline{DES_A}(X)\end{aligned}$$

□

By Theorem 3.7, we can see that either the maximal consistent block based covering or descriptor based covering is selected, the neighborhood system based rough set can narrow the region of upper approximation, and then the accuracy of approximation will be improved.

Example 3.7. Take for instance Table 2.3, since by the decision attribute d, the universe is partitioned into disjoint subsets such that

$$\begin{aligned} U/IND(\{d\}) &= \Big\{\{\text{Poor}\}, \{\text{Good}\}, \{\text{Excel}\}\Big\} \\ &= \Big\{\{x_3\}, \{x_1, x_2, x_4, x_6\}, \{x_5\}\Big\} \end{aligned}$$

then following the neighborhood systems $MNS_{AT}(U)$ and $DNS_{AT}(U)$ we obtained in Example 3.2, we can obtain the following neighborhood systems based rough sets:

$$\begin{aligned} \underline{apr_{AT}}(X_1) &= \{x_3\} \\ \underline{apr_{AT}}(X_2) &= \{x_1, x_2, x_6\} \\ \underline{apr_{AT}}(X_3) &= \emptyset \\ \overline{apr_{AT}}(X_1) &= \{x_3\} \\ \overline{apr_{AT}}(X_2) &= \{x_1, x_2, x_4, x_5, x_6\} \\ \overline{apr_{AT}}(X_3) &= \{x_4, x_5\} \\ \underline{apr'_{AT}}(X_1) &= \{x_3\} \\ \underline{apr'_{AT}}(X_2) &= \{x_1, x_2, x_6\} \\ \underline{apr'_{AT}}(X_3) &= \{x_5\} \\ \overline{apr'_{AT}}(X_1) &= \{x_3\} \\ \overline{apr'_{AT}}(X_2) &= \{x_1, x_2, x_4, x_6\} \\ \overline{apr'_{AT}}(X_3) &= \{x_4, x_5\} \end{aligned}$$

Moreover, if we use the traditional maximal consistent block and descriptor techniques, we obtain the following rough approximations:

$$\begin{aligned} \underline{\mathscr{M}_{AT}}(X_1) &= \{x_3\} \\ \underline{\mathscr{M}_{AT}}(X_2) &= \{x_1, x_2, x_6\} \\ \underline{\mathscr{M}_{AT}}(X_3) &= \emptyset \\ \overline{\mathscr{M}_{AT}}(X_1) &= \{x_3\} \\ \overline{\mathscr{M}_{AT}}(X_2) &= \{x_1, x_2, x_4, x_5, x_6\} \\ \overline{\mathscr{M}_{AT}}(X_3) &= \{x_4, x_5, x_6\} \\ \underline{DES_{AT}}(X_1) &= \{x_3\} \\ \underline{DES_{AT}}(X_2) &= \{\{x_1\}, \{x_2\}, \{x_2, x_6\}\} \\ \underline{DES_{AT}}(X_3) &= \{x_5\} \\ \overline{DES_{AT}}(X_1) &= \{x_3\} \end{aligned}$$

$$\overline{DES_{AT}}(X_2) = \{\{x_1\}, \{x_2\}, \{x_4, x_5\}, \{x_5, x_6\}, \{x_2, x_6\}\}$$
$$\overline{DES_{AT}}(X_3) = \{\{x_5\}, \{x_4, x_5\}, \{x_5, x_6\}\}$$

By the above results of approximations, we can see that

1. $\underline{\mathscr{M}_{AT}}(X_i) = \underline{apr_{AT}}(X_i)$,
   $\cup\underline{DES_{AT}}(X_i) = \underline{apr'_{AT}}(X_i)$ where $i = 1, 2, 3$;
2. $\overline{\mathscr{M}_{AT}}(X_1) = \overline{apr_{AT}}(X_1)$,
   $\overline{\mathscr{M}_{AT}}(X_2) = \overline{apr_{AT}}(X_2)$,
   $\overline{\mathscr{M}_{AT}}(X_3) \supset \overline{apr_{AT}}(X_3)$;
3. $\cup\overline{DES_{AT}}(X_1) = \overline{apr'_{AT}}(X_1)$,
   $\cup\overline{DES_{AT}}(X_2) \supset \overline{apr'_{AT}}(X_2)$,
   $\cup\overline{DES_{AT}}(X_3) \supset \overline{apr'_{AT}}(X_3)$.

These results demonstrate the correctness of Theorem 3.7.

3.3 Reducts Neighborhood Systems

3.3.1 Reducts Neighborhood Systems in Incomplete Information System

Definition 3.9. Let I be an incomplete information system in which $A \subseteq AT$, $\forall x \in U$, define

1. If $MNS_A(x) = MNS_{AT}(x)$, then $MNS_A(x)$ is referred to as a consistent neighborhood system of $MNS_{AT}(x)$ in I; if $MNS_A(x)$ is a consistent neighborhood system of $MNS_{AT}(x)$ in I and $\forall B \subset A$, $MNS_B(x) \neq MNS_{AT}(x)$, then $MNS_A(x)$ is referred to as a reduct neighborhood system of $MNS_{AT}(x)$ in I.
2. If $MNS_A(U) = MNS_{AT}(U)$, then $MNS_A(U)$ is referred to as a consistent neighborhood system of $MNS_{AT}(U)$ in I; if $MNS_A(U)$ is a consistent neighborhood system of $MNS_{AT}(U)$ in I and $\forall B \subset A$, $MNS_B(U) \neq MNS_{AT}(U)$, then $MNS_A(U)$ is referred to as a reduct neighborhood system of $MNS_{AT}(U)$ in I.
3. $DNS_A(x) = DNS_{AT}(x)$, then $DNS_A(x)$ is referred to as a consistent neighborhood system of $DNS_{AT}(x)$ in I; if $DNS_A(x)$ is a consistent neighborhood system of $DNS_{AT}(x)$ in I and $\forall B \subset A$, $DNS_B(x) \neq DNS_{AT}(x)$, then $DNS_A(x)$ is referred to as a reduct neighborhood system of $DNS_{AT}(x)$ in I.
4. $DNS_A(U) = DNS_{AT}(U)$, then $DNS_A(U)$ is referred to as a consistent neighborhood system of $DNS_{AT}(U)$ in I; if $DNS_A(U)$ is a consistent neighborhood system of $DNS_{AT}(U)$ in I and $\forall B \subset A$, $DNS_B(U) \neq DNS_{AT}(U)$, then $DNS_A(U)$ is referred to as a reduct neighborhood system of $DNS_{AT}(U)$ in I.

By Definition 3.9, we can see that

- The consistent neighborhood system of $MNS_{AT}(x)$ in I is constructed on the basis of a subset of the attributes, which preserve the original neighborhood system $MNS_{AT}(x)$ in I; the reduct neighborhood system of $MNS_{AT}(x)$ in I is constructed on the basis of a minimal subset of the attributes, which preserve the original neighborhood system $MNS_{AT}(x)$ in I.
- The consistent neighborhood system of $MNS_{AT}(U)$ in I is constructed on the basis of a subset of the attributes, which preserve the original neighborhood system $MNS_{AT}(U)$ in I; the reduct neighborhood system of $MNS_{AT}(U)$ in I is constructed on the basis of a minimal subset of the attributes, which preserve the original neighborhood system $MNS_{AT}(U)$ in I.
- The consistent neighborhood system of $DNS_{AT}(x)$ in I is constructed on the basis of a subset of the attributes, which preserve the original neighborhood system $DNS_{AT}(x)$ in I; the reduct neighborhood system of $DNS_{AT}(x)$ in I is constructed on the basis of a minimal subset of the attributes, which preserve the original neighborhood system $DNS_{AT}(x)$ in I.
- The consistent neighborhood system of $DNS_{AT}(U)$ in I is constructed on the basis of a subset of the attributes, which preserve the original neighborhood system $DNS_{AT}(U)$ in I; the reduct neighborhood system of $DNS_{AT}(U)$ in I is constructed on the basis of a minimal subset of the attributes, which preserve the original neighborhood system $DNS_{AT}(U)$ in I.

To present the practical approach to obtain the reduct neighborhood systems, we need the following notions.

For each $MN_{AT}(x) \in MNS_{AT}(x)$, $DN_{AT}(x) \in DNS_{AT}(x)$, let us define

$$\mathscr{M}(MN_{AT}(x), y) = \begin{cases} \{a \in AT : z \in MN_{AT}(x), (z, y) \notin TOL(\{a\})\}, & y \notin MN_{AT}(x) \\ \emptyset, & \text{otherwise} \end{cases}$$

$$\mathscr{D}(DN_{AT}(x), y) = \begin{cases} \{a \in AT : DN_{AT}(x) = ||t||, (a, v) \in t, y \notin ||(a, v)||\}, & y \notin ||t|| \\ \emptyset, & \text{otherwise} \end{cases}$$

$\mathscr{M}(MN_{AT}(x), y)$ is referred to as the discernibility attributes set of neighborhood $MN_{AT}(x)$ and object y; $\mathscr{D}(DN_{AT}(x), y)$ is referred to as the discernibility attributes set of neighborhood $DN_{AT}(x)$ and object y.

$$\boldsymbol{M}^{\mathscr{M}} = \{\mathscr{M}(MN_{AT}(x), y) : \mathscr{M}(MN_{AT}(x), y) \neq \emptyset, MN_{AT}(x) \in MNS_{AT}(x)\}$$

and

$$\boldsymbol{M}^{\mathscr{D}} = \{\mathscr{D}(DN_{AT}(x), y) : \mathscr{D}(DN_{AT}(x), y) \neq \emptyset, DN_{AT}(x) \in DNS_{AT}(x)\}$$

are referred to as the maximal consistent block and descriptor based neighborhood systems discernibility matrixes, respectively.

Theorem 3.8. *Let I be an incomplete information system in which* $A \subseteq AT$, $\forall x \in U$,

1. $MNS_A(x)$ *is a consistent neighborhood system of* $MNS_{AT}(x)$ *in I* $\Leftrightarrow$
   $A \cap \mathscr{M}(MN_{AT}(x), y) \neq \varnothing$ *for each* $MN_{AT}(x) \in MNS_{AT}(x)$ *and* $y \notin MN_{AT}(x)$;
2. $MNS_A(U)$ *is a consistent neighborhood system of* $MNS_{AT}(U)$ *in I* $\Leftrightarrow$
   $A \cap \mathscr{M}(MN_{AT}(x), y) \neq \varnothing$ *for each* $\mathscr{M}(MN_{AT}(x), y) \in \mathbf{M}^{\mathscr{M}}$;
3. $DNS_A(x)$ *is a consistent neighborhood system of* $DNS_{AT}(x)$ *in I* $\Leftrightarrow$
   $A \cap \mathscr{D}(DN_{AT}(x), y) \neq \varnothing$ *for each* $||t|| = DN_{AT}(x) \in DNS_{AT}(x)$ *and* $y \notin ||t||$;
4. $DNS_A(U)$ *is a consistent neighborhood system of* $DNS_{AT}(U)$ *in I* $\Leftrightarrow$
   $A \cap \mathscr{D}(DN_{AT}(x), y) \neq \varnothing$ *for each* $\mathscr{D}(DN_{AT}(x), y) \in \mathbf{M}^{\mathscr{D}}$.

Proof.

1. $\Rightarrow$: By condition we have $MNS_A(x) = MNS_{AT}(x)$, thus, $\forall MN_{AT}(x) \in MNS_{AT}(x)$, there must be $MN_A(x) \in MNS_A(x)$ such that $MN_{AT}(x) = MN_A(x)$, it tells us that $\forall MN_{AT}(x) \in MNS_{AT}(x)$, if $y \notin MN_{AT}(x)$, then $y \notin MN_A(x)$ holds. By the basic property of maximal consistent block, there must be $z \in MN_A(x)$ ($z \in MN_{AT}(x)$) such that $(z, y) \notin TOL(A)$, from which we can conclude that $A \cap \mathscr{M}(MN_{AT}(x), y) \neq \varnothing$.

   $\Leftarrow$: $\forall MN_{AT}(x) \in MN_{AT}(x)$, since $A \subseteq AT$, there must be $Y \in \mathscr{M}(A)$ such that $MN_{AT}(x) \subseteq Y$. Obviously, $x \in MN_{AT}(x) \subseteq Y$ holds, then $Y = MN_A(x) \in MNS_A(x)$. From discussion above, we can conclude that $\forall MN_{AT}(x) \in MNS_{AT}(x)$, there must be $MN_A(x) \in MNS_A(x)$ such that $MN_{AT}(x) \subseteq MN_A(x)$.

   $\forall y \notin MN_{AT}(x)$, since $A \cap \mathscr{M}(MN_{AT}(x), y) \neq \varnothing$, then there must be $z \in MN_{AT}(x)$ such that $(z, y) \notin TOL(A)$, it follows that z and y belong to different maximal consistent blocks in terms of the set of attributes A. Obviously, $z \in MN_{AT}(x) \subseteq MN_A(x)$ holds, then $y \notin MN_A(x)$, it follows that $MN_A(x) \subseteq MN_{AT}(x)$. By combining with the results we obtained above, we have $MN_A(x) = MN_{AT}(x)$, that is, $\forall MN_{AT}(x) \in MNS_{AT}(x)$, there must be $MN_A(x) \in MNS_A(x)$ such that $MN_{AT}(x) = MN_A(x)$, i.e. $MNS_{AT}(x) \subseteq MNS_A(x)$.

   In the following, it should be proved that $MNS_A(x) \subseteq MNS_{AT}(x)$. $\forall MN_A(x) \in MNS_A(x)$, there must be $MN_{AT}(x) \in MNS_{AT}(x)$ such that $MN_{AT}(x) \subseteq MN_A(x)$. By what have been discussed above, we know that $MN_A(x) \subseteq MN_{AT}(x)$ still holds, thus, $MN_{AT}(x) = MN_A(x)$. It tells us that for each $MN_A(x) \in MNS_A(x)$, there must be $MN_{AT}(x) \in MNS_{AT}(x)$ such that $MN_{AT}(x) = MN_A(x)$, i.e. $MNS_A(x) \subseteq MNS_{AT}(x)$.
2. It can be derived directly from the proof of 1.
3. $\Rightarrow$: By condition we have $DNS_A(x) = DNS_{AT}(x)$, thus, for each $DN_{AT}(x) \in DNS_{AT}(x)$, we have $||t|| = DN_{AT}(x) = DN_A(x) \in DNS_A(x)$ where $t \in FDES(AT)$. Therefore, $\forall y \notin DN_{AT}(x)$, $y \notin DN_A(x)$ holds. Since $DN_A(x)$ is the support set of a descriptor t' where $t' \in FDES(A)$, there must be $(a, v) \in t'$ ($(a, v) \in t$) such that $y \notin ||(a, v)||$, from which we can conclude that $A \cap \mathscr{D}(DN_{AT}(x), y) \neq \varnothing$.

   $\Leftarrow$: $\forall ||t|| \in DNS_{AT}(x)$ ($t \in FDES(AT), x \in ||t||$), since $A \subseteq AT$, then there must be $||t'|| \in DNS_A(x)$ ($t' \in FDES(A), x \in ||t'||$) such that $||t|| \subseteq ||t'||$, here $t' \geqslant t$.

$\forall y \notin ||t||$, since $A \cap \mathscr{D}(DN_{AT}(x), y) \neq \emptyset$, then there must be $(a, v) \in t'$ such that $y \notin ||(a, v)||$, $a \in A$. Thus, $y \notin ||t'||$, it follows that $||t'|| \subseteq ||t||$. By combining with the result we obtained above, $||t|| = ||t'||$, that is, $\forall ||t|| \in DNS_{AT}(x)$, there must be $||t'|| \in DNS_A(x)$ such that $||t|| = ||t'||$, i.e. $DNS_{AT}(x) \subseteq DNS_A(x)$.

In the following, it should be proved that $DNS_A(x) \subseteq DNS_{AT}(x)$. $\forall ||t'|| \in DNS_A(x)$, there must be $||t|| \in DES_{AT}(x)$ such that $||t|| \subseteq ||t'||$, here $t' \geqslant t$. By the discussion above, we know that $||t'|| \subseteq ||t||$ still holds, thus, $||t|| = ||t'||$. It tells us that $\forall ||t'|| \in DNS_A(x)$, there must be $||t|| \in DNS_{AT}(x)$ such that $||t|| = ||t'||$, i.e. $DNS_A(x) \subseteq DNS_{AT}(x)$.

4. It can be derived directly from the proof of 3. □

Definition 3.10. Let I be an incomplete information system, $\forall x \in U$, define

$$\Delta_{\mathscr{M}}(x) = \bigwedge_{MN_{AT}(x) \in MNS_{AT}(x)} \left(\bigwedge_{y \notin MN_{AT}(x)} \vee \mathscr{M}(MN_{AT}(x), y) \right) \tag{3.20}$$

$$\Delta_{\mathscr{M}} = \bigwedge_{\mathscr{M}(MN_{AT}(x), y) \in \boldsymbol{M}^{\mathscr{M}}} \vee \mathscr{M}(MN_{AT}(x), y) \tag{3.21}$$

$$\Delta_{\mathscr{D}}(x) = \bigwedge_{DN_{AT}(x) \in DNS_{AT}(x)} \left(\bigwedge_{y \notin DN_{AT}(x)} \vee \mathscr{M}(DN_{AT}(x), y) \right) \tag{3.22}$$

$$\Delta_{\mathscr{D}} = \bigwedge_{\mathscr{D}(DN_{AT}(x), y) \in \boldsymbol{M}^{\mathscr{D}}} \vee \mathscr{D}(DN_{AT}(x), y) \tag{3.23}$$

$\Delta_{\mathscr{M}}(x)$, is referred to as the discernibility function for computing reduct neighborhood system of $MNS_{AT}(x)$ in I, $\Delta_{\mathscr{M}}$ is referred to as the discernibility function for computing reduct neighborhood system of $MNS_{AT}(U)$ in I, $\Delta_{\mathscr{D}}(x)$ is referred to as the discernibility function for computing reduct neighborhood system of $DNS_{AT}(x)$ in I, $\Delta_{\mathscr{D}}$ is referred to as the discernibility function for computing reduct neighborhood system of $DNS_{AT}(U)$ in I.

By using the Boolean Reasoning technique, we can obtain the following theorem immediately by Theorem 3.8.

Theorem 3.9. *Let I be an incomplete information system in which $A \subseteq AT$, $\forall x \in U$,*

1. *$MNS_A(x)$ is the reduct neighborhood system of $MNS_{AT}(x)$ if and only if $\wedge A$ is the prime implicant of $\Delta_{\mathscr{M}}(x)$;*
2. *$MNS_A(U)$ is the reduct neighborhood system of $MNS_{AT}(U)$ if and only if $\wedge A$ is the prime implicant of $\Delta_{\mathscr{M}}$;*
3. *$DNS_A(x)$ is the reduct neighborhood system of $DNS_{AT}(x)$ if and only if $\wedge A$ is the prime implicant of $\Delta_{\mathscr{D}}(x)$;*
4. *$DNS_A(U)$ is the reduct neighborhood system of $DNS_{AT}(U)$ if and only if $\wedge A$ is the prime implicant of $\Delta_{\mathscr{D}}$.*

Example 3.8. Take for instance Table 2.1, following Example 3.2, we obtain the maximal consistent block based neighborhood systems discernibility matrix $\boldsymbol{M}^{\mathscr{M}}$ as Table 3.1 shows, the descriptor based neighborhood systems discernibility matrix $\boldsymbol{M}^{\mathscr{D}}$ as Table 3.2 shows.

Table 3.1 Maximal consistent block based neighborhood systems discernibility matrix $M^{\mathscr{M}}$ in Table 2.1

	x_1	x_2	x_3	x_4	x_5	x_6
$MN_{AT}(x_1)$		P	S	MX	MX	P, M
$MN_{AT}(x_2)$	P		S	P, MX	MX	
$MN_{AT}(x_3)$	S	S		S, MX	S, MX	S
$MN_{AT}(x_4)$	MX	P, MX	S, MX			P
$MN^1_{AT}(x_5)$	MX	P, MX	S, MX	P		P
$MN^2_{AT}(x_5)$	P, M, MX	MX	S, MX	P		
$MN^1_{AT}(x_6)$	P, M	MX	S	P, MX	MX	
$MN^2_{AT}(x_6)$	P, M, MX	MX	S, MX	P		

Table 3.2 Descriptor based neighborhood systems discernibility matrix $\boldsymbol{M}^{\mathscr{D}}$ in Table 2.1

	x_1	x_2	x_3	x_4	x_5	x_6
$DN_{AT}(x_1)$		P	S	MX	MX	P
$DN^1_{AT}(x_2)$	P, M		S	P, MX	MX	
$DN^2_{AT}(x_2)$	P		S	P, MX	MX	M
$DN^1_{AT}(x_3)$	M, S	P, S		S, MX	S, MX	P, S
$DN^2_{AT}(x_3)$	S	P, S		S, MX	S, MX	P, M, S
$DN^3_{AT}(x_3)$	P, S	S		P, S, MX	S, MX	M, S
$DN^1_{AT}(x_4)$	M, MX	P, MX	S, MX			P
$DN^2_{AT}(x_4)$	MX	P, MX	S, MX			P, M
$DN^1_{AT}(x_5)$	M, MX	P, MX	S, MX			P
$DN^2_{AT}(x_5)$	MX	P, MX	S, MX			P, M
$DN^3_{AT}(x_5)$	P, M, MX	MX	S, MX	P		
$DN^4_{AT}(x_5)$	P, MX	MX	S, MX	P		M
$DN^1_{AT}(x_6)$	P, M, MX	MX	S, MX	P		
$DN^2_{AT}(x_6)$	P, M		S	P, MX	MX	

Take for instance x_1, since $MNS_{AT}(x_1) = \{\{x_1\}\}$, then by Eq. (3.20), the discernibility function is

$$\begin{aligned}\Delta_{\mathscr{M}}(x_1) &= \bigwedge_{MN_{AT}(x_1)\in MNS_{AT}(x_1)} \left(\bigwedge_{y\notin MN_{AT}(x_1)} \vee \mathscr{M}(MN_{AT}(x_1), y) \right) \\ &= P \wedge S \wedge MX \wedge (P \vee M)\end{aligned}$$

$$= P \wedge S \wedge MX$$

By Theorem 3.9, we know that $MNS_A(x_1)$ ($A = \{P, S, MX\}$) is the reduct neighborhood system of $MNS_{AT}(x_1)$.

Similarly, we obtain the following reduct neighborhood systems:

1. $MNS_A(x_2)$ ($A = \{P, S, MX\}$) is the reduct neighborhood system of $MNS_{AT}(x_2)$;
2. $MNS_A(x_3)$ ($A = \{S\}$) is the reduct neighborhood system of $MNS_{AT}(x_3)$;
3. $MNS_A(x_4)$ ($A = \{P, MX\}$) is the reduct neighborhood system of $MNS_{AT}(x_4)$;
4. $MNS_A(x_5)$ ($A = \{P, MX\}$) is the reduct neighborhood system of $MNS_{AT}(x_5)$;
5. $MNS_A(x_6)$ ($A = \{P, S, MX\}$) is the reduct neighborhood system of $MNS_{AT}(x_6)$;
6. $MNS_A(U)$ ($A = \{P, S, MX\}$) is the reduct neighborhood system of $MNS_{AT}(U)$;
7. $DNS_A(x_1)$ ($A = \{P, S, MX\}$) is the reduct neighborhood system of $DNS_{AT}(x_1)$;
8. $DNS_{AT}(x_2)$ is the reduct neighborhood system of $DNS_{AT}(x_2)$, i.e. no attribute can be deleted for preserving neighborhood system $DNS_{AT}(x_2)$;
9. $DNS_A(x_3)$ ($A = \{S\}$) is the reduct neighborhood system of $DNS_{AT}(x_3)$;
10. $DNS_A(x_4)$ ($A = \{P, MX\}$) is the reduct neighborhood system of $DNS_{AT}(x_4)$;
11. $DNS_A(x_5)$ ($A = \{P, M, MX\}$) is the reduct neighborhood system of $DNS_{AT}(x_5)$;
12. $DNS_A(x_6)$ ($A = \{P, S, MX\}$) is the reduct neighborhood system of $DNS_{AT}(x_6)$;
13. $DNS_{AT}(U)$ is the reduct neighborhood system of $DNS_{AT}(U)$, i.e. no attribute can be deleted for preserving the original neighborhood system $DNS_{AT}(U)$.

3.3.2 Neighborhood Systems Based Approximate Distribution Reducts

Definition 3.11. Let I be an incomplete decision system in which $A \subseteq AT$, the partition determined by the decision attribute d is $U/IND(\{d\}) = \{X_1, X_2, \cdots, X_l\}$, then $\forall x \in U$, let us denote

$$L_{AT}^{MAX}(x) = \left\{X_i \in U/IND(\{d\}) : x \in \underline{apr_{AT}^{M}}(X_i)\right\}$$

$$U_{AT}^{MAX}(x) = \left\{X_i \in U/IND(\{d\}) : x \in \overline{apr_{AT}^{M}}(X_i)\right\}$$

$$L_{AT}^{DES}(x) = \left\{X_i \in U/IND(\{d\}) : x \in \underline{apr_{AT}^{DES}}(X_i)\right\}$$

$$U_{AT}^{DES}(x) = \left\{X_i \in U/IND(\{d\}) : x \in \overline{apr_{AT}^{DES}}(X_i)\right\}$$

define

1. If $L_A^{MAX}(x) = L_{AT}^{MAX}(x)$, then A is referred to as a maximal consistent block based neighborhood systems lower approximate distribution consistent attributes set for $x \in U$ in I; if A is a maximal consistent block based neighborhood systems lower approximate distribution consistent attributes set for $x \in U$ in I and $\forall B \subset A$, B is not the maximal consistent block based neighborhood systems lower approximate distribution consistent attributes set for $x \in U$ in I, then A is referred to as a maximal consistent block based neighborhood systems lower approximate distribution reduct for $x \in U$ in I.

2. If $U_A^{MAX}(x) = U_{AT}^{MAX}(x)$, then A is referred to as a maximal consistent block based neighborhood systems upper approximate distribution consistent attributes set for $x \in U$ in I; if A is a maximal consistent block based neighborhood systems upper approximate distribution consistent attributes set for $x \in U$ in I and $\forall B \subset A$, B is not the maximal consistent block based neighborhood systems upper approximate distribution consistent attributes set for $x \in U$ in I, then A is referred to as a maximal consistent block based neighborhood systems upper approximate distribution reduct for $x \in U$ in I.
3. If $L_A^{DES}(x) = L_{AT}^{DES}(x)$, then A is referred to as a descriptor based neighborhood systems lower approximate distribution consistent attributes set for $x \in U$ in I; if A is a descriptor based neighborhood systems lower approximate distribution consistent attributes set for $x \in U$ in I and $\forall B \subset A$, B is not the descriptor based neighborhood systems lower approximate distribution consistent attributes set for $x \in U$ in I, then A is referred to as a descriptor based neighborhood systems lower approximate distribution reduct for $x \in U$ in I.
4. If $U_A^{DES}(x) = U_{AT}^{DES}(x)$, then A is referred to as a descriptor based neighborhood systems upper approximate distribution consistent attributes set for $x \in U$ in I; if A is a descriptor based neighborhood systems upper approximate distribution consistent attributes set for $x \in U$ in I and $\forall B \subset A$, B is not the descriptor based neighborhood systems upper approximate distribution consistent attributes set for $x \in U$ in I, then A is referred to as a descriptor based neighborhood systems upper approximate distribution reduct for $x \in U$ in I.

By Definition 3.11, we can see that

- A maximal consistent block based neighborhood systems lower approximate distribution consistent attributes set for $x \in U$ in I is the subset of the attributes, which preserves the belongingness of x into the maximal block neighborhood systems based lower approximations of the decision classes; a maximal consistent block based neighborhood systems lower approximate distribution reduct for $x \in U$ in I is the minimal subset of the attributes, which preserves the belongingness of x into the maximal block neighborhood systems based lower approximations of the decision classes.
- A maximal consistent block based neighborhood systems upper approximate distribution consistent attributes set for $x \in U$ in I is the subset of the attributes, which preserves the belongingness of x into the maximal block neighborhood systems based upper approximations of the decision classes; a maximal consistent block based neighborhood systems upper approximate distribution reduct for $x \in U$ in I is the minimal subset of the attributes, which preserves the belongingness of x into the maximal block neighborhood systems based upper approximations of the decision classes.
- A descriptor based neighborhood systems lower approximate distribution consistent attributes set for $x \in U$ in I is the subset of the attributes, which preserves the belongingness of x into the maximal block neighborhood systems based lower

approximations of the decision classes; a descriptor based neighborhood systems lower approximate distribution reduct for $x \in U$ in I is the minimal subset of the attributes, which preserves the belongingness of x into the maximal block neighborhood systems based lower approximations of the decision classes.

- A descriptor based neighborhood systems upper approximate distribution consistent attributes set for $x \in U$ in I is the subset of the attributes, which preserves the belongingness of x into the maximal block neighborhood systems based upper approximations of the decision classes; a descriptor based neighborhood systems upper approximate distribution reduct for $x \in U$ in I is the minimal subset of the attributes, which preserves the belongingness of x into the maximal block neighborhood systems based upper approximations of the decision classes.

The sets of maximal consistent block lower approximate distribution reducts of x, maximal consistent block upper approximate distribution reducts of x, descriptor lower approximate distribution reducts of x, descriptor upper approximate distribution reducts of x are denoted by RED_M^L, RED_M^U, RED_D^L and RED_D^U, respectively.

Theorem 3.10. *Let I be an incomplete decision system,*

1. *If $L_{AT}^{MAX}(x) = \emptyset$, then $\forall a \in AT$, a is the maximal consistent block based neighborhood systems lower approximate distribution reduct of x;*
2. *If $U_{AT}^{MAX}(x) = U/IND(\{d\})$, then $\forall a \in AT$, a is the maximal consistent block based neighborhood systems upper approximate distribution reduct of x;*
3. *If $L_{AT}^{DES}(x) = \emptyset$, then $\forall a \in AT$, a is the descriptor based neighborhood systems lower approximate distribution reduct of x;*
4. *If $U_{AT}^{DES}(x) = U/IND(\{d\})$, then $\forall a \in AT$, a is the descriptor based neighborhood systems upper approximate distribution reduct of x.*

Proof. We only prove 1, others can be proved analogously.

Since $L_{AT}^{MAX}(x) = \emptyset$, then $\forall X_i \in U/IND(\{d\})$, there is no $MN_{AT}(x) \in MNS_{AT}(x)$ such that $MN_{AT}(x) \subseteq X_i$. Moreover, by $a \in AT$, we know that $\forall MN_a(x) \in MNS_a(x)$, there must be $MN_{AT}(x) \in MNS_{AT}(x)$ such that $MN_a(x) \supseteq MN_{AT}(x)$, from which we can conclude that $MN_a(x) \nsubseteq X_i$ for each $X_i \in U/IND(\{d\})$, i.e. $L_a^{MAX}(x) = \emptyset$, a is the maximal consistent block lower approximate distribution reduct of x. □

Example 3.9. Let us consider the data showed in Table 2.3, following Example 3.7, we obtain $L_{AT}^{MAX}(x)$, $U_{AT}^{MAX}(x)$, $L_{AT}^{DES}(x)$ and $U_{AT}^{DES}(x)$ for each object showed in Table 3.3.

Table 3.3 $L_{AT}^{MAX}(x)$, $U_{AT}^{MAX}(x)$, $L_{AT}^{DES}(x)$ and $U_{AT}^{DES}(x)$ for each object in Table 2.3

	x_1	x_2	x_3	x_4	x_5	x_6
$L_{AT}^{MAX}(x_i)$	X_2	X_2	X_1	$\emptyset$	$\emptyset$	X_2
$U_{AT}^{MAX}(x_i)$	X_2	X_2	X_1	$\{X_2, X_3\}$	$\{X_2, X_3\}$	X_2
$L_{AT}^{DES}(x_i)$	X_2	X_2	X_1	$\emptyset$	X_3	X_2
$U_{AT}^{DES}(x_i)$	X_2	X_2	X_1	$\{X_2, X_3\}$	X_3	X_2

Take for instance x_4, since $\underline{L}_{AT}^{MAX}(x_4) = \emptyset$, then $\forall X_i \in U/IND(\{d\})$, $MN_{AT}(x) \subseteq X_i$ does not hold for each $MN_{AT}(x) \in MNS_{AT}(x)$.

Moreover, consider the attribute P in Table 2.3, since

$$\mathscr{M}(P) = \Big\{\{x_1, x_3, x_4, x_5\}, \{x_2, x_3, x_5, x_6\}\Big\}$$

then $MNS_P(x_4) = \{MN_P(x_4)\} = \Big\{\{x_1, x_3, x_4, x_5\}\Big\}$, from which we have $MN_P(x_4) \nsubseteq X_1, X_2, X_3$, i.e. $\underline{L}_P^{MAX}(x_4) = \emptyset$, i.e. $\{P\}$ is the maximal consistent block based neighborhood systems lower approximate distribution reduct of x_4.

Similarly, it is not difficult to obtain that $\{M\}, \{S\}, \{MX\}$ are all maximal consistent block based neighborhood systems lower approximate distribution reducts of x_4.

Moreover, since $\underline{L}_{AT}^{MAX}(x_5) = \emptyset$, then $\{P\}, \{M\}, \{S\}, \{MX\}$ are all maximal consistent block based neighborhood systems lower approximate distribution reducts of x_5.

Since $\underline{L}_{AT}^{DES}(x_4) = \emptyset$, then $\{P\}, \{M\}, \{S\}, \{MX\}$ are all descriptor based neighborhood systems lower approximate distribution reducts of x_4.

Theorem 3.10 can be used to obtain different types of reducts for object in I under some special situations. However, if the condition parts in the above theorem do not hold, we can use the following approach to obtain different types of reducts of I for $x \in U$.

Theorem 3.11. *Let I be an incomplete decision system in which $A \subseteq AT$, then $\forall x \in U$*

1. *A is a maximal consistent block based neighborhood systems lower approximate distribution consistent attributes set for x in I $\Leftrightarrow$ $\forall X_i \in \underline{L}_{AT}^{MAX}(x)$, $\exists MN_{AT}(x) \subseteq X_i$ s.t. $A \cap \mathscr{M}(MN_{AT}(x), y) \neq \emptyset$ where $\forall y \notin X_i$;*
2. *A is a maximal consistent block based neighborhood systems upper approximate distribution consistent attributes set for x in I $\Leftrightarrow$ $\forall X_i \notin \overline{U}_{AT}^{MAX}(x)$, $\exists MN_{AT}(x) \cap X_i = \emptyset$ s.t. $A \cap \mathscr{M}(MN_{AT}(x), y) \neq \emptyset$ where $\forall y \in X_i$;*
3. *A is a descriptor based neighborhood systems lower approximate distribution consistent attributes set for x in I $\Leftrightarrow$ $\forall X_i \in \underline{L}_{AT}^{DES}(x)$, $\exists DN_{AT}(x) \subseteq X_i$ s.t. $A \cap \mathscr{D}(DN_{AT}(x), y) \neq \emptyset$ where $\forall y \notin X_i$;*
4. *A is a descriptor based neighborhood systems upper approximate distribution consistent attributes set for x in I $\Leftrightarrow$ $\forall X_i \notin \overline{U}_{AT}^{DES}(x)$, $\exists DN_{AT}(x) \cap X_i = \emptyset$ s.t. $A \cap \mathscr{D}(DN_{AT}(x), y) \neq \emptyset$ where $\forall y \in X_i$.*

Proof.

1. $\Rightarrow$: $\forall X_i \in \underline{L}_{AT}^{MAX}(x)$, since $\underline{L}_A^{MAX}(x) = \underline{L}_{AT}^{MAX}(x)$, then there must be $MN_A(x) \in MNS_A(x)$ such that $MN_A(x) \subseteq X_i$. By the basic property of maximal consistent block, we know that there must be $MN_{AT}(x) \in MNS_{AT}(x)$ where $MN_{AT}(x) \subseteq MN_A(x)$, it follows that $MN_{AT}(x) \subseteq X_i$. $\forall y \notin X_i$, we know $y \notin MN_{AT}(x)$ and $y \notin MN_A(x)$, it follows that there must be $z \in MN_{AT}(x) \subseteq MN_A(x)$ where $(z, y) \notin TOL(A)$, from which we can conclude that $A \cap \mathscr{M}(MN_{AT}(x), y) \neq \emptyset$.

   $\Leftarrow$: Since $A \subseteq AT$, then $\underline{L}_A^{MAX}(x) \subseteq \underline{L}_{AT}^{MAX}(x)$. Thus, it must be proved that $\underline{L}_A^{MAX}(x) \supseteq \underline{L}_{AT}^{MAX}(x)$. $\forall X_i \in \underline{L}_{AT}^{MAX}(x)$, by condition we know that there must be

$MN_{AT}(x) \in MNS_{AT}(x)$ such that $A \cap \mathcal{M}(MN_{AT}(x), y) \neq \emptyset$ where $\forall y \notin X_i$. Therefore, we know that $(MN_{AT} \cup \{y\})^2 \nsubseteq TOL(A)$. By the basic property of maximal consistent block, there must be $MN_A(x) \in MNS_A(x)$ such that $MN_{AT}(x) \subseteq MN_A(x)$. Since $(MN_{AT} \cup \{y\})^2 \nsubseteq TOL(A)$ for each $y \notin X_i$, then we can conclude that $MN_A(x) \subseteq X_i$ holds, i.e. $X_i \in L_A^{MAX}(x)$, $L_A^{MAX}(x) \supseteq L_{AT}^{MAX}(x)$.

2. $\Rightarrow$: $\forall X_i \notin U_{AT}^{MAX}(x)$, since $L_A^{MAX}(x) = L_{AT}^{MAX}(x)$, then $\exists MN_A(x) \in MNS_A(x)$, $MN_A(x) \cap X_i = \emptyset$ holds. By the basic property of maximal consistent block, we know that there must be $MN_{AT}(x) \in MNS_{AT}(x)$ where $MN_{AT}(x) \subseteq MN_A(x)$, it follows that $MN_{AT}(x) \cap X_i = \emptyset$. $\forall y \in X_i$, we know $y \notin MN_{AT}(x)$ and $y \notin MN_A(x)$, it follows that there must be $z \in MN_{AT}(x) \subseteq MN_A(x)$ where $(z, y) \notin TOL(A)$, from which we can conclude that $A \cap \mathcal{M}(MN_{AT}(x), y) \neq \emptyset$.
$\Leftarrow$: Since $A \subseteq AT$, then $U_A^{MAX}(x) \supseteq U_{AT}^{MAX}(x)$. Thus, it must be proved that $U_A^{MAX}(x) \subseteq U_{AT}^{MAX}(x)$. $\forall X_i \notin U_{AT}^{MAX}(x)$, by condition we know that there must be $MN_{AT}(x) \in MNS_{AT}(x)$ such that $A \cap \mathcal{M}(MN_{AT}(x), y) \neq \emptyset$ where $\forall y \in X_i$. Therefore, we know that $(MN_{AT} \cup \{y\})^2 \nsubseteq TOL(A)$. By the basic property of maximal consistent block, there must be $MN_A(x) \in MNS_A(x)$ such that $MN_{AT}(x) \subseteq MN_A(x)$. Since $(MN_{AT} \cup \{y\})^2 \nsubseteq TOL(A)$ for each $y \notin X_i$, then we can conclude that $MN_A(x) \cap X_i = \emptyset$ holds, i.e. $X_i \notin U_A^{MAX}(x)$, $U_A^{MAX}(x) \subseteq U_{AT}^{MAX}(x)$.
3. The proof of 3 is similar to the proof of 1.
4. The proof of 4 is similar to the proof 2. □

Definition 3.12. Let I be an incomplete information system, $\forall x \in U$, define

$$\Delta_L^{MAX}(x) = \bigwedge_{X_i \in L_{AT}^{M}(x)} \bigvee_{MN_{AT}(x) \subseteq X_i} \left(\bigwedge_{y \notin X_i} \vee \mathcal{M}(MN_{AT}(x), y) \right) \tag{3.24}$$

$$\Delta_U^{MAX}(x) = \bigwedge_{X_i \notin U_{AT}^{M}(x)} \bigvee_{MN_{AT}(x) \cap X_i = \emptyset} \left(\bigwedge_{y \in X_i} \vee \mathcal{M}(MN_{AT}(x), y) \right) \tag{3.25}$$

$$\Delta_L^{DES}(x) = \bigwedge_{X_i \in L_{AT}^{DES}(x)} \bigvee_{DN_{AT}(x) \subseteq X_i} \left(\bigwedge_{y \notin X_i} \vee \mathcal{D}(DN_{AT}(x), y) \right) \tag{3.26}$$

$$\Delta_U^{DES}(x) = \bigwedge_{X_i \notin U_{AT}^{DES}(x)} \bigvee_{DN_{AT}(x) \cap X_i = \emptyset} \left(\bigwedge_{y \in X_i} \vee \mathcal{D}(DN_{AT}(x), y) \right) \tag{3.27}$$

$\Delta_L^{MAX}(x)$ is referred to as the maximal consistent block neighborhood systems based lower approximate discernibility function for x in I, $\Delta_U^{MAX}(x)$ is referred to as the maximal consistent block neighborhood systems based upper approximate discernibility function for x in I, $\Delta_L^{DES}(x)$ is referred to as the descriptor neighborhood systems based lower approximate discernibility function for x in I, $\Delta_U^{DES}(x)$ is referred to as the descriptor neighborhood systems based upper approximate discernibility function for x in I.

By using Boolean Reasoning technique, we can obtain the following theorem immediately by Theorem 3.11.

Theorem 3.12. *Let I be an incomplete information system in which $A \subseteq AT$, $\forall x \in U$,*

1. *A is referred to as a maximal consistent block based neighborhood systems lower approximate distribution reduct for x in I if and only if A is the prime implicant of the maximal consistent block neighborhood systems based lower approximate discernibility function for x in I, i.e. $\Delta_L^{MAX}(x)$;*
2. *A is referred to as a maximal consistent block based neighborhood systems upper approximate distribution reduct for x in I if and only if A is the prime implicant of the maximal consistent block neighborhood systems based upper approximate discernibility function for x in I, i.e. $\Delta_U^{MAX}(x)$;*
3. *A is referred to as a descriptor based neighborhood systems lower approximate distribution reduct for x in I if and only if A is the prime implicant of the descriptor neighborhood systems based lower approximate discernibility function for x in I, i.e. $\Delta_L^{DES}(x)$;*
4. *A is referred to as a descriptor based neighborhood systems upper approximate distribution reduct for x in I if and only if A is the prime implicant of the descriptor neighborhood systems based upper approximate discernibility function for x in I, i.e. $\Delta_U^{DES}(x)$.*

Example 3.10. Take for instance Table 2.3, following Example 3.9, we have $L_{AT}^{MAX}(x_1) = X_2$, thus,

$$\begin{aligned}
\Delta_L^{MAX}(x_1) &= \bigwedge_{X_i \in L_{AT}^{MAX}(x_1)} \bigvee_{MN_{AT}(x_1) \subseteq X_i} \left(\bigwedge_{y \notin X_i} \vee \mathscr{M}(MN_{AT}(x_1), y) \right) \\
&= \bigvee_{MN_{AT}(x_1) \subseteq X_i} \left(\bigwedge_{y \notin X_2} \vee \mathscr{M}(MN_{AT}(x_1), y) \right) \\
&= (\vee \mathscr{M}(MN_{AT}(x_1), x_3)) \wedge (\vee \mathscr{M}(MN_{AT}(x_1), x_5)) \\
&= S \wedge MX
\end{aligned}$$

By Theorem 3.12, we know that $\{S, MX\}$ is the maximal consistent block based neighborhood systems lower approximate distribution reduct for x_1. Similarly, we obtain the four types of the neighborhood systems based approximate distribution reducts in Table 2.3 as Table 3.4 shows.

Table 3.4 Four types of neighborhood systems based approximate distribution reducts for x in Table 2.3

	x_1	x_2	x_3	x_4	x_5	x_6
maximal consistent block based neighborhood systems lower approximate distribution reduct for x	$\{\{S, MX\}\}$	$\{\{S, MX\}\}$	$\{\{S\}\}$	$\{\{P\},\{M\},\{S\},\{MX\}\}$	$\{\{P\},\{M\},\{S\},\{MX\}\}$	$\{\{S, MX\}\}$
maximal consistent block based neighborhood systems upper approximate distribution reduct for x	$\{\{S, MX\}\}$	$\{\{S, MX\}\}$	$\{\{S\}\}$	$\{\{S\},\{MX\}\}$	$\{\{S\},\{MX\}\}$	$\{\{MX\}\}$

Continued

	x_1	x_2	x_3	x_4	x_5	x_6
descriptor based neighborhood systems lower approximate distribution reduct for x	$\{\{S, MX\}\}$	$\{\{S, MX\}\}$	$\{\{S\}, \{P, M, MX\}\}$	$\{\{P\}, \{M\}, \{S\}, \{MX\}\}$	$\{\{P, M, MX\}\}$	$\{\{S, MX\}\}$
descriptor based neighborhood systems upper approximate distribution reduct for x	$\{\{S, MX\}\}$	$\{\{S, MX\}\}$	$\{\{S\}, \{P, M, MX\}\}$	$\{\{S\}, \{MX\}\}$	$\{\{P, M, MX\}\}$	$\{\{MX\}\}$

3.4 Conclusions

Neighborhood system granular computing model, is also referred to as the local granular computing model, plays a very important role in the development of the theory of granular computing. In this chapter, the following results can be concluded.

1. A new operation is defined on the neighborhood system, from which new knowledge, i.e. neighborhoods, can be derived from the original neighborhood system. By comparing with the original neighborhood system, the expanded neighborhood system not only contains more knowledge, but also can induce a topological space on the universe because it satisfies with the intersection condition.
2. The mathematical structure of the neighborhood system is studied. The expanded neighborhood system can induce a semigroup. Such result is consistent to Pawlak's knowledge representation system and then it provides a way to the discussion of representation of neighborhood system.
3. The maximal consistent block and descriptor techniques, the neighborhood system based rough set approach is introduced into the incomplete information system. By comparing with the traditional maximal consistent blocks and descriptors based rough sets, we can obtain the smaller upper approximations if the neighborhood systems based rough sets are employed. From this point of view, neighborhood system based rough set may be more suitable for the dealing with incomplete information systems with "*" unknown values.

References

[1] Lin, T. Y.: Granular computing: Practices, theories, and future directions. Encyclopedia on Complexity of Systems Science 4339–4355 (2009)

[2] Lin, T. Y.: Neighborhood systems and approximation in relational databases and knowledge bases. In: Proceedings of the 4th International Symposium on Methodologies of Intelligent Systems, pp. 75–86 (1989)

[3] Lin, T. Y.: Neighborhood Systems: A qualitative theory for fuzzy and rough sets. In: Wang, P. (eds.), Advances in Machine Intelligence and Soft Computing, pp. 132–155 (1997)

[4] Lin, T. Y.: Granular computing: From rough sets and neighborhood systems to information granulation and computing in words. In: European Congress on Intelligent Techniques and Soft Computing, pp. 1602–1606 (1997)

[5] Lin, T. Y.: Granular computing on binary relations I: Data mining and neighborhood systems. In: Skowron, A., Polkowski, L. (eds.), Rough Sets in Knowledge Discovery, pp. 107–121 (1998)

[6] Lin, T. Y.: Granular computing on binary relations II: Rough set representations and belief functions. In: Skowron, A., Polkowski, L. (eds.), Rough Sets in Knowledge Discovery, pp. 121–140 (1998)

[7] Lin, T. Y.: Neighborhood systems: Mathematical models of information granulations. In: 2003 IEEE International Conference on Systems, Man and Cybernetics, pp. 3188–3193 (2003)

[8] Lin, T. Y.: Granular computing on partitions, coverings and neighborhood systems. In: Proceedings of the International Forum on Theory of GrC from Rough Set Perspective, pp. 1–7 (2006)

[9] Lin, T. Y.: Granular computing I: The concept of granulation and its formal model, International Journal of Granular Computing, Rough Sets and Intelligent Systems **1**, 21–42 (2009)

[10] Lin, T. Y., Zadeh, L. A.: Foreword to the special section on granular computing. IEEE T. Fuzzy Syst. **16**, 282–284 (2008)

[11] Wu, W. Z., Zhang, W. X.: Neighborhood operator systems and approximations. Inform. Sci. **144**, 201–217 (2002)

[12] Yao, Y. Y.: Information granulation and rough set approximation. Int. J. Intell. Syst. **16**, 87–104 (2001)

[13] Yao, Y. Y.: Rough sets, neighborhood systems, and granular computing. In: Proceedings of the 1999 IEEE Canadian Conference on Electrical and Computer Engineering, pp. 1553–1558 (1999)

[14] Yao, Y. Y.: Granular computing using neighborhood systems. In: The 3rd On–line World Conference on Soft Computing, pp. 539–553 (1999)

[15] Yao, Y. Y.: Information tables with neighborhood semantics. In: Data Mining and Knowledge Discovery: Theory, Tools, and Technology II, pp. 108–116 (2000)

[16] Yao, Y. Y.: Neighborhood systems and approximate retrieval. Inform. Sci. **176**, 3431–3452 (2006)

[17] Qian, Y. H., Liang, J. Y., Dang, C. Y.: Knowledge structure, knowledge granulation and knowledge distance in a knowledge base. Int. J. Approx. Reason. **50**, 174–188 (2009)

[18] Qian, Y. H., Liang, J. Y., Wu, W. Z.: Information granularity in fuzzy binary GrC model. IEEE T. Fuzzy Syst. **19**, 253–264 (2011)

[19] Yang, X. B., Lin, T. Y.: Knowledge operations in neighborhood system. In: 2010 IEEE International Conference on Granular Computing, pp. 822–825 (2010)

Part III
Dominance-based Rough Sets and Incomplete Information Systems

Chapter 4
Dominance-based Rough Sets in "∗" Incomplete Information System

Abstract Dominance-based rough set approach is a very important expansion of Pawlak's rough set approach since the former takes the preference-ordered domains of the attributes into account. In this chapter, the dominance-based rough set approach is introduced into the incomplete information system, in which all unknown values can be compared with any other values in the domains of the corresponding attributes. The "↑" and "↓" descriptors are employed to generate all certain rules from the incomplete information system. Moreover, the expanded dominance relation is also compared with the limited dominance relation, from which we can conclude that the limited dominance-based rough set approach is more suitable than the expanded dominance-based rough set approach when dealing with the incomplete information system.

Pawlak's rough set model, was constructed on the basis of an indiscernibility relation. Such relation can be used to deal with the regular attributes such as symptoms, colors, and textural features. Nevertheless, the indiscernibility relation based rough set approach is not able to discover inconsistencies coming from the consideration of criteria, that is, attributes with preference-ordered domains (scales), such as product quality, market share, and debt ratio. To solve such problem, Greco et al. [1–7] have proposed an extension of Pawlak's rough set, which is referred to as the dominance-based Rough Set Approach (DRSA). Presently, work on the dominance-based rough set progressing rapidly. For example, by considering three types of monotonicity properties in rough set, Błaszczyńki et al. [8–11] proposed the monotonic variable consistency dominance-based rough set. Dembczyński et al. [12, 13], Qian et al. [14] and Yang et al. [15] introduced the dominance-based rough set into interval-valued environment, respectively. Fan et al. [16] proposed some decision logic languages for rule representation from the viewpoint of the dominance-based rough set. Greco et al. [1, 6] and Hu et al. [17–19] generalized the dominance-based rough set into fuzzy environment, respectively. Błaszczyńki et al. proposed the variable consistency dominance-based rough set approach [8, 9, 11]. Inuiguchi [20] et al. proposed the variable precision dominance-based rough set. Kottwski et al. [21] proposed the stochastic dominance-based rough set.

In this chapter, we are mainly focusing on the dominance-based rough set approaches to incomplete information system, in which all unknown values are "do not care" conditions. Following Chapter 2, such unknown values are still denoted by "*" in this chapter.

4.1 Dominance-based Rough Set

Let $\succeq_a$ be a weak preference relation on U (often called outranking), which can be used to represent a preference on the set of objects with respect to criterion a; $x \succeq_a y$ means "x is at least as good as y with respect to criterion a". We say that x dominates y with respect to $A \subseteq AT$, (or, x A-dominates y), denoted by $(x, y) \in DOM(A)$, if $x \succeq_a y$ for each $a \in A$. Assuming, without loss of generality, that domains of all criteria are ordered such that preference increases with the values, in other words, $(x, y) \in DOM(A)$ is equivalent to: $f(x, a) \geqslant f(y, a)$ for each $a \in A$. From this point of view, Greco et al. defined the following two sets for each $x \in U$:

- A set of objects dominating x, called A-dominating set, $DOM_A^+(x) = \{y \in U : (y, x) \in DOM(A)\}$;
- A set of objects dominated by x, called A-dominated set, $DOM_A^-(x) = \{y \in U : (x, y) \in DOM(A)\}$.

Moreover, assume that the decision attribute d makes a partition of U into a finite number of classes; let $\boldsymbol{CL} = \{CL_n, n \in N\}$, $N = \{1, 2, \cdots, m\}$, be a set of these classes that are ordered, that is, $\forall r_1, r_2 \in N$ such that $r_1 > r_2$, the objects from CL_{r_1} are preferred to the objects from CL_{r_2}. Therefore, in dominance-based rough set approach, the sets to be approximated are upward unions and downward unions of those classes, which are defined as $CL_n^{\geqslant} = \bigcup\limits_{n' \geqslant n} CL_{n'}$ and $CL_n^{\leqslant} = \bigcup\limits_{n' \leqslant n} CL_{n'}$, respectively where $n, n' \in N$. The statement $x \in CL_n^{\geqslant}$ means "x belongs to at least class CL_n", where $x \in CL_n^{\leqslant}$ means "x belongs to at most class CL_n".

Definition 4.1. Let I be a complete decision system in which $A \subseteq AT$, $n \in N$, the lower and upper approximations of $CL_n^{\geqslant}$ are defined as:

$$\underline{DOM_A}(CL_n^{\geqslant}) = \{x \in U : DOM_A^+(x) \subseteq CL_n^{\geqslant}\} \tag{4.1}$$

$$\overline{DOM_A}(CL_n^{\geqslant}) = \{x \in U : DOM_A^-(x) \cap CL_n^{\geqslant} \neq \emptyset\} \tag{4.2}$$

the lower and upper approximations of $CL_n^{\leqslant}$ are defined as:

$$\underline{DOM_A}(CL_n^{\leqslant}) = \{x \in U : DOM_A^-(x) \subseteq CL_n^{\leqslant}\} \tag{4.3}$$

$$\overline{DOM_A}(CL_n^{\leqslant}) = \{x \in U : DOM_A^+(x) \cap CL_n^{\leqslant} \neq \emptyset\} \tag{4.4}$$

The pairs $[\underline{DOM_A}(CL_n^{\geqslant}), \overline{DOM_A}(CL_n^{\geqslant})]$ and $[\underline{DOM_A}(CL_n^{\leqslant}), \overline{DOM_A}(CL_n^{\leqslant})]$ are referred to as the dominance-based rough sets of $CL_n^{\geqslant}$ and $CL_n^{\leqslant}$, respectively. Similar to the classical rough set approach, the boundary regions of the unions of decision classes $CL_n^{\geqslant}$ and $CL_n^{\leqslant}$ are:

$$BN_A(CL_n^{\geqslant}) = \overline{DOM_A}(CL_n^{\geqslant}) - \overline{DOM_A}(CL_n^{\geqslant}) \tag{4.5}$$

$$BN_A(CL_n^{\leqslant}) = \overline{DOM_A}(CL_n^{\leqslant}) - \overline{DOM_A}(CL_n^{\leqslant}) \tag{4.6}$$

Definition 4.2. Let I be a complete decision system in which $A \subseteq AT$, the quality of approximation of multi-criteria classification $\boldsymbol{CL}$ by A is defined as

$$\gamma_A(\boldsymbol{CL}) = \frac{\left|U - \bigcup_{n \in N} BN_A(CL_n^{\geqslant})\right|}{|U|} = \frac{\left|U - \bigcup_{n \in N} BN_A(CL_n^{\leqslant})\right|}{|U|} \tag{4.7}$$

By Definition 4.2, we can see that the quality of approximation of multi-criteria classification $\boldsymbol{CL}$ is defined as the ratio between the number of A-correctly classified objects and the number of all the objects in the data table.

Proposition 4.1. *Let I be a complete decision system in which $A \subseteq AT$, then we have following properties about the dominance-based lower and upper approximations:*

1. $\forall n \in N$,

$$\underline{DOM_A}(CL_n^{\geqslant}) \subseteq CL_n^{\geqslant} \subseteq \overline{DOM_A}(CL_n^{\geqslant}) \tag{4.8}$$

$$\underline{DOM_A}(CL_n^{\leqslant}) \subseteq CL_n^{\leqslant} \subseteq \overline{DOM_A}(CL_n^{\leqslant}) \tag{4.9}$$

2. $\forall n \in N$,

$$\overline{DOM_A}(CL_n^{\geqslant}) = \bigcup_{x \in CL_n^{\geqslant}} DOM_A^{+}(x) \tag{4.10}$$

$$\overline{DOM_A}(CL_n^{\leqslant}) = \bigcup_{x \in CL_n^{\leqslant}} DOM_A^{-}(x) \tag{4.11}$$

3. a. $\underline{DOM_A}(CL_n^{\geqslant}) = U - \overline{DOM_A}(CL_{n-1}^{\leqslant})$ $(n = 2, \cdots, m)$,
   b. $\underline{DOM_A}(CL_n^{\leqslant}) = U - \overline{DOM_A}(CL_{n+1}^{\geqslant})$ $(n = 1, \cdots, m-1)$,
   c. $\overline{DOM_A}(CL_n^{\geqslant}) = U - \underline{DOM_A}(CL_{n-1}^{\leqslant})$ $(n = 2, \cdots, m)$,
   d. $\overline{DOM_A}(CL_n^{\leqslant}) = U - \underline{DOM_A}(CL_{n+1}^{\geqslant})$ $(n = 1, \cdots, m-1)$,
   e. $BN_A(CL_n^{\geqslant}) = BN_A(CL_{n-1}^{\leqslant})$ $(n = 2, \cdots, m)$,
   f. $BN_A(CL_n^{\leqslant}) = BN_A(CL_{n+1}^{\geqslant})$ $(n = 1, \cdots, m-1)$;

4. $\forall n \in N$,

   a. $\underline{DOM_A}(CL_n^{\geqslant}) \subseteq \underline{DOM_{AT}}(CL_n^{\geqslant})$,
   b. $\overline{DOM_A}(CL_n^{\geqslant}) \supseteq \overline{DOM_{AT}}(CL_n^{\geqslant})$,
   c. $\underline{DOM_A}(CL_n^{\leqslant}) \subseteq \underline{DOM_{AT}}(CL_n^{\leqslant})$,
   d. $\overline{DOM_A}(CL_n^{\leqslant}) \supseteq \overline{DOM_{AT}}(CL_n^{\leqslant})$,

e. $BN_A(CL_n^{\geqslant}) \supseteq BN_{AT}(CL_n^{\geqslant})$,

f. $BN_A(CL_n^{\leqslant}) \supseteq BN_{AT}(CL_n^{\leqslant})$.

The end result of the dominance-based rough set is a representation of the information contained in the data system considered in terms of "if$\cdots$then$\cdots$" decision rules. Intuitively, this is linked to the interpretation of the rough approximations in terms of the following implications:

- $\underline{DOM_A}(CL_n^{\geqslant}) = \{x \in U : (y, x) \in DOM(A) \Rightarrow y \in CL_n^{\geqslant}\}$;
- $\overline{DOM_A}(CL_n^{\geqslant}) = \{x \in U :$ if $(y, x) \in DOM(A)$, then y could belong $CL_n^{\geqslant}\}$;
- $\underline{DOM_A}(CL_n^{\leqslant}) = \{x \in U : (x, y) \in DOM(A) \Rightarrow y \in CL_n^{\leqslant}\}$;
- $\overline{DOM_A}(CL_n^{\leqslant}) = \{x \in U :$ if $(x, y) \in DOM(A)$, then y could belong $CL_n^{\leqslant}\}$.

Therefore, by using the dominance-based rough approximations, the knowledge hidden in a decision system I may be unravelled and expressed in the form of decision rules ("at least" and "at most" rules) from every training example x in I such that:

- Certain "at least" rule: if $f(x, a_1) \geqslant r_{a_1}$ and $f(x, a_2) \geqslant r_{a_2}$ and $\cdots$ and $f(x, a_k) \geqslant r_{a_k}$, then $x \in CL_n^{\geqslant}$;
- Possible "at least" rule: if $f(x, a_1) \geqslant r_{a_1}$ and $f(x, a_2) \geqslant r_{a_2}$ and $\cdots$ and $f(x, a_k) \geqslant r_{a_k}$, then x could belongs to $CL_n^{\geqslant}$;
- Certain "at most" rule: if $f(x, a_1) \leqslant r_{a_1}$ and $f(x, a_2) \leqslant r_{a_2}$ and $\cdots$ and $f(x, a_k) \leqslant r_{a_k}$, then $x \in CL_n^{\leqslant}$;
- Possible "at most" rule: if $f(x, a_1) \leqslant r_{a_1}$ and $f(x, a_2) \leqslant r_{a_2}$ and $\cdots$ and $f(x, a_k) \leqslant r_{a_k}$, then x could belongs to $CL_n^{\leqslant}$.

where $A = \{a_1, a_2, \cdots, a_k\} \subseteq AT$, $(r_{a_1}, r_{a_2}, \cdots, r_{a_k}) \in V_{a_1} \times V_{a_2} \times \cdots \times V_{a_k}$ and $n \in N$.

Remark 4.1. It should be noticed that since $CL_1^{\geqslant} = U$, then for each $x \in U$, there must be $x \in CL_1^{\geqslant}$ regardless of the attributes values of x are. For such reason, it is meaningless to obtain the "at least" rule if $n = 1$. Similarly, it is meaningless to obtain the "at most" rule if $n = m$. From discussions above, what rules will be explored in the following are all based on the upward unions of decision classes $CL_n^{\geqslant}(2 \leqslant n \leqslant m)$ and downward unions of decision classes $CL_n^{\leqslant}(1 \leqslant n \leqslant m-1)$.

4.2 Expanded Dominance-based Rough Set in Incomplete Information System with "*" Unknown Values

The classical dominance-based rough set was firstly proposed for dealing with the complete information systems. In Chapter 2, we have known that incomplete information systems can be seen everywhere in the real world applications. Therefore, it is important to introduce the dominance-based rough set approach into incomplete environment. In Ref. [22], Shao and Zhang have successfully expanded the dominance relation and the dominance-based rough set into incomplete information system, in which the unknown values can be compared with any other values in the domains of the corresponding attributes. Such expansion of the dominance relation is showed in Definition 4.3.

Definition 4.3. Let I be an incomplete decision system in which $A \subseteq AT$, the expanded dominance relation in terms of A is defined as:

$$EDOM(A)=\{(x,y) \in U^2 : \forall a \in A, f(x,a) \geqslant f(y,a) \vee f(x,a)=* \vee f(y,a) = *\} \quad (4.12)$$

Obviously, $EDOM(A)$ is reflexive, but not necessarily symmetric and transitive. By $EDOM(A)$, we can also define the following two sets for each $x \in U$:

- $EDOM_A^+(x)=\{y \in U : (y,x) \in EDOM(A)\}$ is the set of objects that may dominate x in terms of A;
- $EDOM_A^-(x)=\{y \in U : (x,y) \in EDOM(A)$ is the set of objects that may be dominated by x in terms of A.

Theorem 4.1. *Let I be an incomplete decision system in which $A \subseteq AT$, we have*

$$EDOM(A) = \bigcap_{a \in A} EDOM(\{a\}) \quad (4.13)$$

$$EDOM(AT) \subseteq EDOM(A) \quad (4.14)$$

Proof. It can be derived directly from Definition 4.3. □

Definition 4.4. Let I be an incomplete complete decision system in which $A \subseteq AT$, $n \in N$, the lower and upper approximations of $CL_n^{\geqslant}$ in terms of the expanded dominance relation $EDOM(A)$ are defined as:

$$\underline{EDOM_A}(CL_n^{\geqslant})=\{x \in U : EDOM_A^+(x) \subseteq CL_n^{\geqslant}\} \quad (4.15)$$

$$\overline{EDOM_A}(CL_n^{\geqslant})=\{x \in U : EDOM_A^-(x) \cap CL_n^{\geqslant} \neq \varnothing\} \quad (4.16)$$

the lower and upper approximations of $CL_n^{\leqslant}$ in terms of the expanded dominance relation $EDOM(A)$ are defined as:

$$\underline{EDOM_A}(CL_n^{\leqslant})=\{x \in U : EDOM_A^-(x) \subseteq CL_n^{\leqslant}\} \quad (4.17)$$

$$\overline{EDOM_A}(CL_n^{\leqslant})=\{x \in U : EDOM_A^+(x) \cap CL_n^{\leqslant} \neq \varnothing\} \quad (4.18)$$

By Definitioin 4.4 the boundary regions of the unions of decision classes $CL_n^{\geqslant}$ and $CL_n^{\leqslant}$ are:

$$BN_A^E(CL_n^{\geqslant}) = \overline{EDOM_A}(CL_n^{\geqslant}) - \overline{EDOM_A}(CL_n^{\geqslant}) \quad (4.19)$$

$$BN_A^E(CL_n^{\leqslant}) = \overline{EDOM_A}(CL_n^{\leqslant}) - \overline{EDOM_A}(CL_n^{\leqslant}) \quad (4.20)$$

Definition 4.5. Let I be an incomplete complete decision system in which $A \subseteq AT$, the quality of approximation of multi-criteria classification $\boldsymbol{CL}$ in terms of the expanded dominance relation $EDOM(A)$ is defined as:

$$\gamma_A^E(CL) = \frac{\left|U - \bigcup_{n\in N} BN_A^E(CL_n^{\geqslant})\right|}{|U|} = \frac{\left|U - \bigcup_{n\in N} BN_A^E(CL_n^{\leqslant})\right|}{|U|} \tag{4.21}$$

Proposition 4.2. *Let I be an incomplete complete decision system in which $A \subseteq AT$, then then we have following properties about the expanded dominance-based lower and upper approximations:*

1. $\forall n \in N$,

$$\underline{EDOM_A}(CL_n^{\geqslant}) \subseteq CL_n^{\geqslant} \subseteq \overline{EDOM_A}(CL_n^{\geqslant}) \tag{4.22}$$

$$\underline{EDOM_A}(CL_n^{\leqslant}) \subseteq CL_n^{\leqslant} \subseteq \overline{EDOM_A}(CL_n^{\leqslant}) \tag{4.23}$$

2. $\forall n \in N$,

$$\overline{EDOM_A}(CL_n^{\geqslant}) = \bigcup_{x\in CL_n^{\geqslant}} EDOM_A^{+}(x) \tag{4.24}$$

$$\overline{EDOM_A}(CL_n^{\leqslant}) = \bigcup_{x\in CL_n^{\leqslant}} EDOM_A^{-}(x) \tag{4.25}$$

3. a. $\underline{EDOM_A}(CL_n^{\geqslant}) = U - \overline{EDOM_A}(CL_{n-1}^{\leqslant})$ $(n = 2, \cdots, m)$,
   b. $\underline{EDOM_A}(CL_n^{\leqslant}) = U - \overline{EDOM_A}(CL_{n+1}^{\geqslant})$ $(n = 1, \cdots, m-1)$,
   c. $\overline{EDOM_A}(CL_n^{\geqslant}) = U - \underline{EDOM_A}(CL_{n-1}^{\leqslant})$ $(n = 2, \cdots, m)$,
   d. $\overline{EDOM_A}(CL_n^{\leqslant}) = U - \underline{EDOM_A}(CL_{n+1}^{\geqslant})$ $(n = 1, \cdots, m-1)$,
   e. $BN_A^E(CL_n^{\geqslant}) = BN_A^E(CL_{n-1}^{\leqslant})$ $(n = 2, \cdots, m)$,
   f. $BN_A^E(CL_n^{\leqslant}) = BN_A^E(CL_{n+1}^{\geqslant})$ $(n = 1, \cdots, m-1)$;

4. $\forall n \in N$,

   a. $\underline{EDOM_A}(CL_n^{\geqslant}) \subseteq \underline{EDOM_{AT}}(CL_n^{\geqslant})$,
   b. $\overline{EDOM_A}(CL_n^{\geqslant}) \supseteq \overline{EDOM_{AT}}(CL_n^{\geqslant})$,
   c. $\underline{EDOM_A}(CL_n^{\leqslant}) \subseteq \underline{EDOM_{AT}}(CL_n^{\leqslant})$,
   d. $\overline{EDOM_A}(CL_n^{\leqslant}) \supseteq \overline{EDOM_{AT}}(CL_n^{\leqslant})$,
   e. $BN_A^E(CL_n^{\geqslant}) \supseteq BN_{AT}^E(CL_n^{\geqslant})$,
   f. $BN_A^E(CL_n^{\leqslant}) \supseteq BN_{AT}^E(CL_n^{\leqslant})$.

Proof.

1. $\forall x \in \underline{EDOM_A}(CL_n^{\geqslant})$, we have $EDOM_A^{+}(x) \subseteq CL_n^{\geqslant}$. Since the expanded dominance relation is reflexive, then $x \in EDOM_A^{+}(x)$, it follows that $x \in CL_n^{\geqslant}$, from which we can conclude that $\underline{EDOM_A}(CL_n^{\geqslant}) \subseteq CL_n^{\geqslant}$.

$\forall x \in CL_n^{\geqslant}$, since the expanded dominance relation is reflexive, then we have $x \in EDOM_A^-(x)$, it follows that $EDOM_A^-(x) \cap CL_n^{\geqslant} \neq \emptyset$, from which we can conclude that $CL_n^{\geqslant} \subseteq \overline{EDOM_A}(CL_n^{\geqslant})$.

From discussions above, we have $\underline{EDOM_A}(CL_n^{\geqslant}) \subseteq CL_n^{\geqslant} \subseteq \overline{EDOM_A}(CL_n^{\geqslant})$. Similarly, it is not difficult to prove $\underline{EDOM_A}(CL_n^{\leqslant}) \subseteq CL_n^{\leqslant} \subseteq \overline{EDOM_A}(CL_n^{\leqslant})$.

2. $\forall y \in \overline{EDOM_A}(CL_n^{\geqslant})$, we have $EDOM_A^-(y) \cap CL_n^{\geqslant} \neq \emptyset$, then there must be $x \in U$ such that $x \in EDOM_A^-(y)$ and $x \in CL_n^{\geqslant}$. Since $x \in EDOM_A^-(y)$, then by the definition of the expanded dominance relation, we have $y \in EDOM_A^+(x)$, it follows that $y \in \bigcup_{x \in CL_n^{\geqslant}} EDOM_A^+(x)$, i.e. $\overline{EDOM_A}(CL_n^{\geqslant}) \subseteq \bigcup_{x \in CL_n^{\geqslant}} EDOM_A^+(x)$.

   $\forall y \in \bigcup_{x \in CL_n^{\geqslant}} EDOM_A^+(x)$, we have $y \in EDOM_A^+(x)$ where $x \in CL_n^{\geqslant}$. By the definition of the expanded dominance relation, we have $x \in EDOM_A^-(y)$, i.e. $EDOM_A^-(y) \cap CL_n^{\geqslant} \neq \emptyset$, from which we can conclude that $x \in \overline{EDOM_A}(CL_n^{\geqslant})$, i.e. $\bigcup_{x \in CL_n^{\geqslant}} EDOM_A^+(x) \subseteq \overline{EDOM_A}(CL_n^{\geqslant})$.

   From discussions above, we have $\overline{EDOM_A}(CL_n^{\geqslant}) = \bigcup_{x \in CL_n^{\geqslant}} EDOM_A^+(x)$. Similarly, it is not difficult to prove $\overline{EDOM_A}(CL_n^{\leqslant}) = \bigcup_{x \in CL_n^{\leqslant}} EDOM_A^-(x)$.

3. $\forall n \in \{2, \cdots, m\}$, $\forall x \in \underline{EDOM_A}(CL_n^{\geqslant})$, we have $EDOM_A^+(x) \subseteq CL_n^{\geqslant}$, it follows that $EDOM_A^+(x) \cap CL_{n-1}^{\leqslant} = \emptyset$, i.e. $x \notin \overline{EDOM_A}(CL_{n-1}^{\leqslant})$.

   $\forall n \in \{2, \cdots, m\}$, $\forall x \in U - \overline{EDOM_A}(CL_{n-1}^{\leqslant})$, we have $EDOM_A^+(x) \cap CL_{n-1}^{\leqslant} = \emptyset$, it follows that $EDOM_A^+(x) \subseteq CL_n^{\geqslant}$, i.e. $x \in \underline{EDOM_A}(CL_n^{\geqslant})$.

   From discussions above, we have $\underline{EDOM_A}(CL_n^{\geqslant}) = U - \overline{EDOM_A}(CL_{n-1}^{\leqslant})$ where $n = 2, \cdots, m$. Similarly, it is not difficult to prove other equations.

4. Since $A \subseteq AT$, then by Theorem 4.1, we know that $EDOM_{AT}^+(x) \subseteq EDOM_A^+(x)$ holds for each $x \in U$. $\forall x \in \underline{EDOM_A}(CL_n^{\geqslant})$, we have $EDOM_A^+(x) \subseteq CL_n^{\geqslant}$, it follows that $EDOM_{AT}^+(x) \subseteq CL_n^{\geqslant}$, from which we can conclude that $x \in \underline{EDOM_{AT}}(CL_n^{\geqslant})$.

   Similarly, it is not difficult to prove other formulas. □

Example 4.1. Table 4.1 is the incomplete decision system that was analyzed by Shao and Zhang in Ref. [22].

In Table 4.1,

- $U = \{x_1, x_2, \cdots, x_{12}\}$ is the universe;
- $AT = \{a_1, a_2, a_3\}$ is the set of condition attributes, d is the decision attribute, $V_{\{a_1\}} = \{1, 2, 3\}$, $V_{\{a_2\}} = V_{\{a_3\}} = V_{\{d\}} = \{1, 2\}$.

Table 4.1 Incomplete decision system

U	a_1	a_2	a_3	d
x_1	3	2	2	2
x_2	2	1	1	1
x_3	2	1	*	2
x_4	1	1	1	1
x_5	1	*	1	1
x_6	3	1	1	2
x_7	2	1	1	2
x_8	3	2	2	2
x_9	2	1	2	2
x_{10}	1	1	2	1
x_{11}	2	1	2	1
x_{12}	3	*	2	2

The decision attributes d makes a partition of U into a finite number of classes such that $\boldsymbol{CL} = \{CL_1, CL_2\}$, where

$$CL_1 = \{x_2, x_4, x_5, x_{10}\}$$
$$CL_2 = \{x_1, x_3, x_6, x_7, x_8, x_9, x_{11}, x_{12}\}$$

then the sets to be approximated are:

$$CL_1^{\geqslant} = U, \quad CL_1^{\leqslant} = CL_1, \quad CL_2^{\geqslant} = CL_2, \quad CL_2^{\leqslant} = U$$

By Definition 4.4, we have:

$$\underline{EDOM_{AT}}(CL_1^{\geqslant}) = U$$
$$\underline{EDOM_{AT}}(CL_2^{\geqslant}) = \{x_1, x_6, x_8, x_9, x_{11}, x_{12}\}$$
$$\overline{EDOM_{AT}}(CL_1^{\geqslant}) = U$$
$$\overline{EDOM_{AT}}(CL_2^{\geqslant}) = \{x_1, x_2, x_3, x_6, x_7, x_8, x_9, x_{11}, x_{12}\}$$
$$\underline{EDOM_{AT}}(CL_1^{\leqslant}) = \{x_4, x_5, x_{10}\}$$
$$\underline{EDOM_{AT}}(CL_2^{\leqslant}) = U$$
$$\overline{EDOM_{AT}}(CL_1^{\leqslant}) = \{x_2, x_3, x_4, x_5, x_7, x_{10}\}$$
$$\overline{EDOM_{AT}}(CL_2^{\leqslant}) = U$$

Similar to the syntaxes of certain rules in complete decision system, we can generate the following certain rules from the lower approximations of $CL_2^{\geqslant}$ and $CL_1^{\leqslant}$ in Table 4.1:

R_1: $f(x, a_1) \geqslant 3 \wedge f(x, a_2) \geqslant 2 \wedge f(x, a_3) \geqslant 2 \rightarrow f(x, a_4) \geqslant 2$ // supported by objects x_1, x_8 in $\underline{EDOM_{AT}}(CL_2^{\geqslant})$;

R_2: $f(x, a_1) \geqslant 3 \wedge f(x, a_2) \geqslant 1 \wedge f(x, a_3) \geqslant 1 \rightarrow f(x, a_4) \geqslant 2$ // supported by object x_6 in $\underline{EDOM_{AT}}(CL_2^{\geqslant})$;

R_3: $f(x, a_1) \geqslant 2 \wedge f(x, a_2) \geqslant 1 \wedge f(x, a_3) \geqslant 2 \rightarrow f(x, a_4) \geqslant 2$ // supported by objects x_9, x_{11} in $\underline{EDOM_{AT}}(CL_2^{\geqslant})$;

R_4: $f(x, a_1) \geqslant 3 \wedge f(x, a_2) \geqslant * \wedge f(x, a_3) \geqslant 2 \rightarrow f(x, a_4) \geqslant 2$ // supported by object x_{12} in $\underline{EDOM_{AT}}(CL_2^{\geqslant})$;

R_5: $f(x, a_1) \leqslant 1 \wedge f(x, a_2) \leqslant 1 \wedge f(x, a_3) \leqslant 1 \rightarrow f(x, a_4) \leqslant 1$ // supported by object x_4 in $\underline{EDOM_{AT}}(CL_1^{\leqslant})$;

R_6: $f(x, a_1) \leqslant 1 \wedge f(x, a_2) \leqslant * \wedge f(x, a_3) \leqslant 1 \rightarrow f(x, a_4) \leqslant 1$ // supported by object x_5 in $\underline{EDOM_{AT}}(CL_1^{\leqslant})$;

R_7: $f(x, a_1) \leqslant 1 \wedge f(x, a_2) \leqslant 1 \wedge f(x, a_3) \leqslant 2 \rightarrow f(x, a_4) \leqslant 1$ // supported by object x_{10} in $\underline{EDOM_{AT}}(CL_1^{\leqslant})$.

4.3 Valued Dominance-based Fuzzy Rough Set Approach

4.3.1 Valued Dominance Relation

The dominance relation showed in Definition 4.3 only represents whether an object is possibly dominating another one because the unknown values "$*$" is considered as comparable with any other values in the domains of the corresponding attributes.

Let us consider Table 4.2, which is also a student evaluation table.

In Table 4.2,

- $U = \{x_1, x_2, x_3, x_4, x_5, x_6\}$ is the set of six students, i.e., universe;
- AT = {Mathematics, Physics, Literature} is the set of the condition attributes, d = Global evaluation, is the decision attribute.

Following Definition 4.3, it is not difficult to observe that $(x_6, x_1) \in EDOM(AT)$ and $(x_6, x_2) \in EDOM(AT)$, i.e. student x_6 may dominate students x_1 and x_2 by considering the evaluations of all three courses. However, we can see that x_6 is certainly dominating x_1 because both of these two students have certain values on all of the these courses, while x_6 is only possibly dominating x_2 because the evaluation of student x_2 on "Literature" is unknown. If the value of x_2 on "Literature" is "medium" or better, then x_6 does not dominate x_2. On the other hand, if the value of x_2 on "Literature" is "bad", then we have x_6 dominates x_2.

Table 4.2 Student evaluations with unknown values

U	Mathematics	Physics	Literature	Global Evaluation
x_1	medium	bad	bad	bad
x_2	good	medium	*	good
x_3	medium	*	medium	medium
x_4	*	medium	medium	medium
x_5	bad	good	bad	bad
x_6	good	medium	bad	good

From discussions above, we may desire to express the intuitive idea that x_6 is "more dominating" x_1 than x_2. This is due to the fact that in the case of x_2 there is one attribute value is unknown and the reset all are dominated by the values of x_6 on the corresponding attributes, while in the case of x_1 all values are not only known but also be dominated by the values of x_6 on the corresponding attributes. From this point of view, the valued dominance relation [23] can be used to show the probability that x is dominating y.

Definition 4.6. Let I be an incomplete decision system in which $A \subseteq AT$, the valued dominance relation in terms of A is defined as

$$VDOM(A) = \left\{VDOM_A(x,y) : x, y \in U\right\} \tag{4.26}$$

$VDOM_A(x,y)$ is the probability that x is dominating y such that

$$VDOM_A(x,y) = \begin{cases} 1, & x = y \\ \prod_{a \in A} VDOM_{(\{a\})}(x,y), & x \neq y \end{cases}$$

where $\forall a \in A$,

$$VDOM_{(\{a\})}(x,y) = \begin{cases} \dfrac{|\{(v_1,v_2) : v_1, v_2 \in V_a \wedge v_1 \geqslant v_2\}|}{|V_a|^2}, & f(x,a) = * \wedge f(y,a) = * \\ \dfrac{|\{v : v \in V_a \wedge v \leqslant f(x,a)\}|}{|V_a|}, & f(x,a) \neq * \wedge f(y,a) = * \\ \dfrac{|\{v : v \in V_a \wedge v \geqslant f(y,a)\}|}{|V_a|}, & f(x,a) = * \wedge f(y,a) \neq * \\ 0, & f(x,a) \neq * \wedge f(y,a) \neq * \\ & \wedge f(x,a) < f(y,a) \\ 1, & f(x,a) \neq * \wedge f(y,a) \neq * \\ & \wedge f(x,a) \geqslant f(y,a) \end{cases}$$

The above definition of valued dominance relation is based on the hypothesis that there exists a uniform probability distribution among attributes values, i.e. all values have the same probability to be associated to an element $x \in U$. It is same the assumption in valued tolerance relation.

Example 4.2. In Table 4.2, $V_a = V_b = V_c = \{\text{Bad, Medium, Good}\}$.

Take for instance x_6 and x_2, by Definition 4.6, we have $VDOM_{(\{a\})}(x_6, x_2) = VDOM_{(\{b\})}(x_6, x_2) = 1$. Moreover, since $\{v : v \in V_c \wedge v \leqslant f(x_6, c)\} = \{\text{Bad}\}$, we have $VDOM_{(\{c\})}(x_6, x_2) = \dfrac{1}{3}$, from which we can conclude that $VDOM_{AT}(x_6, x_2) = \dfrac{1}{3}$.

Similarly, we obtain the following Table 4.3 for representing the valued dominance relation in Table 4.2.

Table 4.3 Valued dominance relation in Table 4.2

	x_1	x_2	x_3	x_4	x_5	x_6
x_1	1	0	0	0	0	0
x_2	1	1	$\frac{4}{9}$	$\frac{2}{3}$	0	1
x_3	1	0	1	$\frac{4}{9}$	$\frac{1}{3}$	0
x_4	$\frac{2}{3}$	$\frac{2}{9}$	$\frac{4}{9}$	1	0	$\frac{1}{3}$
x_5	0	0	0	0	1	0
x_6	1	$\frac{1}{3}$	0	0	0	1

4.3.2 Fuzzy Rough Approximations

Similar to the valued tolerance relation (see Section 2.2), we also use the fuzzy rough technique to construct rough approximation through the valued dominance relation.

Definition 4.7. Let I be an incomplete decision system in which $A \subseteq AT$, $n \in N$, the lower and upper approximations of $CL_n^{\geqslant}$ are denoted by $\underline{VDOM_A}(CL_n^{\geqslant})$ and $\overline{VDOM_A}(CL_n^{\geqslant})$, respectively, whose memberships for each $x \in U$, are defined as:

$$\begin{aligned}\mu_{\underline{VDOM_A}(CL_n^{\geqslant})}(x) &= \wedge_{y\in U}\left(\mu_{CL_n^{\geqslant}}(y) \vee (1 - VDOM_A(y,x))\right)\\ &= \wedge\{1 - VDOM_A(y,x) : y \notin CL_n^{\geqslant}\} \qquad (4.27)\end{aligned}$$

$$\begin{aligned}\mu_{\overline{VDOM_A}(CL_n^{\geqslant})}(x) &= \vee_{y\in U}\left(\mu_{CL_n^{\geqslant}}(y) \wedge VDOM_A(x,y)\right)\\ &= \vee\{VDOM_A(x,y) : y \in CL_n^{\geqslant}\} \qquad (4.28)\end{aligned}$$

the lower and upper approximations of $CL_n^{\leqslant}$ are denoted by $\underline{VDOM_A}(CL_n^{\leqslant})$ and $\overline{VDOM_A}(CL_n^{\leqslant})$, respectively, whose memberships for each $x \in U$, are defined as:

$$\begin{aligned}\mu_{\underline{VDOM_A}(CL_n^{\leqslant})}(x) &= \wedge_{y\in U}\left(\mu_{(CL_n^{\leqslant}}(y) \vee (1 - VDOM_A(x,y))\right)\\ &= \wedge\{1 - VDOM_A(x,y) : y \notin CL_n^{\leqslant}\} \qquad (4.29)\end{aligned}$$

$$\begin{aligned}\mu_{\overline{VDOM_A}(CL_n^{\leqslant})}(x) &= \vee_{y\in U}\left(\mu_{CL_n^{\leqslant}}(y) \wedge VDOM_A(y,x)\right)\\ &= \vee\{VDOM_A(y,x) : y \in CL_n^{\leqslant}\} \qquad (4.30)\end{aligned}$$

Obviously, the fuzzy rough approximations in Definition 4.7 are fuzzy generalization of dominance-based rough set model presented in Definition 4.4. Moreover, if we do not consider the probability an object is dominating another, i.e. $VDOM_A(x,y) = \{0,1\}$ instead of $[0,1]$, then the pairs $\left[\underline{VDOM_A}(CL_n^{\geqslant}), \overline{VDOM_A}(CL_n^{\geqslant})\right]$, $\left[\underline{VDOM_A}(CL_n^{\leqslant}), \overline{VDOM_A}(CL_n^{\leqslant})\right]$ will degenerate to be $\left[\underline{EDOM_A}(CL_n^{\geqslant}), \overline{EDOM_A}(CL_n^{\geqslant})\right]$ and $\left[\underline{EDOM_A}(CL_n^{\leqslant}), \overline{EDOM_A}(CL_n^{\leqslant})\right]$, respectively.

Example 4.3. Take for instance Table 4.2, the decision attributes d makes a partition of U into a finite number of classes such that $\boldsymbol{CL} = \{CL_1, CL_2, CL_3\}$ where $CL_1 = \{\text{bad}\} = \{x_1, x_5\}$, $CL_2 = \{\text{medium}\} = \{x_3, x_4\}$, $CL_3 = \{\text{good}\} = \{x_2, x_6\}$. Thus, the set to be approximated are:

- $CL_1^{\leqslant} = CL_1$, i.e. the class of (at most) bad students;
- $CL_2^{\leqslant} = CL_1 \bigcup CL_2$, i.e. the class of at most medium students;
- $CL_3^{\leqslant} = CL_1 \bigcup CL_2 \bigcup CL_3$, i.e. the class of at most good students;
- $CL_1^{\geqslant} = CL_1 \bigcup CL_2 \bigcup CL_3$, i.e. the class of at least bad students;
- $CL_2^{\geqslant} = CL_2 \bigcup CL_3$, i.e. the class of at least medium students;
- $CL_3^{\geqslant} = CL_3$, i.e. the class of (at least) good students.

By Definition 4.7 we can compute the valued dominance–based lower and upper approximate memberships for $x \in U$ as Table 4.4 shows.

Table 4.4 Lower and upper approximations in Table 4.2

	x_1	x_2	x_3	x_4	x_5	x_6
$\mu_{\underline{VDOM_{AT}}(CL_1^{\geqslant})}(x)$	1	1	1	1	1	1
$\mu_{\underline{VDOM_{AT}}(CL_2^{\geqslant})}(x)$	0	1	1	1	0	1
$\mu_{\underline{VDOM_{AT}}(CL_3^{\geqslant})}(x)$	0	$\frac{7}{9}$	0	0	0	$\frac{2}{3}$
$\mu_{\overline{VDOM_{AT}}(CL_1^{\geqslant})}(x)$	1	1	1	1	1	1
$\mu_{\overline{VDOM_{AT}}(CL_2^{\geqslant})}(x)$	0	1	1	1	0	1
$\mu_{\overline{VDOM_{AT}}(CL_3^{\geqslant})}(x)$	0	1	0	$\frac{1}{3}$	0	1
$\mu_{\underline{VDOM_{AT}}(CL_1^{\leqslant})}(x)$	1	0	0	0	1	0
$\mu_{\underline{VDOM_{AT}}(CL_2^{\leqslant})}(x)$	1	0	1	$\frac{2}{3}$	1	0
$\mu_{\underline{VDOM_{AT}}(CL_3^{\leqslant})}(x)$	1	1	1	1	1	1
$\mu_{\overline{VDOM_{AT}}(CL_1^{\leqslant})}(x)$	1	0	0	0	1	0
$\mu_{\overline{VDOM_{AT}}(CL_2^{\leqslant})}(x)$	1	$\frac{2}{9}$	1	1	1	$\frac{1}{3}$
$\mu_{\overline{VDOM_{AT}}(CL_3^{\leqslant})}(x)$	1	1	1	1	1	1

Proposition 4.3. *Let I be an incomplete decision system in which $A \subseteq AT$, the valued dominance-based fuzzy rough sets have the following properties:*

1. *for* $1 \leqslant n \leqslant m$,

$$\underline{VDOM_A}(CL_n^{\geqslant}) \subseteq CL_n^{\geqslant} \subseteq \overline{VDOM_A}(CL_n^{\geqslant}) \tag{4.31}$$

$$\underline{VDOM_A}(CL_n^{\leqslant}) \subseteq CL_n^{\leqslant} \subseteq \overline{VDOM_A}(CL_n^{\leqslant}) \tag{4.32}$$

2.

$$\underline{VDOM_A}(CL_n^{\geqslant}) = U - \overline{VDOM_A}(CL_{n-1}^{\leqslant}), \quad n = 2, \cdots, m \tag{4.33}$$

$$\underline{VDOM_A}(CL_n^{\leqslant}) = U - \overline{VDOM_A}(CL_{n+1}^{\geqslant}), \quad n = 1, \cdots, m-1 \tag{4.34}$$

$$\overline{VDOM_A}(CL_n^{\geqslant}) = U - \underline{VDOM_A}(CL_{n-1}^{\leqslant}), \quad n = 2, \cdots, m \tag{4.35}$$

$$\overline{VDOM_A}(CL_n^{\leqslant}) = U - \underline{VDOM_A}(CL_{n+1}^{\leqslant}), \quad n = 1, \cdots, m-1 \tag{4.36}$$

3.

$$\underline{VDOM_A}(CL_n^{\geqslant}) \subseteq \underline{VDOM_{AT}}(CL_n^{\geqslant}) \tag{4.37}$$

$$\overline{VDOM_A}(CL_n^{\geqslant}) \supseteq \overline{VDOM_{AT}}(CL_n^{\geqslant}) \tag{4.38}$$

$$\underline{VDOM_A}(CL_n^{\leqslant}) \subseteq \underline{VDOM_{AT}}(CL_n^{\leqslant}) \tag{4.39}$$

$$\overline{VDOM_A}(CL_n^{\leqslant}) \supseteq \overline{VDOM_{AT}}(CL_n^{\leqslant}) \tag{4.40}$$

Proof.
1. $\forall x \in U$, if $x \in CL_n^{\geqslant}$, i.e. $\mu_{CL_n^{\geqslant}}(x) = 1$, then

$$\mu_{\underline{VDOM_A}(CL_n^{\geqslant})}(x) = \wedge_{y \in U}\left(\mu_{CL_n^{\geqslant}}(y) \vee (1 - VDOM_A(y, x))\right)$$
$$\leqslant \left(\mu_{CL_n^{\geqslant}}(x) \vee (1 - VDOM_A(x, x))\right)$$
$$= 1 = \mu_{CL_n^{\geqslant}}(x)$$

On other hand, if $x \notin CL_n^{\geqslant}$, i.e. $\mu_{CL_n^{\geqslant}}(x) = 0$, then

$$\mu_{\underline{VDOM_A}(CL_n^{\geqslant})}(x) = \wedge_{y \in U}(\mu_{CL_n^{\geqslant}}(y) \vee (1 - VDOM_A(y, x)))$$
$$\leqslant \left(\mu_{CL_n^{\geqslant}}(x) \vee (1 - VDOM_A(x, x))\right)$$
$$= 0 = \mu_{CL_n^{\geqslant}}(x)$$

From discussions above, we can conclude that $\underline{VDOM_A}(CL_n^{\geqslant}) \subseteq CL_n^{\geqslant}$.

$\forall x \in U$, if $x \in CL_n^{\geqslant}$, i.e. $\mu_{CL_n^{\geqslant}}(x) = 1$, then

$$\mu_{\overline{VDOM_A}(CL_n^{\geqslant})}(x) = \vee_{y \in U}\left(\mu_{CL_n^{\geqslant}}(y) \wedge VDOM_A(y, x)\right)$$
$$\geqslant (\mu_{CL_n^{\geqslant}}(x) \wedge VDOM_A(x, x))$$
$$= 1 = \mu_{CL_n^{\geqslant}}(x)$$

On other hand, if $x \notin CL_n^{\geqslant}$, i.e. $\mu_{CL_n^{\geqslant}}(x) = 0$, then

$$\mu_{\overline{VDOM_A}(CL_n^{\geqslant})}(x) = \vee_{y \in U}\left(\mu_{CL_n^{\geqslant}}(y) \wedge VDOM_A(y, x)\right)$$
$$\geqslant \left(\mu_{CL_n^{\geqslant}}(x) \wedge VDOM_A(x, x)\right)$$
$$= 0 = \mu_{CL_n^{\geqslant}}(x).$$

From discussions above, we can conclude that $CL_n^{\geqslant} \subseteq \overline{VDOM_A}(CL_n^{\geqslant})$. Similarly, it is not difficult to prove Eq. (4.32).

2. $\forall x \in U$, since $1 - \mu_{CL_n^{\geqslant}}(x) = \mu_{CL_{n-1}^{\leqslant}}(x)$ where $n = 2, \cdots, m$, we have

$$\mu_{\underline{VDOM_A}(CL_n^{\geqslant})}(x) = \wedge_{y \in U}\left(\mu_{CL_n^{\geqslant}}(y) \vee (1 - VDOM_A(y, x))\right)$$
$$= 1 - \vee_{y \in U}\left((1 - \mu_{CL_{n-1}^{\leqslant}}(y)) \wedge VDOM_A(y, x)\right)$$
$$= 1 - \vee_{y \in U}\left(\mu_{CL_{n-1}^{\leqslant}}(y) \wedge VDOM_A(y, x)\right)$$
$$= 1 - \mu_{\overline{VDOM_A}(CL_{n-1}^{\leqslant})}(x)$$

From discussions above, we can conclude that $\underline{VDOM_A}(CL_n^{\geqslant}) = U - \overline{VDOM_A}(CL_{n-1}^{\leqslant})$. Similarly, it is not difficult to prove Eqs. (4.34), (4.35) and (4.36).

3. $\forall y, x \in U$, by Definition 4.6 we have $VDOM_A(y, x) \geqslant VDOM_{AT}(y, x)$. Thus,

$$\mu_{\underline{VDOM_A}(CL_n^{\geqslant})}(x) = \wedge_{y \in U}\left(\mu_{CL_n^{\geqslant}}(y) \vee (1 - VDOM_A(y, x))\right)$$
$$\leqslant \wedge_{y \in U}\left(\mu_{CL_n^{\geqslant}}(y) \vee (1 - VDOM_{AT}(y, x))\right)$$
$$= \mu_{\underline{VDOM_{AT}}(CL_n^{\geqslant})}(x)$$

$$\begin{aligned}\mu_{\overline{VDOM_A}(CL_n^{\geqslant})}(x) &= \vee_{y\in U}\left(\mu_{CL_n^{\geqslant}}(y) \wedge VDOM_A(y,x)\right)\\ &\geqslant \vee_{y\in U}\left(\mu_{CL_n^{\geqslant}}(y) \wedge VDOM_{AT}(y,x)\right)\\ &= \mu_{\overline{VDOM_{AT}}(CL_n^{\geqslant})}(x)\end{aligned}$$

Thus, Eqs. (4.37) and (4.38) are proved. Similarly, it is not difficult to prove Eqs. (4.39) and (4.40). □

In Proposition 4.3, 1 says that the upward (downward) unions of decision classes include their fuzzy rough lower approximations and are included in their fuzzy rough upper approximation; 2 represents complementarity properties of the proposed fuzzy rough approximations; 3 expresses monotonicity of the proposed fuzzy rough set in terms of the monotonous varieties of condition attributes.

4.3.3 Extraction of Decision Rules

Similar to the classical dominance-based rough set approach, let I be an incomplete decision system, the knowledge hidden in I may be unravelled and expressed in the form of decision rules ("at least" and "at most" rules) from every training example x such that:

- "At least" rule:
  $r^{\geqslant}(x)$: If $f(y,a) \geqslant f(x,a)$ for each $a \in AT$, then $f(y,d) \geqslant f(x,d)$ with certainty factor $c_{AT}^{\geqslant}(x)$;
- "At most" rule:
  $r^{\leqslant}(x)$: If $f(y,a) \leqslant f(x,a)$ for each $a \in AT$, then $f(y,d) \leqslant f(x,d)$ with certainty factor $c_{AT}^{\leqslant}(x)$.

Certain factors $c_{AT}^{\geqslant}(x)$ and $c_{AT}^{\leqslant}(x)$ are used to express the credibility degrees of "at most" and "at least" decision rules, which are generated from x, respectively. They are defined as:

$$c_{AT}^{\geqslant}(x) = \frac{\sum\limits_{y\in[x]_d^{\geqslant}} VDOM_{AT}(y,x)}{\sum\limits_{y\in U} VDOM_{AT}(y,x)} \tag{4.41}$$

$$c_{AT}^{\leqslant}(x) = \frac{\sum\limits_{y\in[x]_d^{\leqslant}} VDOM_{AT}(x,y)}{\sum\limits_{y\in U} VDOM_{AT}(x,y)} \tag{4.42}$$

where $[x]_d^{\geqslant} = \{y \in U : f(y,d) \geqslant f(x,d)\}$ and $[x]_d^{\leqslant} = \{y \in U : f(y,d) \leqslant f(x,d)\}$.

If $c_{AT}^{\geqslant}(x) = 1$ ($c_{AT}^{\leqslant}(x) = 1$), then $r^{\geqslant}(x)$ ($r^{\leqslant}(x)$) are referred to as certain, while if $c_{AT}^{\geqslant}(x) > 0$ ($c_{AT}^{\leqslant}(x) > 0$), then $r^{\geqslant}(x)$ ($r^{\leqslant}(x)$) are referred to as possible.

Theorem 4.2. *Let S be an incomplete decision system, $\forall x \in U$, we have:*

1. $r^{\geqslant}(x)$ *is certain* $\Leftrightarrow \mu_{\underline{VDOM_{AT}}([x]_d^{\geqslant})}(x) = 1$;
2. $r^{\geqslant}(x)$ *is possible* $\Leftrightarrow \mu_{\overline{VDOM_{AT}}([x]_d^{\geqslant})}(x) = 1$;
3. $r^{\leqslant}(x)$ *is certain* $\Leftrightarrow \mu_{\underline{VDOM_{AT}}([x]_d^{\leqslant})}(x) = 1$;
4. $r^{\leqslant}(x)$ *is possible* $\Leftrightarrow \mu_{\overline{VDOM_{AT}}([x]_d^{\leqslant})}(x) = 1$.

Proof. We only prove 1 and 2, the proofs of 3 and 4 are similar to the proofs of 1 and 2.

1.

$$\begin{aligned} c_{AT}^{\geqslant}(x) = 1 &\Leftrightarrow \sum_{y\in[x]_d^{\geqslant}} VDOM_{AT}(y,x) = \sum_{y\in U} VDOM_{AT}(y,x) \\ &\Leftrightarrow VDOM_{AT}(y,x) = 0 \text{ for } \forall y \notin [x]_d^{\geqslant} \\ &\Leftrightarrow \wedge\{1 - VDOM_{AT}(y,x) : y \notin [x]_d^{\geqslant}\} = 1 \\ &\Leftrightarrow \mu_{\underline{VDOM_{AT}}([x]_d^{\geqslant})}(x) = 1 \end{aligned}$$

2. Since $x \in [x]_d^{\geqslant}$, we have

$$\begin{aligned} c_{AT}^{\geqslant}(x) > 0 &\Leftrightarrow \sum_{y\in[x]_d^{\geqslant}} VDOM_{AT}(y,x) > 0 \\ &\Leftrightarrow VDOM_{AT}(x,x) = 1 \\ &\Leftrightarrow \vee\{VDOM_{AT}(x,y) : y \in [x]_d^{\geqslant}\} = 1 \\ &\Leftrightarrow \mu_{\overline{VDOM_{AT}}([x]_d^{\geqslant})}(x) = 1 \end{aligned}$$

□

Example 4.4. In Table 4.2, take for instance student x_1, we can obtain the "at least" rule such that:

$r^{\geqslant}(x_1)$: If $f(y,\text{Mathematics}) \geqslant$ medium and $f(y,\text{Physics}) \geqslant$ bad and $f(y,\text{Literature}) \geqslant$ bad, then y is at least a bad student;

Since $\mu_{\underline{VDOM_{AT}}([x_1]_d^{\geqslant})}(x_1) = \mu_{\underline{VDOM_{AT}}(CL_1^{\geqslant})}(x_1) = 1$, by Theorem 4.2 we can see that rule $r^{\geqslant}(x_1)$ is a certain rule.

Similarly, we can generate all "at least" and "at most" decision rules from Table 1 as followings:

"at least" rules:

$r^{\geqslant}(x_2)$: if $f(y,\text{Mathematics}) \geqslant$ Good and $f(y,\text{Physics}) \geqslant$ Medium and $f(y,\text{Literature}) \geqslant$ *, then y is at least a Good student //

$$\begin{aligned} c_{AT}^{\geqslant}(x_2) &= \frac{\sum_{y\in[x]_d^{\geqslant}} VDOM_{AT}(y,x_2)}{\sum_{y\in U} VDOM_{AT}(y,x_2)} \\ &= \frac{VDOM_{AT}(x_2,x_2) + VDOM_{AT}(x_6,x_2)}{\sum_{y\in U} VDOM_{AT}(y,x_2)} \end{aligned}$$

$$= \frac{1 + \frac{1}{3}}{1 + \frac{2}{9} + \frac{1}{3}} = 0.86$$

$r^{\geqslant}(x_3)$: if $f(y, \text{Mathematics}) \geqslant$medium and $f(y, \text{Physics}) \geqslant *$ and $f(y, \text{Literature}) \geqslant$ medium, then y is at least a medium student // $c^{\geqslant}_{AT}(x_3) = 1$;

$r^{\geqslant}(x_4)$: if $f(y, \text{Mathematics}) \geqslant *$ and $f(y, \text{Physics}) \geqslant$medium and $f(y, \text{Literature}) \geqslant$ medium, then y is at least a medium student // $c^{\geqslant}_{AT}(x_4) = 1$;

$r^{\geqslant}(x_5)$: if $f(y, \text{Mathematics}) \geqslant$bad and $f(y, \text{Physics}) \geqslant$ good and $f(y, \text{Literature}) \geqslant$ bad, then y is at least a bad student // $c^{\geqslant}_{AT}(x_5) = 1$;

$r^{\geqslant}(x_6)$: if $f(y, \text{Mathematics}) \geqslant$ good and $f(y, \text{Physics}) \geqslant$ medium and $f(y, \text{Literature}) \geqslant$ bad, then y is at least a good student // $c^{\geqslant}_{AT}(x_6) = 0.86$.

"at most" rules:

$r^{\leqslant}(x_1)$: if $f(y, \text{Mathematics}) \leqslant$ medium and $f(y, \text{Physics}) \leqslant$ bad and $f(y, \text{Literature}) \leqslant$ bad, then y is at most a bad student // $c^{\leqslant}_{AT}(x_1) = 1$;

$r^{\leqslant}(x_2)$: if $f(y, \text{Mathematics}) \leqslant$ good and $f(y, \text{Physics}) \leqslant$ medium and $f(y, \text{Literature}) \leqslant *$, then y is at most a good student // $c^{\leqslant}_{AT}(x_2) = 1$;

$r^{\leqslant}(x_3)$: if $f(y, \text{Mathematics}) \leqslant$medium and $f(y, \text{Physics}) \leqslant *$ and $f(y, \text{Literature}) \leqslant$ medium, then y is at most a medium student // $c^{\leqslant}_{AT}(x_3) = 1$;

$r^{\leqslant}(x_4)$: if $f(y, \text{Mathematics}) \leqslant *$ and $f(y, \text{Physics}) \leqslant$medium and $f(y, \text{Literature}) \leqslant$ medium, then y is at most a medium student // $c^{\leqslant}_{AT}(x_4) = 0.79$;

$r^{\leqslant}(x_5)$: if $f(y, \text{Mathematics}) \leqslant$ bad and $f(y, \text{Physics}) \leqslant$ good and $f(y, \text{Literature}) \leqslant$ bad, then y is at most a bad student // $c^{\leqslant}_{AT}(x_5) = 1$;

$r^{\leqslant}(x_6)$: if $f(y, \text{Mathematics}) \leqslant$ good and $f(y, \text{Physics}) \leqslant$ medium and $f(y, \text{Literature}) \leqslant$ bad, then y is at most a good student // $c^{\leqslant}_{AT}(x_6) = 1$.

4.4 ↑ and ↓ Descriptors and Certain Rules

In Example 4.1, it is not difficult to observe that the generated rules have following limitations:

1. Unknown values $*$ may exist in some rules. For example, R_4 and R_6;
2. Not all certain rules have been obtained.

For instance, let us consider the following subset of universe:

$$P = \{x \in U : f(x, a_1) \geqslant 2 \wedge f(x, a_2) \geqslant 2 \wedge f(x, a_3) \geqslant 1\} = \{x_1, x_8, x_{12}\}$$

Therefore, we have $P \subseteq CL_2^{\geqslant}$, that is,

$$f(x, a_1) \geqslant 2 \wedge f(x, a_2) \geqslant 2 \wedge f(x, a_3) \geqslant 1 \rightarrow f(x, a_4) \geqslant 2$$

is a certain rule which should be derived from Table 4.1. However, no object in $\underline{EDOM_{AT}}(CL_2^{\geqslant})$ is corresponding to this rule. Moreover, it must be noticed that this rule can not be classified into any one of the rules of R_1, R_2, R_3 and R_4. In other words, it is a new "at least" certain rule.

To solve the above two problems, Yang et al. [24] proposed a descriptors based approach to derive certain rules.

4.4.1 Definition of ↑ and ↓ Descriptors

Similar to the descriptor showed in Chapter 2, the symbols $\wedge$ and $\vee$ denote the logical connectives "and" (conjunction) and "or" (disjunction), respectively. In an incomplete decision system I, if $A \subseteq AT$, then any attribute-value pair $(\geqslant_a, v)$ is referred to as a $\uparrow$ A-atomic property, while any attribute-value pair $(\leqslant_a, v)$ is referred to as a $\downarrow$ A-atomic property, where $a \in A$ and $v \in V_a$. Any $\uparrow$ ($\downarrow$) A-atomic properties or conjunction of different $\uparrow$ ($\downarrow$) A-atomic properties are referred to as the $\uparrow$ ($\downarrow$) A-descriptors. To distinguish with these these two different descriptors, let us denote $\uparrow$ and $\downarrow$ A-descriptors by $t^{\uparrow}$ and $t^{\downarrow}$, respectively.

Let $t^{\uparrow}$ ($t^{\downarrow}$) are $\uparrow$ ($\downarrow$) A-descriptors, the sets of the attributes occurring in $t^{\uparrow}$ ($t^{\downarrow}$) are denoted by $A(t^{\uparrow})$ ($A(t^{\downarrow})$). If $(\geqslant_a, v)((\leqslant_a, v))$ are $\uparrow$ ($\downarrow$) atomic properties occurring in $t^{\uparrow}$ ($t^{\downarrow}$) respectively, we simply say that $(\geqslant_a, v) \in t^{\uparrow}$ ($(\leqslant_a, v) \in t^{\downarrow}$).

Let $t^{\uparrow}$ and $s^{\uparrow}$ are two $\uparrow$ A-descriptors, if for all $(\geqslant_a, v) \in t^{\uparrow}$, we have $(\geqslant_a, v) \in s^{\uparrow}$, that is, $t^{\uparrow}$ is constructed from a subset of atomic properties occurring in $s^{\uparrow}$, then we say $t^{\uparrow}$ is coarser than $s^{\uparrow}$ or $s^{\uparrow}$ is finer than $t^{\uparrow}$ and is denoted by $t^{\uparrow} \geq s^{\uparrow}$ or $s^{\uparrow} \leq t^{\uparrow}$. If $t^{\uparrow}$ is constructed from a proper subset of atomic properties occurring in $s^{\uparrow}$, then we say $t^{\uparrow}$ is proper coarser than $s^{\uparrow}$ and is denoted by $t^{\uparrow} > s^{\uparrow}$ or $s^{\uparrow} < t^{\uparrow}$. Similarly, it is not difficult to define $t^{\downarrow} \geq s^{\downarrow}$ and $t^{\downarrow} > s^{\downarrow}$.

Moreover, if $t^{\uparrow}$ ($t^{\downarrow}$) are $\uparrow$ ($\downarrow$) A-descriptors and $A(t^{\uparrow}) = A$ ($A(t^{\downarrow}) = A$), then $t^{\uparrow}$ ($t^{\downarrow}$) are referred to as full $\uparrow$ ($\downarrow$) A-descriptors. Here, suppose that $\wedge_{a_i \in A}(\geqslant_{a_i}, v_i)$ ($\wedge_{a_i \in A}(\leqslant_{a_i}, v_i)$) are full $\uparrow$ ($\downarrow$) A-descriptors, we denote

$$\| \wedge_{a_i \in A} (\geqslant_{a_i}, v_i)\| = \{x \in U : \forall a_i \in A, f(x, a_i) = * \vee f(x, a_i) \geqslant v_i\}$$
$$\| \wedge_{a_i \in A} (\leqslant_{a_i}, v_i)\| = \{x \in U : \forall a_i \in A, f(x, a_i) = * \vee f(x, a_i) \leqslant v_i\}$$

then $\|\wedge_{a_i \in A}(\geqslant_{a_i}, v_i)\|$ ($\|\wedge_{a_i \in A}(\leqslant_{a_i}, v_i)\|$) are referred to as the support sets of $\wedge_{a_i \in A}(\geqslant_{a_i}, v_i)$ ($\wedge_{a_i \in A}(\leqslant_{a_i}, v_i)$).

Here, let us denote

$$DES^{\uparrow}(A) = \{t^{\uparrow} : t^{\uparrow} \text{ is a } \uparrow A\text{-descriptor}, \|t^{\uparrow}\| \neq \emptyset\}$$
$$DES^{\downarrow}(A) = \{t^{\downarrow} : t^{\downarrow} \text{is a } \downarrow A\text{-descriptor}, \|t^{\downarrow}\| \neq \emptyset\}$$
$$FDES^{\uparrow}(A) = \{t^{\uparrow} : t^{\uparrow} \in DES^{\uparrow}(A), A(t^{\uparrow}) = A\}$$
$$FDES^{\downarrow}(A) = \{t^{\downarrow} : t^{\downarrow} \in DES^{\downarrow}(A), A(t^{\downarrow}) = A\}$$

From the viewpoint of two different descriptors, the universe U could be partitioned into several subsets that may overlap, and the result is denoted by $U/A^{\uparrow}$ or $U/A^{\downarrow}$ such that $U/A^{\uparrow} = \{\|t^{\uparrow}\| : t^{\uparrow} \in FDES^{\uparrow}(A)\}$, $U/A^{\downarrow} = \{\|t^{\downarrow}\| : t^{\downarrow} \in FDES^{\downarrow}(A)\}$.

Example 4.5. Take for instance Table 4.1, we have

1. $FDES^{\uparrow}(AT) = \{t_1^{\uparrow}, t_2^{\uparrow}, t_3^{\uparrow}, t_4^{\uparrow}, t_5^{\uparrow}, t_6^{\uparrow}, t_7^{\uparrow}, t_8^{\uparrow}, t_9^{\uparrow}, t_{10}^{\uparrow}, t_{11}^{\uparrow}, t_{12}^{\uparrow}\}$, in which

$$t_1^{\uparrow} = (\geqslant_{a_1}, 3) \wedge (\geqslant_{a_2}, 2) \wedge (\geqslant_{a_3}, 2)$$

$$
\begin{aligned}
t_2^{\uparrow} &= (\geqslant_{a_1} 3) \wedge (\geqslant_{a_2}, 2) \wedge (\geqslant_{a_3}, 1) \\
t_3^{\uparrow} &= (\geqslant_{a_1} 3) \wedge (\geqslant_{a_2}, 1) \wedge (\geqslant_{a_3}, 2) \\
t_4^{\uparrow} &= (\geqslant_{a_1} 3) \wedge (\geqslant_{a_2}, 1) \wedge (\geqslant_{a_3}, 1) \\
t_5^{\uparrow} &= (\geqslant_{a_1} 2) \wedge (\geqslant_{a_2}, 2) \wedge (\geqslant_{a_3}, 2) \\
t_6^{\uparrow} &= (\geqslant_{a_1} 2) \wedge (\geqslant_{a_2}, 2) \wedge (\geqslant_{a_3}, 1) \\
t_7^{\uparrow} &= (\geqslant_{a_1} 2) \wedge (\geqslant_{a_2}, 1) \wedge (\geqslant_{a_3}, 2) \\
t_8^{\uparrow} &= (\geqslant_{a_1} 2) \wedge (\geqslant_{a_2}, 1) \wedge (\geqslant_{a_3}, 1) \\
t_9^{\uparrow} &= (\geqslant_{a_1} 1) \wedge (\geqslant_{a_2}, 2) \wedge (\geqslant_{a_3}, 2) \\
t_{10}^{\uparrow} &= (\geqslant_{a_1} 1) \wedge (\geqslant_{a_2}, 2) \wedge (\geqslant_{a_3}, 1) \\
t_{11}^{\uparrow} &= (\geqslant_{a_1} 1) \wedge (\geqslant_{a_2}, 1) \wedge (\geqslant_{a_3}, 2) \\
t_{12}^{\uparrow} &= (\geqslant_{a_1} 1) \wedge (\geqslant_{a_2}, 1) \wedge (\geqslant_{a_3}, 1)\}
\end{aligned}
$$

$$
\begin{aligned}
U/AT^{\uparrow} = &\{||t_1^{\uparrow}||, ||t_2^{\uparrow}||, ||t_3^{\uparrow}||, ||t_4^{\uparrow}||, ||t_5^{\uparrow}||, ||t_6^{\uparrow}||, ||t_7^{\uparrow}||, ||t_8^{\uparrow}||, ||t_9^{\uparrow}||, ||t_{10}^{\uparrow}||, ||t_{11}^{\uparrow}||, ||t_{12}^{\uparrow}||\} \\
= &\Big\{\{x_1, x_8, x_{12}\}, \{x_1, x_8, x_{12}\}, \{x_1, x_8, x_{12}\}, \{x_1, x_6, x_8, x_{12}\}, \{x_1, x_8, x_{12}\}, \\
&\{x_1, x_8, x_{12}\}, \{x_1, x_3, x_8, x_9, x_{11}, x_{12}\}, \{x_1, x_2, x_3, x_6, x_7, x_8, x_9, x_{11}, x_{12}\}, \\
&\{x_1, x_8, x_{12}\}, \{x_1, x_5, x_8, x_{12}\}, \{x_1, x_3, x_8, x_9, x_{10}, x_{11}, x_{12}\}, U\Big\}
\end{aligned}
$$

2. $FDES^{\downarrow}(AT) = \{t_1^{\downarrow}, t_2^{\downarrow}, t_3^{\downarrow}, t_4^{\downarrow}, t_5^{\downarrow}, t_6^{\downarrow}, t_7^{\downarrow}, t_8^{\downarrow}, t_9^{\downarrow}, t_{10}^{\downarrow}, t_{11}^{\downarrow}, t_{12}^{\downarrow}\}$, in which

$$
\begin{aligned}
t_1^{\downarrow} &= \{(\leqslant_{a_1} 3) \wedge (\leqslant_{a_2}, 2) \wedge (\leqslant_{a_3}, 2) \\
t_2^{\downarrow} &= (\leqslant_{a_1} 3) \wedge (\leqslant_{a_2}, 2) \wedge (\leqslant_{a_3}, 1) \\
t_3^{\downarrow} &= (\leqslant_{a_1} 3) \wedge (\leqslant_{a_2}, 1) \wedge (\leqslant_{a_3}, 2) \\
t_4^{\downarrow} &= (\leqslant_{a_1} 3) \wedge (\leqslant_{a_2}, 1) \wedge (\leqslant_{a_3}, 1) \\
t_5^{\downarrow} &= (\leqslant_{a_1} 2) \wedge (\leqslant_{a_2}, 2) \wedge (\leqslant_{a_3}, 2) \\
t_6^{\downarrow} &= (\leqslant_{a_1} 2) \wedge (\leqslant_{a_2}, 2) \wedge (\leqslant_{a_3}, 1) \\
t_7^{\downarrow} &= (\leqslant_{a_1} 2) \wedge (\leqslant_{a_2}, 1) \wedge (\leqslant_{a_3}, 2) \\
t_8^{\downarrow} &= (\leqslant_{a_1} 2) \wedge (\leqslant_{a_2}, 1) \wedge (\leqslant_{a_3}, 1) \\
t_9^{\downarrow} &= (\leqslant_{a_1} 1) \wedge (\leqslant_{a_2}, 2) \wedge (\leqslant_{a_3}, 2) \\
t_{10}^{\downarrow} &= (\leqslant_{a_1} 1) \wedge (\leqslant_{a_2}, 2) \wedge (\leqslant_{a_3}, 1) \\
t_{11}^{\downarrow} &= (\leqslant_{a_1} 1) \wedge (\leqslant_{a_2}, 1) \wedge (\leqslant_{a_3}, 2) \\
t_{12}^{\downarrow} &= (\leqslant_{a_1} 1) \wedge (\leqslant_{a_2}, 1) \wedge (\leqslant_{a_3}, 1)\}
\end{aligned}
$$

$$
\begin{aligned}
U/A^{\downarrow} = &\{||t_1^{\downarrow}||, ||t_2^{\downarrow}||, ||t_3^{\downarrow}||, ||t_4^{\downarrow}||, ||t_5^{\downarrow}||, ||t_6^{\downarrow}||, ||t_7^{\downarrow}||, ||t_8^{\downarrow}||, ||t_9^{\downarrow}||, ||t_{10}^{\downarrow}||, ||t_{11}^{\downarrow}||, ||t_{12}^{\downarrow}||\} \\
= &\Big\{U, \{x_2, x_3, x_4, x_5, x_6, x_7\}, \{x_2, x_3, x_4, x_5, x_6, x_7, x_9, x_{10}, x_{11}, x_{12}\}, \\
&\{x_2, x_3, x_4, x_5, x_6, x_7\}, \{x_2, x_3, x_4, x_5, x_7, x_9, x_{10}, x_{11}\}, \{x_2, x_3, x_4, x_5, x_7\}, \\
&\{x_2, x_3, x_4, x_5, x_7, x_9, x_{10}, x_{11}\}, \{x_2, x_3, x_4, x_5, x_7\}, \{x_4, x_5, x_{10}\}, \{x_4, x_5\}, \\
&\{x_4, x_5, x_{10}\}, \{x_4, x_5\}\Big\}
\end{aligned}
$$

4.4.2 Reduct of ↑ and ↓ Descriptors

In Chapter 2, we know that the reducts of descriptors are minimal conjunction of the atomic properties, which preserves the support set of the original descriptor. Following such idea, we can define the following reducts for ↑ and ↓ descriptors.

Definition 4.8. Let I be an incomplete decision system

1. if $t^{\uparrow} \in FDES^{\uparrow}(AT)$, then $t'^{\uparrow} \in DES^{\uparrow}(AT)$ is referred to as a reduct descriptor of $t^{\uparrow}$ if and only if the following two conditions hold:

   a. $t'^{\uparrow} \geq t^{\uparrow}, ||t'^{\uparrow}|| = ||t^{\uparrow}||$;
   b. $||t''^{\uparrow}|| \neq ||t^{\uparrow}||$ for each $t''^{\uparrow} > t'^{\uparrow}$.

2. if $t^{\downarrow} \in FDES^{\downarrow}(AT)$, then $t'^{\downarrow} \in DES^{\downarrow}(AT)$ is referred to as a reduct descriptor of $t^{\downarrow}$ if and only if the following two conditions hold:

   a. $t'^{\downarrow} \geq t^{\downarrow}, ||t'^{\downarrow}|| = ||t^{\downarrow}||$;
   b. $||t''^{\downarrow}|| \neq ||t^{\downarrow}||$ for each $t''^{\downarrow} > t'^{\downarrow}$.

A reduct of a descriptor allows us to classify objects with a smaller number of required atomic properties. It can be observed that reducts of $t^{\uparrow}$ ($t^{\downarrow}$) coarser than $t^{\uparrow}$ ($t^{\downarrow}$) while preserving the same support sets of $t^{\uparrow}$ ($t^{\downarrow}$). The sets of all reducts descriptors of $t^{\uparrow}$ ($t^{\downarrow}$) are denoted by $redes(t^{\uparrow})$ ($redes(t^{\downarrow})$).

In the following, we will present the practical approach to compute reducts of ↑ and ↓ descriptors. Firstly, $\forall x \in U$ and $\forall t^{\uparrow} \in FDES^{\uparrow}(AT)$ ($\forall t^{\downarrow} \in FDES^{\downarrow}(AT)$), suppose that $t^{\uparrow} = \wedge_{a_i \in AT}(\geqslant_{a_i}, v_i)$ ($t^{\downarrow} = \wedge_{a_i \in AT}(\leqslant_{a_i}, v_i)$), let us denote by

$$DIS_{AT}(t^{\uparrow}, x) = \begin{cases} \{a_i \in AT : f(x, a_i) \text{ is known and } f(x, a_i) < v_i\}, & x \notin ||t^{\uparrow}|| \\ AT, & x \in ||t^{\uparrow}|| \end{cases}$$

$$DIS_{AT}(t^{\downarrow}, x) = \begin{cases} \{a_i \in AT : f(x, a_i) \text{ is known and } f(x, a_i) > v_i\}, & x \notin ||t^{\downarrow}|| \\ AT, & x \in ||t^{\downarrow}|| \end{cases}$$

where $v_i \in V_{a_i}$.

$DIS_{AT}(t^{\uparrow}, x)$ ($DIS_{AT}(t^{\downarrow}, x)$) are referred to as the ↑ (↓)-discernibility attributes sets of $t^{\uparrow}$ ($t^{\downarrow}$) and x, $DIS_{AT}(t^{\uparrow}) = \{D_{AT}(t^{\uparrow}, x) : x \in U, t^{\uparrow} \in FDES^{\uparrow}(AT)\}$ and $DIS_{AT}(t^{\downarrow}) = \{D_{AT}(t^{\downarrow}, x) : x \in U, t^{\downarrow} \in FDES^{\downarrow}(AT)\}$, are referred to as the ↑ (↓)-discernibility matrixes of $t^{\uparrow}$ ($t^{\downarrow}$) in I.

Theorem 4.3. *Let I be an incomplete decision system,*

1. *if $t^{\uparrow} \in FDES^{\uparrow}(AT)$ and $t'^{\uparrow} \in DES^{\uparrow}(AT)$ such that $t'^{\uparrow} \geq t^{\uparrow}$, then we have*

$$||t^{\uparrow}|| = ||t'^{\uparrow}|| \Leftrightarrow DIS_{AT}(t^{\uparrow}, x) \cap AT(t'^{\uparrow}) \neq \emptyset \text{ for each } \forall x \in U \text{ and } x \notin ||t^{\uparrow}||;$$

2. *if $t^{\downarrow} \in FDES^{\downarrow}(AT)$ and $t'^{\downarrow} \in DES^{\downarrow}(AT)$ such that $t'^{\downarrow} \geq t^{\downarrow}$, then we have*

$$||t^{\downarrow}|| = ||t'^{\downarrow}|| \Leftrightarrow DIS_{AT}(t^{\downarrow}, x) \cap AT(t'^{\downarrow}) \neq \emptyset \text{ for each} \forall x \in U \text{ and } x \notin ||t^{\downarrow}||.$$

Proof.

1. "⇒": $\forall x \in U$, if $x \notin \|t^{\uparrow}\|$, then $x \notin \|t'^{\uparrow}\|$ by $\|t^{\uparrow}\| = \|t'^{\uparrow}\|$. Thus, there must be $a_j \in AT(t'^{\uparrow}) \subseteq AT$ such that $f(x, a_j) < v_j$ where $v_j \in V_{a_j}$ and $(\geqslant_{a_j}, v_j) \in \|t'^{\uparrow}\|$. It should be noticed that $t'^{\uparrow} \geq t^{\uparrow}$, then we have $a_j \in DIS_{AT}(t^{\uparrow}, x)$, it follows that $DIS_{AT}(t^{\uparrow}, x) \cap AT(t'^{\uparrow}) \neq \emptyset$.

   "⇐": It is obvious that $\|t'^{\uparrow}\| \supseteq \|t^{\uparrow}\|$ by $t'^{\uparrow} \geq t^{\uparrow}$. Therefore, it must be proved that $\|t'^{\uparrow}\| \subseteq \|t^{\uparrow}\|$. $\forall x \in U$, $x \notin \|t^{\uparrow}\|$, since by assumption $DIS_{AT}(t^{\uparrow}, x) \cap AT(t'^{\uparrow}) \neq \emptyset$, there must be $a_j \in AT(t'^{\uparrow})$ such that $f(x, a_j) < v_j$ where $v_j \in V_{a_j}$ and $(\geqslant_{a_j}, v_j) \in \|t'^{\uparrow}\|$, it follows that $x \notin \|t'^{\uparrow}\|$, i.e., $\|t'^{\uparrow}\| \subseteq \|t^{\uparrow}\|$.

2. The proof of 2 is similar to the proof of 1. □

Definition 4.9. Let I be an incomplete decision system, $t^{\uparrow} \in FDES^{\uparrow}(AT)$, $t^{\downarrow} \in FDES^{\downarrow}(AT)$, define

$$\Delta(t^{\uparrow}) = \bigwedge_{x \in U} \vee DIS_{AT}(t^{\uparrow}, x) \tag{4.43}$$

$$\Delta(t^{\downarrow}) = \bigwedge_{x \in U} \vee DIS_{AT}(t^{\downarrow}, x) \tag{4.44}$$

$\Delta(t^{\uparrow})$ ($\Delta(t^{\downarrow})$) are referred to as the $t^{\uparrow}$ ($t^{\downarrow}$)-discernibility functions in I.

By using Boolean reasoning techniques, we can obtain the following theorem from Theorem 4.3.

Theorem 4.4. *Let I be an incomplete decision system,*

1. *if $t^{\uparrow} \in FDES^{\uparrow}(AT)$ and $t'^{\uparrow} \in DES^{\uparrow}(AT)$ such that $t'^{\uparrow} \geq t^{\uparrow}$, then $t'^{\uparrow}$ is a reduct descriptor of $t^{\uparrow}$ if and only if $AT(t')$ is the prime implicant of the $t^{\uparrow}$ discernibility function, i.e. $\Delta(t^{\uparrow})$;*
2. *if $t^{\downarrow} \in FDES^{\downarrow}(AT)$ and $t'^{\downarrow} \in DES^{\downarrow}(AT)$ such that $t'^{\downarrow} \geq t^{\downarrow}$, then $t'^{\downarrow}$ is a reduct descriptor of $t^{\downarrow}$ if and only if $AT(t')$ is the prime implicant of the $t^{\downarrow}$ discernibility function, i.e. $\Delta(t^{\downarrow})$.*

Example 4.6. Following Example 4.5, Table 4.5 is the ↑-discernibility matrix of full ↑ AT-descriptors in Table 4.1.

By Definition 4.9, we have

$$\begin{aligned}\Delta(t_1^{\uparrow}) &= \bigwedge_{x \in U} \vee D_{AT}(t_1^{\uparrow}, x)\\ &= (a_1 \vee a_2) \wedge (a_1 \vee a_3) \wedge (a_2 \vee a_3) \wedge AT\\ &= (a_1 \wedge a_2) \vee (a_1 \wedge a_3) \vee (a_2 \wedge a_3)\end{aligned}$$

By Theorem 4.4, we have
$redes(t_1^{\uparrow}) = \{(\geqslant_{a_1}, 3) \wedge (\geqslant_{a_2}, 2), (\geqslant_{a_2}, 2) \wedge (\geqslant_{a_3}, 2), (\geqslant_{a_1}, 3) \wedge (\geqslant_{a_3}, 2)\}$

Table 4.5 ↑-discernibility matrix in Table 4.1

U	x_1	x_2	x_3	x_4	x_5	x_6	x_7	x_8	x_9	x_{10}	x_{11}	x_{12}
$t_1^{\uparrow}$	AT	AT	a_1, a_2	AT	a_1, a_3	a_2, a_3	a_2, a_3	AT	a_1, a_2	a_1, a_2	a_1, a_2	AT
$t_2^{\uparrow}$	AT	a_1, a_2	a_1, a_2	a_1, a_2	a_1	a_2	a_1, a_2	AT	a_1, a_2	a_1, a_2	a_1, a_2	AT
$t_3^{\uparrow}$	AT	a_1, a_3	a_1	a_1, a_3	a_1, a_3	a_3	a_1, a_3	AT	a_1	a_1	a_1	AT
$t_4^{\uparrow}$	AT	a_1	a_1	a_1	a_1	AT	a_1	AT	a_1	a_1	a_1	a_1
$t_5^{\uparrow}$	AT	a_2, a_3	a_2	AT	a_1, a_3	a_2, a_3	a_2, a_3	AT	a_2	a_1, a_2	a_2	AT
$t_6^{\uparrow}$	AT	a_2	a_2	a_1, a_2	a_1	AT	a_2	AT	a_2	a_1, a_2	a_2	AT
$t_7^{\uparrow}$	AT	a_3	AT	a_1, a_3	a_1, a_3	a_3	a_3	AT	AT	a_1	AT	AT
$t_8^{\uparrow}$	AT	AT	AT	a_1	a_1	AT	AT	AT	AT	a_1	AT	AT
$t_9^{\uparrow}$	AT	a_2, a_3	a_2	a_2, a_3	a_3	a_2, a_3	a_2, a_3	AT	a_2	a_2	a_2	AT
$t_{10}^{\uparrow}$	AT	a_2	a_2	a_2	AT	a_2	a_2	AT	a_2	a_2	a_2	AT
$t_{11}^{\uparrow}$	AT	a_3	AT	a_3	a_3	a_3	a_3	AT	AT	AT	AT	AT
$t_{12}^{\uparrow}$	AT	AT	AT	AT	AT	AT	AT	AT	AT	AT	AT	AT

Similarly, it is not difficult to compute all reduct descriptors of all full ↑ AT-descriptors in Table 4.1 as following:

$$redes(t_2^{\uparrow}) = \{(\geqslant_{a_1}, 3) \wedge (\geqslant_{a_2}, 2)\}$$
$$redes(t_3^{\uparrow}) = \{(\geqslant_{a_1}, 3) \wedge (\geqslant_{a_3}, 2)\}$$
$$redes(t_4^{\uparrow}) = \{(\geqslant_{a_1}, 3)\}$$
$$redes(t_5^{\uparrow}) = \{(\geqslant_{a_1}, 2) \wedge (\geqslant_{a_2}, 2), (\geqslant_{a_2}, 2) \wedge (\geqslant_{a_3}, 2)\}$$
$$redes(t_6^{\uparrow}) = \{(\geqslant_{a_1}, 2) \wedge (\geqslant_{a_2}, 2)\}$$
$$redes(t_7^{\uparrow}) = \{(\geqslant_{a_1}, 2) \wedge (\geqslant_{a_3}, 2)\}$$
$$redes(t_8^{\uparrow}) = \{(\geqslant_{a_1}, 2)\}$$
$$redes(t_9^{\uparrow}) = \{(\geqslant_{a_2}, 2) \wedge (\geqslant_{a_3}, 2)\}$$
$$redes(t_{10}^{\uparrow}) = \{(\geqslant_{a_2}, 2)\}$$
$$redes(t_{11}^{\uparrow}) = \{(\geqslant_{a_3}, 2)\}$$
$$redes(t_{12}^{\uparrow}) = \{(\geqslant_{a_1}, 1), (\geqslant_{a_2}, 1), (\geqslant_{a_3}, 1)\}$$

For all ↓ full AT-descriptors in Table 4.1, we have

$$redes(t_1^{\downarrow}) = \{(\leqslant_{a_1}, 3), (\leqslant_{a_2}, 2), (\leqslant_{a_3}, 2)\}$$
$$redes(t_2^{\downarrow}) = \{(\leqslant_{a_3}, 1)\}$$
$$redes(t_3^{\downarrow}) = \{(\leqslant_{a_2}, 1)\}$$
$$redes(t_4^{\downarrow}) = \{(\leqslant_{a_3}, 1)\}$$
$$redes(t_5^{\downarrow}) = \{(\leqslant_{a_1}, 2)\}$$
$$redes(t_6^{\downarrow}) = \{(\leqslant_{a_1}, 2), (\leqslant_{a_3}, 1)\}$$
$$redes(t_7^{\downarrow}) = \{(\leqslant_{a_1}, 2)\}$$
$$redes(t_8^{\downarrow}) = \{(\leqslant_{a_1}, 2), (\leqslant_{a_3}, 1)\}$$
$$redes(t_9^{\downarrow}) = \{(\leqslant_{a_1}, 1)\}$$
$$redes(t_{10}^{\downarrow}) = \{(\leqslant_{a_1}, 1)\}$$

$$redes(t_{11}^{\downarrow}) = \{(\leqslant_{a_1}, 1)\}$$
$$redes(t_{12}^{\downarrow}) = \{(\leqslant_{a_1}, 1), (\leqslant_{a_3}, 1)\}$$

4.4.3 ↑ and ↓ Certain Rules

Definition 4.10. Let I be an incomplete decision system, suppose that $\wedge_{a_i \in A}(\geqslant_{a_i}, v_i) \in FDES^{\uparrow}(A)$ and $\wedge_{a_i \in A}(\leqslant_{a_i}, v_i) \in FDES^{\downarrow}(A)$, define the following:

1. if $\| \wedge_{a_i \in A} (\geqslant_{a_i}, v_i)\| \subseteq CL_n^{\geqslant}$ $(n = 2, \cdots, m)$, then
$$f(x, a_1) \geqslant v_1 \wedge f(x, a_2) \geqslant v_2 \wedge \cdots \wedge f(x, a_k) \geqslant v_k \rightarrow x \in CL_n^{\geqslant}$$
is referred to as a ↑ certain rule, it can be denoted by $\wedge_{a_i \in A}(\geqslant_{a_i}, v_i) \rightarrow x \in CL_n^{\geqslant}$;

2. if $\| \wedge_{a_i \in A} (\leqslant_{a_i}, v_i)\| \subseteq CL_n^{\leqslant}$ $(n = 1, \cdots, m-1)$, then
$$f(x, a_1) \leqslant v_1 \wedge f(x, a_2) \leqslant v_2 \wedge \cdots \wedge f(x, a_k) \leqslant v_k \rightarrow x \in CL_n^{\leqslant}$$
is referred to as a ↓ certain rule, it can be denoted by $\wedge_{a_i \in A}(\leqslant_{a_i}, v_i) \rightarrow x \in CL_n^{\leqslant}$.

where $A = \{a_1, a_2, \cdots, a_k\}$, $(v_1, v_2, \cdots, v_k) \in V_{a_1} \times V_{a_2} \times \cdots \times V_{a_k}$, $1 \leqslant i \leqslant k$.

Example 4.7. Following Example 4.5, we have $\|t_1^{\uparrow}\|, \|t_2^{\uparrow}\|, \|t_3^{\uparrow}\|, \|t_4^{\uparrow}\|, \|t_5^{\uparrow}\|, \|t_6^{\uparrow}\|, \|t_7^{\uparrow}\|, \|t_9^{\uparrow}\| \subseteq CL_2^{\geqslant}$. Therefore, we can obtain the following ↑ certain rules:

$r_1 : f(x, a_1) \geqslant 3 \wedge f(x, a_2) \geqslant 2 \wedge f(x, a_3) \geqslant 2 \rightarrow f(x, a_4) \geqslant 2$//supported by $t_1^{\uparrow}$
$r_2 : f(x, a_1) \geqslant 3 \wedge f(x, a_2) \geqslant 2 \wedge f(x, a_3) \geqslant 1 \rightarrow f(x, a_4) \geqslant 2$//supported by $t_2^{\uparrow}$
$r_3 : f(x, a_1) \geqslant 3 \wedge f(x, a_2) \geqslant 1 \wedge f(x, a_3) \geqslant 2 \rightarrow f(x, a_4) \geqslant 2$//supported by $t_3^{\uparrow}$
$r_4 : f(x, a_1) \geqslant 3 \wedge f(x, a_2) \geqslant 1 \wedge f(x, a_3) \geqslant 1 \rightarrow f(x, a_4) \geqslant 2$//supported by $t_4^{\uparrow}$
$r_5 : f(x, a_1) \geqslant 2 \wedge f(x, a_2) \geqslant 2 \wedge f(x, a_3) \geqslant 2 \rightarrow f(x, a_4) \geqslant 2$//supported by $t_5^{\uparrow}$
$r_6 : f(x, a_1) \geqslant 2 \wedge f(x, a_2) \geqslant 2 \wedge f(x, a_3) \geqslant 1 \rightarrow f(x, a_4) \geqslant 2$//supported by $t_6^{\uparrow}$
$r_7 : f(x, a_1) \geqslant 2 \wedge f(x, a_2) \geqslant 1 \wedge f(x, a_3) \geqslant 2 \rightarrow f(x, a_4) \geqslant 2$//supported by $t_7^{\uparrow}$
$r_8 : f(x, a_1) \geqslant 1 \wedge f(x, a_2) \geqslant 2 \wedge f(x, a_3) \geqslant 2 \rightarrow f(x, a_4) \geqslant 2$//supported by $t_9^{\uparrow}$

Obviously, rules r_1, r_2 and r_3 can be classified into r_4, rule r_5 can be classified into r_6 and r_7, respectively.

Moreover, we have $\|t_9^{\downarrow}\|, \|t_{10}^{\downarrow}\|, \|t_{11}^{\downarrow}\|, \|t_{12}^{\downarrow}\| \subseteq CL_1^{\leqslant}$. Therefore, we can obtain the following ↓ certain rules:

$r_1' : f(x, a_1) \leqslant 1 \wedge f(x, a_2) \leqslant 2 \wedge f(x, a_3) \leqslant 2 \rightarrow f(x, a_4) \leqslant 1$//supported by $t_9^{\downarrow}$
$r_2' : f(x, a_1) \leqslant 1 \wedge f(x, a_2) \leqslant 2 \wedge f(x, a_3) \leqslant 1 \rightarrow f(x, a_4) \leqslant 1$//supported by $t_{10}^{\downarrow}$
$r_3' : f(x, a_1) \leqslant 1 \wedge f(x, a_2) \leqslant 1 \wedge f(x, a_3) \leqslant 2 \rightarrow f(x, a_4) \leqslant 1$//supported by $t_{11}^{\downarrow}$
$r_4' : f(x, a_1) \leqslant 1 \wedge f(x, a_2) \leqslant 1 \wedge f(x, a_3) \leqslant 1 \rightarrow f(x, a_4) \leqslant 1$//supported by $t_{12}^{\downarrow}$

Obviously, rules r_2', r_3' and r_4' can be classified into rule r_1'.

By comparing with Example 4.1, we can see that there is no unknown value "$*$" in any one of the ↑ (↓) certain rules. Moreover, r_6 and r_8 are new rules, which can not be derived from the lower approximation of $CL_2^{\geqslant}$; rules R_5, R_6 and R_7 in Example 4.1 only reflect partial information of rule r_1', i.e., r_1' is more generalized than R_5, R_6 and R_7.

From discussions above, we can see that descriptor-based approach to knowledge acquisition is better than the lower approximation-based approach to knowledge acquisition in dominance principle because it not only eliminates the rules with unknown values but also generates more certain rules from the incomplete decision system.

4.4.4 Optimal ↑ and ↓ Certain Rules

The reduct of descriptor can also be used to simplify the certain rules.

Definition 4.11. Let I be an incomplete decision system, $t^{\uparrow} \in FDES^{\uparrow}(AT)$, $t^{\downarrow} \in FDES^{\downarrow}(AT)$:

1. if $\exists CL_n^{\geqslant}$ $(2 \leqslant n \leqslant m)$ such that $||t^{\uparrow}|| \subseteq CL_n^{\geqslant}$, then $t^{\uparrow}$ is referred to as consistent in $CL_n^{\geqslant}$;
2. if $\exists CL_n^{\leqslant}$ $(1 \leqslant n \leqslant m-1)$ such that $||t^{\downarrow}|| \subseteq CL_n^{\leqslant}$, then $t^{\downarrow}$ is referred to as consistent in $CL_n^{\leqslant}$.

Definition 4.12. Let I be an incomplete decision system,

1. Suppose that $t^{\uparrow} \in FDES^{\uparrow}(AT)$ and $t^{\uparrow}$ is consistent in $CL_n^{\geqslant}$, if $t'^{\uparrow} \succeq t^{\uparrow}$ and $t'^{\uparrow}$ is consistent in $CL_n^{\geqslant}$, then $t'^{\uparrow}$ is referred to as a relative consistent descriptor of $t^{\uparrow}$; if $t'^{\uparrow}$ is a relative consistent descriptor of $t^{\uparrow}$ and for $\forall t''^{\uparrow} \succ t'^{\uparrow}$, $t''^{\uparrow}$ is not the relative consistent descriptor of $t^{\uparrow}$, then $t'^{\uparrow}$ is referred to as a relative reduct descriptor of $t^{\uparrow}$.
2. Suppose that $t^{\downarrow} \in FDES^{\downarrow}(AT)$ and $t^{\downarrow}$ is consistent in $CL_n^{\leqslant}$, if $t'^{\downarrow} \succeq t^{\downarrow}$ and $t'^{\downarrow}$ is consistent in $CL_n^{\leqslant}$, then $t'^{\downarrow}$ is referred to as a relative consistent descriptor of $t^{\downarrow}$; if $t'^{\downarrow}$ is a relative consistent descriptor of $t^{\downarrow}$ and for $\forall t''^{\downarrow} \succ t'^{\downarrow}$, $t''^{\downarrow}$ is not the relative consistent descriptor of $t^{\downarrow}$, then $t'^{\downarrow}$ is referred to as a relative reduct descriptor of $t^{\downarrow}$.

A relative reduct descriptor of $t^{\uparrow}$ allows us to preserve the consistent characteristic of $t^{\uparrow}$ in $CL_n^{\geqslant}$ $(2 \leqslant n \leqslant m)$ with the *minimal* number of required atomic properties. A relative reduct descriptor of $t^{\downarrow}$ allows us to preserve the consistent characteristic of $t^{\downarrow}$ in $CL_n^{\leqslant}$ $(1 \leqslant n \leqslant m-1)$ with the *minimal* number of required atomic properties.

Suppose that $t^{\uparrow} \in FDES^{\uparrow}(AT)$, $t^{\uparrow}$ is consistent in CL_n $(2 \leqslant n \leqslant m)$, $t^{\downarrow} \in FDES^{\downarrow}(AT)$, $t^{\downarrow}$ is consistent in CL_n $(1 \leqslant n \leqslant m-1)$, it is not difficult to obtain the following conclusions:

1. If $t'^{\uparrow}$ is a relative consistent descriptor of $t^{\uparrow}$, then $t'^{\uparrow} \to CL_n^{\geqslant}$ is a simplified ↑ credible rule for $t^{\uparrow} \to CL_n^{\geqslant}$, here $t^{\uparrow} \to CL_n^{\geqslant}$ is referred to as the initial ↑ credible rule induced by $t^{\uparrow}$; if $t'^{\uparrow}$ is a relative reduct descriptor of $t^{\uparrow}$, then $t^{\uparrow} \to CL_n^{\geqslant}$ is an optimal ↑ credible rule.
2. If $t'^{\downarrow}$ is a relative consistent descriptor of $t^{\downarrow}$, then $t'^{\downarrow} \to CL_n^{\leqslant}$ is a simplified ↓ credible rule for $t^{\downarrow} \to CL_n^{\leqslant}$, here $t^{\downarrow} \to CL_n^{\leqslant}$ is referred to as the initial ↓ credible rule induced by $t^{\downarrow}$; if $t'^{\downarrow}$ is a relative reduct descriptor of $t^{\downarrow}$, then $t'^{\downarrow} \to CL_n^{\leqslant}$ is an optimal ↓ credible rule.

From discussions above, to obtain the optimal ↑ (↓) credible rules in incomplete decision system, the computation of relative reducts of ↑ (↓) descriptors has become a necessity. In the following, we will present the practical approach to compute two notions of relative reduct descriptors.

Theorem 4.5. *Let I be an incomplete decision system,*

1. *Suppose that* $t^{\uparrow} \in FDES^{\uparrow}(AT)$ *and* $t^{\uparrow}$ *is consistent in* $CL_n^{\geqslant}$ $(2 \leqslant n \leqslant m)$, *for* $\forall t'^{\uparrow} \in DES^{\uparrow}(AT)$, *we have the following:*
   $t'^{\uparrow}$ *is a relative consistent descriptor of* $t^{\uparrow} \Leftrightarrow t'^{\uparrow} \succeq t^{\uparrow}, \forall x \in U - CL_n^{\geqslant}, DIS_{AT}(t^{\uparrow}, x) \cap AT(t'^{\uparrow}) \neq \varnothing$.
2. *Suppose that* $t^{\downarrow} \in FDES^{\downarrow}(AT)$ *and* $t^{\downarrow}$ *is consistent in* $CL_n^{\leqslant}$ $(1 \leqslant n \leqslant m-1)$, *for* $\forall t'^{\downarrow} \in DES^{\downarrow}(AT)$, *we have the following:*
   $t'^{\downarrow}$ *is a relative consistent descriptor of* $t^{\downarrow} \Leftrightarrow t'^{\downarrow} \succeq t^{\downarrow}, \forall x \in U - CL_n^{\leqslant}, DIS_{AT}(t^{\downarrow}, x) \cap AT(t'^{\downarrow}) \neq \varnothing$.

Proof.
1. "⇒": It is clear that $t'^{\uparrow} \succeq t^{\uparrow}$ because $t'^{\uparrow}$ is a relative consistent descriptor of $t^{\uparrow}$. Suppose that $\exists x \in U - CL_n^{\geqslant}$ such that $DIS_{AT}(t^{\uparrow}, x) \cap AT(t'^{\uparrow}) = \varnothing$, then we have $x \in ||t'^{\uparrow}||$. By condition we have $t'^{\uparrow}$ is consistent in $CL_n^{\geqslant}$, $||t'^{\uparrow}|| \subseteq CL_n^{\geqslant}$, it follows that $x \in CL_n^{\geqslant}$. This is contradictive to $x \in U - CL_n^{\geqslant}$.
   "⇐": Suppose that $t'^{\uparrow}$ is not consistent in $CL_n^{\geqslant}$, then there must be $y \in U$ such that $y \in ||t'^{\uparrow}||$ and $y \notin CL_n^{\geqslant}$. By condition we have $DIS_{AT}(t^{\uparrow}, y) \cap AT(t'^{\uparrow}) \neq \varnothing$ because $y \in U - CL_n^{\geqslant}$. This is contradictive to $DIS_{AT}(t^{\uparrow}, y) \cap AT(t'^{\uparrow}) = \varnothing$ because $y \in ||t'^{\uparrow}||$.
2. The proof of 2 is similar to the proof of 1. □

Definition 4.13. Let I be an incomplete decision system, $\forall t^{\uparrow} \in FDES^{\uparrow}(AT)$ and $\forall t^{\downarrow} \in FDES^{\downarrow}(AT)$:

1. if $t^{\uparrow}$ is consistent in $CL_n^{\geqslant}$ $(2 \leqslant n \leqslant m)$, then define

$$\Delta_n(t^{\uparrow}) = \bigwedge_{x \in U - CL_n^{\geqslant}} \vee DIS_{AT}(t^{\uparrow}, x) \tag{4.45}$$

2. if $t^{\downarrow}$ is consistent in $CL_n^{\leqslant}$ $(1 \leqslant n \leqslant m-1)$, then define

$$\Delta_n(t^{\downarrow}) = \bigwedge_{x \in U - CL_n^{\leqslant}} \vee DIS_{AT}(t^{\downarrow}, x) \tag{4.46}$$

$\Delta_n(t^{\uparrow})$ $(\Delta_n(t^{\downarrow}))$ are referred to as the $t^{\uparrow}$ $(t^{\downarrow})$ relative discernibility functions.

By using the Boolean reasoning technique, we can also obtain the following theorem by Theorem 4.5.

Theorem 4.6. *Let I be an incomplete decision system,*

1. *Suppose that* $t^{\uparrow} \in FDES^{\uparrow}(AT)$ *and* $t^{\uparrow}$ *is consistent in* $CL_n^{\geqslant}$ $(2 \leqslant n \leqslant m)$, $t'^{\uparrow} \in DES^{\uparrow}(AT)$, $t'^{\uparrow} \succeq t^{\uparrow}$, *then:*
   $t'^{\uparrow}$ *is a relative reduct descriptor of* $t^{\uparrow}$ *if and only if* $AT(t')$ *is the prime implicant of the* $t^{\uparrow}$ *relative discernibility function, i.e.* $\Delta_n(t^{\uparrow})$;
2. *Suppose that* $t^{\downarrow} \in FDES^{\downarrow}(AT)$ *and* $t^{\downarrow}$ *is consistent in* $CL_n^{\leqslant}$ $(1 \leqslant n \leqslant m-1)$, $t'^{\downarrow} \in DES^{\downarrow}(AT)$, $t'^{\downarrow} \succeq t^{\downarrow}$, *then:*
   $t'^{\downarrow}$ *is a relative reduct descriptor of* $t^{\downarrow}$ *if and only if* $AT(t')$ *is the prime implicant of the* $t^{\downarrow}$ *relative discernibility function, i.e.* $\Delta_n(t^{\downarrow})$.

4.4.5 An Illustrative Example

We use an example for evaluation to illustrate the concept of relative reduct descriptor and optimal certain rules.

Suppose that the director of the school must give a global evaluation to some students. This evaluation should be based on the level in Mathematics, Physics and Literature. However, there are some missing values. The director gave the examples of evaluation as shown in Table 4.6. The example concern fifteen students described by means of four attributes:

- $a \Rightarrow$ level in Mathematics;
- $b \Rightarrow$ level in Physics;
- $c \Rightarrow$ level of Literature;
- $d \Rightarrow$ global evaluation (decision class).

Table 4.6 Student evaluations with unknown values

U	Mathematics	Physics	Literature	Global Evaluation
x_1	medium	medium	bad	bad
x_2	good	medium	bad	medium
x_3	bad	good	*	bad
x_4	medium	good	bad	medium
x_5	*	good	bad	medium
x_6	bad	bad	medium	bad
x_7	good	bad	medium	bad
x_8	medium	*	medium	medium
x_9	good	medium	medium	good
x_{10}	medium	good	medium	good
x_{11}	good	bad	good	bad
x_{12}	*	medium	*	medium
x_{13}	good	medium	good	good
x_{14}	bad	good	good	bad
x_{15}	medium	good	good	good

By the global evaluation, we have $\boldsymbol{CL} = \{CL_3, CL_2, CL_1\}$, where
$CL_3 = \{\text{good students}\} = \{x_9, x_{10}, x_{13}, x_{15}\}$,
$CL_2 = \{\text{medium students}\} = \{x_2, x_4, x_5, x_8, x_{12}\}$,
$CL_1 = \{\text{bad students}\} = \{x_1, x_3, x_6, x_7, x_{11}, x_{14}\}$.
Thus, the following unions of classes should be approximated:

- $CL_2^{\geqslant} = CL_2 \cup CL_3$, i.e. the class of at least medium students,

- $CL_3^{\geqslant} = CL_3$, i.e. the class of (at least) good students.
- $CL_2^{\leqslant} = CL_1 \cup CL_2$, i.e. the class of at most medium students.
- $CL_1^{\leqslant} = CL_1$, i.e. the class of (at most) bad students.

Firstly, similar to Example 4.7, all initial $\uparrow$ certain rules can be derived from Table 4.6:

r_1: if Mathematics $\geqslant$ medium and Physics $\geqslant$ good and Literature $\geqslant$ good, then student $\geqslant$ good // supported by descriptor $t_1^{\uparrow} = (\geqslant_{\text{Mathematics}}, \text{medium}) \wedge (\geqslant_{\text{Physcis}}, \text{good}) \wedge (\geqslant_{\text{Literature}}, \text{good})$;

r_2: if Mathematics $\geqslant$ good and Physics $\geqslant$ good and Literature $\geqslant$ bad, then student $\geqslant$ medium // supported by descriptor $t_2^{\uparrow} = (\geqslant_{\text{Mathematics}}, \text{good}) \wedge (\geqslant_{\text{Physcis}}, \text{good}) \wedge (\geqslant_{\text{Literature}}, \text{bad})$;

r_3: if Mathematics $\geqslant$ good and Physics $\geqslant$ medium and Literature $\geqslant$ good, then student $\geqslant$ medium // supported by descriptor $t_3^{\uparrow} = (\geqslant_{\text{Mathematics}}, \text{good}) \wedge (\geqslant_{\text{Physcis}}, \text{medium}) \wedge (\geqslant_{\text{Literature}}, \text{good})$;

r_4: if Mathematics $\geqslant$ good and Physics $\geqslant$ medium and Literature $\geqslant$ medium, then student $\geqslant$ medium // supported by descriptor $t_4^{\uparrow} = (\geqslant_{\text{Mathematics}}, \text{good}) \wedge (\geqslant_{\text{Physcis}}, \text{medium}) \wedge (\geqslant_{\text{Literature}}, \text{medium})$;

r_5: if Mathematics $\geqslant$ good and Physics $\geqslant$ medium and Literature $\geqslant$ bad, then student $\geqslant$ medium // supported by descriptor $t_5^{\uparrow} = (\geqslant_{\text{Mathematics}}, \text{good}) \wedge (\geqslant_{\text{Physcis}}, \text{medium}) \wedge (\geqslant_{\text{Literature}}, \text{bad})$;

r_6: if Mathematics $\geqslant$ medium and Physics $\geqslant$ good and Literature $\geqslant$ good, then student $\geqslant$ medium // supported by descriptor $t_6^{\uparrow} = (\geqslant_{\text{Mathematics}}, \text{medium}) \wedge (\geqslant_{\text{Physcis}}, \text{good}) \wedge (\geqslant_{\text{Literature}}, \text{good})$;

r_7: if Mathematics $\geqslant$ medium and Physics $\geqslant$ good and Literature $\geqslant$ medium, then student $\geqslant$ medium // supported by descriptor $t_7^{\uparrow} = (\geqslant_{\text{Mathematics}}, \text{medium}) \wedge (\geqslant_{\text{Physcis}}, \text{good}) \wedge (\geqslant_{\text{Literature}}, \text{medium})$;

r_8: if Mathematics $\geqslant$ medium and Physics $\geqslant$ good and Literature $\geqslant$ bad, then student $\geqslant$ medium // supported by descriptor $t_8^{\uparrow} = (\geqslant_{\text{Mathematics}}, \text{medium}) \wedge (\geqslant_{\text{Physcis}}, \text{good}) \wedge (\geqslant_{\text{Literature}}, \text{bad})$;

r_9: if Mathematics $\geqslant$ medium and Physics $\geqslant$ medium and Literature $\geqslant$ good, then student $\geqslant$ medium // supported by descriptor $t_9^{\uparrow} = (\geqslant_{\text{Mathematics}}, \text{medium}) \wedge (\geqslant_{\text{Physcis}}, \text{medium}) \wedge (\geqslant_{\text{Literature}}, \text{good})$;

r_{10}: if Mathematics $\geqslant$ medium and Physics $\geqslant$ medium and Literature $\geqslant$ medium, then student $\geqslant$ medium // supported by descriptor $t_{10}^{\uparrow} = (\geqslant_{\text{Mathematics}}, \text{medium}) \wedge (\geqslant_{\text{Physcis}}, \text{medium}) \wedge (\geqslant_{\text{Literature}}, \text{medium})$.

All initial $\downarrow$ credible rules can be derived from Table 4.6:

r_1': if Mathematics $\leqslant$ bad and Physics $\leqslant$ bad and Literature $\leqslant$ good, then student $\leqslant$ bad // supported by descriptor $s_1^{\downarrow} = (\leqslant_{\text{Mathematics}}, \text{bad}) \wedge (\leqslant_{\text{Physcis}}, \text{bad}) \wedge (\leqslant_{\text{Literature}}, \text{good})$;

r_2': if Mathematics $\leqslant$ bad and Physics $\leqslant$ bad and Literature $\leqslant$ medium, then student $\leqslant$ bad // supported by descriptor $s_2^{\downarrow} = (\leqslant_{\text{Mathematics}}, \text{bad}) \wedge (\leqslant_{\text{Physcis}}, \text{bad}) \wedge (\leqslant_{\text{Literature}}, \text{medium})$;

r'_3: if Mathematics ⩽ good and Physics ⩽ good and Literature ⩽ bad, then student ⩽ medium // supported by descriptor $s_3^{\downarrow} = (\leqslant_{\text{Mathematics}}, \text{good}) \wedge (\leqslant_{\text{Physcis}}, \text{good}) \wedge (\leqslant_{\text{Literature}}, \text{bad})$;

r'_4: if Mathematics ⩽ good and Physics ⩽ medium and Literature ⩽ bad, then student ⩽ medium // supported by descriptor $s_4^{\downarrow} = (\leqslant_{\text{Mathematics}}, \text{good}) \wedge (\leqslant_{\text{Physcis}}, \text{medium}) \wedge (\leqslant_{\text{Literature}}, \text{bad})$;

r'_5: if Mathematics ⩽ good and Physics ⩽ bad and Literature ⩽ good, then student ⩽ medium // supported by descriptor $s_5^{\downarrow} = (\leqslant_{\text{Mathematics}}, \text{good}) \wedge (\leqslant_{\text{Physcis}}, \text{bad}) \wedge (\leqslant_{\text{Literature}}, \text{good})$;

r'_6: if Mathematics ⩽ good and Physics ⩽ bad and Literature ⩽ medium, then student ⩽ medium // supported by descriptor $s_6^{\downarrow} = (\leqslant_{\text{Mathematics}}, \text{good}) \wedge (\leqslant_{\text{Physcis}}, \text{bad}) \wedge (\leqslant_{\text{Literature}}, \text{medium})$;

r'_7: if Mathematics ⩽ medium and Physics ⩽ good and Literature ⩽ bad, then student ⩽ medium // supported by descriptor $s_7^{\downarrow} = (\leqslant_{\text{Mathematics}}, \text{medium}) \wedge (\leqslant_{\text{Physcis}}, \text{good}) \wedge (\leqslant_{\text{Literature}}, \text{bad})$;

r'_8: If Mathematics ⩽ medium and Physics ⩽ medium and Literature ⩽ good, then student ⩽ medium // supported by descriptor $s_8^{\downarrow} = (\leqslant_{\text{Mathematics}}, \text{medium}) \wedge (\leqslant_{\text{Physcis}}, \text{medium}) \wedge (\leqslant_{\text{Literature}}, \text{good})$;

r'_9: if Mathematics ⩽ medium and Physics ⩽ medium and Literature ⩽ medium, then student ⩽ medium // supported by descriptor $s_9^{\downarrow} = (\leqslant_{\text{Mathematics}}, \text{medium}) \wedge (\leqslant_{\text{Physcis}}, \text{medium}) \wedge (\leqslant_{\text{Literature}}, \text{medium})$;

r'_{10}: if Mathematics ⩽ medium and Physics ⩽ medium and Literature ⩽ bad, then student ⩽ medium // supported by descriptor $s_{10}^{\downarrow} = (\leqslant_{\text{Mathematics}}, \text{medium}) \wedge (\leqslant_{\text{Physcis}}, \text{medium}) \wedge (\leqslant_{\text{Literature}}, \text{bad})$;

r'_{11}: if Mathematics ⩽ medium and Physics ⩽ bad and Literature ⩽ good, then student ⩽ medium // supported by descriptor $s_{11}^{\downarrow} = (\leqslant_{\text{Mathematics}}, \text{medium}) \wedge (\leqslant_{\text{Physcis}}, \text{bad}) \wedge (\leqslant_{\text{Literature}}, \text{good})$;

r'_{12}: if Mathematics ⩽ medium and Physics ⩽ bad and Literature ⩽ medium, then student ⩽ medium; // supported by descriptor $s_{12}^{\downarrow} = (\leqslant_{\text{Mathematics}}, \text{medium}) \wedge (\leqslant_{\text{Physcis}}, \text{bad}) \wedge (\leqslant_{\text{Literature}}, \text{medium})$;

r'_{13}: if Mathematics ⩽ bad and Physics ⩽ good and Literature ⩽ good, then student ⩽ medium // supported by descriptor $s_{13}^{\downarrow} = (\leqslant_{\text{Mathematics}}, \text{bad}) \wedge (\leqslant_{\text{Physcis}}, \text{good}) \wedge (\leqslant_{\text{Literature}}, \text{good})$;

r'_{14}: if Mathematics ⩽ bad and Physics ⩽ good and Literature ⩽ medium, then student ⩽ medium // supported by descriptor $s_{14}^{\downarrow} = (\leqslant_{\text{Mathematics}}, \text{bad}) \wedge (\leqslant_{\text{Physcis}}, \text{good}) \wedge (\leqslant_{\text{Literature}}, \text{medium})$;

r'_{15}: if Mathematics ⩽ bad and Physics ⩽ good and Literature ⩽ bad, then student ⩽ medium // supported by descriptor $s_{15}^{\downarrow} = (\leqslant_{\text{Mathematics}}, \text{bad}) \wedge (\leqslant_{\text{Physcis}}, \text{good}) \wedge (\leqslant_{\text{Literature}}, \text{bad})$;

r'_{16}: if Mathematics ⩽ bad and Physics ⩽ medium and Literature ⩽ good, then student ⩽ medium // supported by descriptor $s_{16}^{\downarrow} = (\leqslant_{\text{Mathematics}}, \text{bad}) \wedge (\leqslant_{\text{Physcis}},$

medium) $\wedge$ ($\leqslant_{\text{Literature}}$, good);

r'_{17}: if Mathematics $\leqslant$ bad and Physics $\leqslant$ medium and Literature $\leqslant$ medium, then student $\leqslant$ medium // supported by descriptor $s^{\downarrow}_{17}$ = ($\leqslant_{\text{Mathematics}}$, bad) $\wedge$ ($\leqslant_{\text{Physcis}}$, medium) $\wedge$ ($\leqslant_{\text{Literature}}$, medium);

r'_{18}: if Mathematics $\leqslant$ bad and Physics $\leqslant$ medium and Literature $\leqslant$ bad, then student $\leqslant$ medium // supported by descriptor $s^{\downarrow}_{18}$ = ($\leqslant_{\text{Mathematics}}$, bad) $\wedge$ ($\leqslant_{\text{Physcis}}$, medium) $\wedge$ ($\leqslant_{\text{Literature}}$, bad);

r'_{19}: if Mathematics $\leqslant$ bad and Physics $\leqslant$ bad and Literature $\leqslant$ good, then student $\leqslant$ medium // supported by descriptor $s^{\downarrow}_{19}$ = ($\leqslant_{\text{Mathematics}}$, bad) $\wedge$ ($\leqslant_{\text{Physcis}}$, bad) $\wedge$ ($\leqslant_{\text{Literature}}$, good);

r'_{20}: if Mathematics $\leqslant$ bad and Physics $\leqslant$ bad and Literature $\leqslant$ medium, then student $\leqslant$ medium // supported by descriptor $s^{\downarrow}_{20}$ = ($\leqslant_{\text{Mathematics}}$, bad) $\wedge$ ($\leqslant_{\text{Physcis}}$, bad) $\wedge$ ($\leqslant_{\text{Literature}}$, medium).

For any one of the initial $\uparrow$ and $\downarrow$ certain rules, there is a descriptor supporting this rule. Compute the relative reduct descriptor of such descriptor.

Take for instance r_1, r_1 is supported by the $\uparrow$ descriptor $t^{\uparrow}_1$. By Definition 4.13, we have

$$\begin{aligned}\Delta_3(t^{\uparrow}_1) &= \bigwedge_{x \in U - CL^{\geqslant}_3} \vee DIS_{AT}(t^{\uparrow}_1, x)\\ &= a \wedge b \wedge c\end{aligned}$$

Therefore, no atomic property can be eliminated from $t^{\uparrow}_1$ to preserve the property such that $\|t^{\uparrow}_1\| \subseteq CL^{\geqslant}_3$, i.e., the initial $\uparrow$ certain rule r_1 cannot be simplified. Similarly, it is not difficult to observe that initial $\uparrow$ certain rules r_9 and r_{10} also cannot be simplified.

Take for instance r_2, r_2 is supported by the $\uparrow$ descriptor $t^{\uparrow}_2$. By Definition 4.13, we have

$$\begin{aligned}\Delta_2(t^{\uparrow}_2) &= \bigwedge_{x \in U - CL^{\geqslant}_2} \vee DIS_{AT}(t^{\uparrow}_3, x)\\ &= a \wedge b\end{aligned}$$

By Theorem 4.6, ($\geqslant_{\text{Mathematics}}$, good) $\wedge$ ($\geqslant_{\text{Physcis}}$, good) is the relative reduct descriptor of $t^{\uparrow}_2$, from which we can obtain an optimal $\uparrow$ certain rule such that

Opt_r_1: if Mathematics $\geqslant$ good, and Physics $\geqslant$ good then student $\geqslant$ medium.

Similarly, we can get the following optimal $\uparrow$ certain rules:

Opt_r_2: if Mathematics $\geqslant$ good, and Physics $\geqslant$ medium, then student $\geqslant$ medium // supported by the relative reduct descriptors of $t^{\uparrow}_3$, $t^{\uparrow}_4$, $t^{\uparrow}_5$;

Opt_r_3: if Mathematics $\geqslant$ medium, and Physics $\geqslant$ good, then student $\geqslant$ medium // supported by the relative reduct descriptors of $t^{\uparrow}_6$, $t^{\uparrow}_7$, $t^{\uparrow}_8$.

Obviously, rule Opt_r_1 can be classified into rules Opt_r_2 and Opt_r_3, respectively.

Take for instance r'_1, r'_1 is supported by the $\downarrow$ descriptor $s^{\downarrow}_1$. By Definition 4.13, we have

$$\begin{aligned}\varDelta_1(s_1^{\downarrow}) &= \bigwedge_{x \in U - CL_3^{\geqslant}} \vee DIS_{AT}(s_1^{\downarrow}, x)\\ &= a \wedge b\end{aligned}$$

By Theorem 4.6, $(\leqslant_{\text{Mathematics}}, \text{bad}) \wedge (\leqslant_{\text{Physcis}}, \text{bad})$ is the relative reduct descriptor of $s_1^{\downarrow}$, from which we can obtain an optimal $\downarrow$ certain rule such that

$Opt\_r'_1$: if Mathematics $\leqslant$ bad and Physics $\leqslant$ bad, then student $\leqslant$ bad.

This rule is also supported by the relative reduct descriptor of $s_2^{\downarrow}$.

Similarly, we can get the following optimal $\downarrow$ certain rules:

$Opt\_r'_2$: if Literature $\leqslant$ bad, then student $\leqslant$ medium // supported by the relative reduct descriptors of $s_3^{\downarrow}$, $s_4^{\downarrow}$, $s_7^{\downarrow}$, $s_{10}^{\downarrow}$;

$Opt\_r'_3$: if Physics $\leqslant$ bad, then student $\leqslant$ medium // supported by the relative reduct descriptors of $s_5^{\downarrow}$, $s_6^{\downarrow}$, $s_{11}^{\downarrow}$, $s_{12}^{\downarrow}$;

$Opt\_r'_4$: if Mathematics $\leqslant$ medium and Physics $\leqslant$ medium, then student $\leqslant$ medium // supported by the relative reduct descriptors of $s_8^{\downarrow}$, $s_9^{\downarrow}$, $s_{10}^{\downarrow}$;

$Opt\_r'_5$: if Mathematics $\leqslant$ bad, then student $\leqslant$ medium // supported by the relative reduct descriptors of $s_{13}^{\downarrow}$, $s_{14}^{\downarrow}$, $s_{15}^{\downarrow}$, $s_{16}^{\downarrow}$, $s_{17}^{\downarrow}$;

$Opt\_r'_6$: if Mathematics $\leqslant$ bad and Literature $\leqslant$ bad, then student $\leqslant$ medium // supported by the relative reduct descriptors of $s_{18}^{\downarrow}$;

$Opt\_r'_7$: if Mathematics $\leqslant$ bad and Physics $\leqslant$ bad, then student $\leqslant$ medium // supported by the relative reduct descriptors of $s_{19}^{\downarrow}$, $s_{20}^{\downarrow}$.

Obviously, rule $Opt\_r'_7$ can be classified into rule $Opt\_r'_4$.

4.5 Limited Dominance-based Rough Set Approach

Following Shao and Zhang's work, Luo and Yang proposed another expansion of the dominance relation, which is referred to as the limited dominance relation [25].

4.5.1 Limited Dominance-based Rough Set

Definition 4.14. [25] Let I be an incomplete decision system in which $A \subseteq AT$, then the limited dominance relation in terms of A is denoted by $LDOM(A)$ such that

$$\begin{aligned}LDOM(A) = \{(x,y) \in U^2 : \forall a \in A, f(x,a) \geqslant f(y,a) \vee (f(x,a) = \max V_a \wedge f(y,a) = *) \vee \\ (f(x,a) = * \wedge f(y,a) = \min V_a)\} \cup I_U \qquad (4.47)\end{aligned}$$

where $\max V_a = \{v \in V_a : \forall v' \in V_a, v \geqslant v'\}$, $\min V_a = \{v \in V_a : \forall v' \in V_a, v \leqslant v'\}$, I_U is the an identical relation such that $I_U = \{(x,x) : x \in U\}$

Obviously, $\forall (x,y) \in LDOM(\{a\})$, we have the following three different cases:

1. $f(x,a)$ and $f(y,a)$ are all known $\Rightarrow$ $f(x,a) \geqslant f(y,a)$, i.e. x dominates y on a definitely;

2. $f(y, a)$ is unknown while $f(x, a)$ is known $\Rightarrow$ $f(x, a)$ is the maximal value in the domain of attribute a, i.e. x dominates y on a definitely;
3. $f(x, a)$ is unknown while $f(y, a)$ is known $\Rightarrow$ $f(y, a)$ is the minimal value in the domain of attribute a, i.e. x dominates y on a definitely.

By the above analysis, we can see that different from the expanded dominance relation showed in Definition 4.3, in the limited dominance relation, x dominates y definitely instead of possibly. Moreover, $LDOM(A)$ is reflexive and transitive but not necessarily symmetric. Similar to the classical dominance-based rough set, we can also define the following two sets for each $x \in U$ through the limited dominance relation:

- $LDOM_A^+(x) = \{y \in U : (y, x) \in LDOM(A)\}$ is the set of objects that dominating x in terms of limited dominance relation $LDOM(A)$;
- $LDOM_A^-(x) = \{y \in U : (x, y) \in LDOM(A)\}$ is the set of objects that dominated by x in terms of limited dominance relation $LDOM(A)$.

Definition 4.15. Let I be an incomplete complete decision system in which $A \subseteq AT$, $n \in N$, the lower and upper approximations of $CL_n^{\geqslant}$ in terms of the limited dominance relation $LDOM(A)$ are defined as:

$$\underline{LDOM_A}(CL_n^{\geqslant}) = \{x \in U : LDOM_A^+(x) \subseteq CL_n^{\geqslant}\} \tag{4.48}$$

$$\overline{LDOM_A}(CL_n^{\geqslant}) = \{x \in U : LDOM_A^-(x) \cap CL_n^{\geqslant} \neq \varnothing\} \tag{4.49}$$

the lower and upper approximations of $CL_n^{\leqslant}$ in terms of the limited dominance relation $LDOM(A)$ are defined as:

$$\underline{LDOM_A}(CL_n^{\leqslant}) = \{x \in U : LDOM_A^-(x) \subseteq CL_n^{\leqslant}\} \tag{4.50}$$

$$\overline{LDOM_A}(CL_n^{\leqslant}) = \{x \in U : LDOM_A^+(x) \cap CL_n^{\leqslant} \neq \varnothing\} \tag{4.51}$$

By Definition 4.15, the boundary regions of the unions of decision classes $CL_n^{\geqslant}$ and $CL_n^{\leqslant}$ are:

$$BN_A^L(CL_n^{\geqslant}) = \overline{LDOM_A}(CL_n^{\geqslant}) - \overline{LDOM_A}(CL_n^{\geqslant}) \tag{4.52}$$

$$BN_A^L(CL_n^{\leqslant}) = \overline{LDOM_A}(CL_n^{\leqslant}) - \overline{LDOM_A}(CL_n^{\leqslant}) \tag{4.53}$$

Proposition 4.4. *Let I be an incomplete complete decision system in which $A \subseteq AT$, then the limited dominance-based lower and upper approximations have the following properties:*

1. $\forall n \in N$,

$$\underline{LDOM_A}(CL_n^{\geqslant}) \subseteq CL_n^{\geqslant} \subseteq \overline{LDOM_A}(CL_n^{\geqslant}) \tag{4.54}$$

$$\underline{LDOM_A}(CL_n^{\leqslant}) \subseteq CL_n^{\leqslant} \subseteq \overline{LDOM_A}(CL_n^{\leqslant}) \tag{4.55}$$

2. $\forall n \in N$,

$$\overline{LDOM_A}(CL_n^{\geqslant}) = \bigcup_{x \in CL_n^{\geqslant}} LDOM_A^+(x) \tag{4.56}$$

$$\overline{LDOM_A}(CL_n^{\leqslant}) = \bigcup_{x \in CL_n^{\leqslant}} LDOM_A^-(x) \tag{4.57}$$

3. a. $\underline{LDOM_A}(CL_n^{\geqslant}) = U - \overline{LDOM_A}(CL_{n-1}^{\leqslant})$ $(n = 2, \cdots, m)$,
   b. $\underline{LDOM_A}(CL_n^{\leqslant}) = U - \overline{LDOM_A}(CL_{n+1}^{\geqslant})$ $(n = 1, \cdots, m-1)$,
   c. $\overline{LDOM_A}(CL_n^{\geqslant}) = U - \underline{LDOM_A}(CL_{n-1}^{\leqslant})$ $(n = 2, \cdots, m)$,
   d. $\overline{LDOM_A}(CL_n^{\leqslant}) = U - \underline{LDOM_A}(CL_{n+1}^{\geqslant})$ $(n = 1, \cdots, m-1)$,
   e. $BN_A^L(CL_n^{\geqslant}) = BN_A^L(CL_{n-1}^{\leqslant})$ $(n = 2, \cdots, m)$,
   f. $BN_A^L(CL_n^{\leqslant}) = BN_A^L(CL_{n+1}^{\geqslant})$ $(n = 1, \cdots, m-1)$;
4. $\forall n \in N$,
   a. $\underline{LDOM_A}(CL_n^{\geqslant}) \subseteq \underline{LDOM_{AT}}(CL_n^{\geqslant})$,
   b. $\overline{LDOM_A}(CL_n^{\geqslant}) \supseteq \overline{LDOM_{AT}}(CL_n^{\geqslant})$,
   c. $\underline{LDOM_A}(CL_n^{\leqslant}) \subseteq \underline{LDOM_{AT}}(CL_n^{\leqslant})$,
   d. $\overline{LDOM_A}(CL_n^{\leqslant}) \supseteq \overline{LDOM_{AT}}(CL_n^{\leqslant})$,
   e. $BN_A^L(CL_n^{\geqslant}) \supseteq BN_{AT}^L(CL_n^{\geqslant})$,
   f. $BN_A^L(CL_n^{\leqslant}) \supseteq BN_{AT}^L(CL_n^{\leqslant})$.

Proof. The proof of Proposition 4.4 is similar to the proof of Proposition 4.2. □

Definition 4.16. Let I be an incomplete complete decision system in which $A \subseteq AT$, the quality of approximation of multi-criteria classification $\boldsymbol{CL}$ in terms of the limited dominance relation $LDOM(A)$ is defined as:

$$\gamma_A^L(\boldsymbol{CL}) = \frac{\left| U - \bigcup_{n \in N} BN_A^L(CL_n^{\geqslant}) \right|}{|U|} = \frac{\left| U - \bigcup_{n \in N} BN_A^L(CL_n^{\leqslant}) \right|}{|U|} \tag{4.58}$$

4.5.2 Comparisons Between Expanded and Limited Dominance-based Rough Sets

Theorem 4.7. *Let I be an incomplete complete decision system in which $A \subseteq AT$, then we have*

$$LDOM(A) \subseteq EDOM(A) \tag{4.59}$$

Proof. It can be derived directly from Definition 4.3 and Definition 4.14. □

Theorem 4.8. *Let I be an incomplete complete decision system in which $A \subseteq AT$, then $\forall n \in N$, we have*

$$\underline{EDOM_A}(CL_n^{\geqslant}) \subseteq \underline{LDOM_A}(CL_n^{\geqslant}) \tag{4.60}$$

$$\overline{LDOM_A}(CL_n^{\geqslant}) \subseteq \overline{EDOM_A}(CL_n^{\geqslant}) \tag{4.61}$$

$$\underline{EDOM_A}(CL_n^{\leqslant}) \subseteq \underline{LDOM_A}(CL_n^{\leqslant}) \tag{4.62}$$

$$\overline{LDOM_A}(CL_n^{\leqslant}) \subseteq \overline{EDOM_A}(CL_n^{\leqslant}) \tag{4.63}$$

Proof. By results of Theorem 4.7, since $LDOM(A) \subseteq EDOM(A)$, then $LDOM_A^+(x) \subseteq EDOM_A^+(x)$ and $LDOM_A^-(x) \subseteq EDOM_A^-(x)$ hold obviously. Therefore, it is a trivial to prove the above formulas. □

Theorem 4.8 tells us that by using the limited dominance relation in the incomplete decision system, we can obtain greater lower approximations and smaller upper approximations than using Shao and Zhang's expanded dominance relation. Thus, the uncertain regions (boundary regions) of the unions of decision classes are reduced.

Theorem 4.9. *Let I be an incomplete complete decision system in which $A \subseteq AT$, then we have*

$$\gamma_A^L(\boldsymbol{CL}) \geqslant \gamma_A^E(\boldsymbol{CL}) \tag{4.64}$$

Proof. It can be derived directly from the results of Theorem 4.8. □

Theorem 4.9 tells us that the limited dominance relation can improve the accuracies of approximations of unions of decision classes and quality of approximation of the partition $\boldsymbol{CL}$ in the incomplete decision system.

Example 4.8. Assuming that Table 4.7 is a summary of reviewers' reports for 10 papers submitted to a journal. This table details 10 papers evaluated by means of four attributes:

- a_1: originality;
- a_2: presentation;
- a_3: technical soundness;
- d: overall evaluation.

Table 4.7 A summary of reviewers' reports for 10 papers

U	a_1	a_2	a_3	d
x_1	excellent	excellent	good	accept
x_2	good	fair	good	revision
x_3	excellent	*	good	revision
x_4	fair	excellent	fair	reject
x_5	fair	poor	fair	reject
x_6	good	excellent	fair	reject
x_7	*	poor	good	reject
x_8	excellent	fair	fair	revision
x_9	good	*	fair	revision
x_{10}	excellent	good	good	revision

The components of the data table are:

- $U = \{x_1, x_2, \cdots, x_{10}\}$ is the universe;

- $AT = \{a_1, a_2, a_3\}$ is the set of all condition attributes, d is the decision attribute, $V_{a_1} = V_{a_2} = V_{a_3} =$ {excellent, good, fair, poor} where "excellent" > "good" > "fair" > "poor";
- d is the decision attribute such that $V_d =$ {accept, revision, reject} where "accept" > "revision" > "reject".

The decision d determines the partition on the universe such that $\boldsymbol{CL} = \{CL_1, CL_2, CL_3\}$ where

1. $CL_1 = \{x \in U : f(x,d) = \text{reject}\} = \{x_4, x_5, x_6, x_7\}$;
2. $CL_2 = \{x \in U : f(x,d) = \text{revision}\} = \{x_2, x_3, x_8, x_9, x_{10}\}$;
3. $CL_3 = \{x \in U : f(x,d) = \text{accept}\} = \{x_1\}$.

By the expanded dominance relation-based rough sets showed in Definition 4.3, we have

$$\underline{EDOM_A}(CL_1^{\geqslant}) = U$$
$$\underline{EDOM_A}(CL_2^{\geqslant}) = \{x_1, x_2, x_3, x_8, x_{10}\}$$
$$\underline{EDOM_A}(CL_3^{\geqslant}) = \varnothing$$
$$\underline{EDOM_A}(CL_1^{\leqslant}) = \{x_4, x_5\}$$
$$\underline{EDOM_A}(CL_2^{\leqslant}) = \{x_2, x_4, x_5, x_6, x_7, x_8, x_9, x_{10}\}$$
$$\underline{EDOM_A}(CL_3^{\leqslant}) = U$$
$$\overline{EDOM_A}(CL_1^{\geqslant}) = U$$
$$\overline{EDOM_A}(CL_2^{\geqslant}) = \{x_1, x_2, x_3, x_6, x_7, x_8, x_{10}\}$$
$$\overline{EDOM_A}(CL_3^{\geqslant}) = \{x_1, x_3\}$$
$$\overline{EDOM_A}(CL_1^{\leqslant}) = \{x_3, x_4, x_5, x_6, x_7, x_9\}$$
$$\overline{EDOM_A}(CL_2^{\leqslant}) = U$$
$$\overline{EDOM_A}(CL_3^{\leqslant}) = U$$

If we use the limited dominance-based rough sets showed in Definition 4.15, then we obtain

$$\underline{LDOM_A}(CL_1^{\geqslant}) = U$$
$$\underline{LDOM_A}(CL_2^{\geqslant}) = \{x_1, x_2, x_3, x_8, x_{10}\}$$
$$\underline{LDOM_A}(CL_3^{\geqslant}) = \{x_1\}$$
$$\underline{LDOM_A}(CL_1^{\leqslant}) = \{x_4, x_5, x_7\}$$
$$\underline{LDOM_A}(CL_2^{\leqslant}) = \{x_2, x_3, x_4, x_5, x_6, x_7, x_8, x_9, x_{10}\}$$
$$\underline{LDOM_A}(CL_3^{\leqslant}) = U$$
$$\overline{LDOM_A}(CL_1^{\geqslant}) = U$$
$$\overline{LDOM_A}(CL_2^{\geqslant}) = \{x_1, x_2, x_3, x_6, x_8, x_9, x_{10}\}$$
$$\overline{LDOM_A}(CL_3^{\geqslant}) = \{x_1\}$$
$$\overline{LDOM_A}(CL_1^{\leqslant}) = \{x_4, x_5, x_6, x_7, x_9\}$$
$$\overline{LDOM_A}(CL_2^{\leqslant}) = \{x_2, x_3, x_4, x_5, x_6, x_7, x_8, x_9, x_{10}\}$$
$$\overline{LDOM_A}(CL_3^{\leqslant}) = U$$

The above results show the correctness of Theorem 4.8.

The qualities of approximations of the partition CL in terms of two different dominance-based rough set models are: $\gamma^{E}_{AT}(\boldsymbol{CL}) = 0.6$, $\gamma^{L}_{AT}(\boldsymbol{CL})$=0.8. It shows the correctness of Theorem 4.9.

4.6 Conclusions

In this chapter, several approaches to generalize the dominance-based rough set are presented in incomplete information system with "*" unknown values. They are summarized as following:

- Firstly, Shao and Zhang proposed an expanded dominance relation, which is based on the assumption that the unknown values can be compared with any other values in the domains of the corresponding attributes, such idea is consistent to that of the tolerance relation;
- Secondly, to measure the probability of dominance principle between two objects, the concept of the valued dominance relation is presented, such idea is consistent to that of the valued tolerance relation;
- Thirdly, by analyzing the limitation of the decision rules induced from the expanded dominance-based rough set approach, a descriptor approach is used to generate the certain decision rules from the incomplete decision system, by comparing with the expanded dominance-based rough set approach, the descriptor approach can not only eliminate those rules with unknown values, but also generate more certain rules.
- Finally, by analyzing the limitation of the classification of the expanded dominance relation, the concept of the limited dominance relation is presented, limited dominance relation is different from the expanded dominance relation since the former is based on the assumption that the unknown values can only be compared with the maximal and minimal values in the domain of the corresponding attribute, therefore, by using the limited dominance relation, we can obtain greater lower approximation and smaller upper approximation than the expanded dominance relation can do, from this point of view, the limited dominance relation may be more suitable than the expanded dominance relation in dealing with the incomplete information system with "*" unknown values.

References

[1] Greco, S., Inuiguchi, M., Słowiński, R.: Fuzzy rough sets and multiple-premise gradual decision rules. Int. J. Approx. Reason. **41**, 179–211 (2006)

[2] Greco, S., Matarazzo, B., Słowiński, R.: Handing missing values in rough set analysis of mutiattribute and muti-criteria decision problems. In: 7th International Workshop on New Directions in Rough Sets, Data Mining and Granular–Soft Computing, pp. 146–157 (1999)

[3] Greco, S., Matarazzo, B., Słowiński, R.: Rough approximation by dominance relations. Int. J. Intell. Syst. **17**, 153–171 (2002)

[4] Greco, S., Matarazzo, B., Słowiński, R.: Rough sets theory for multicriteria decision analysis. Eur. J. Oper. Res. **129**, 1–47 (2002)

[5] Greco, S., Matarazzo, B., Słowiński, R.: Dominance-based rough set approach to case–Based reasoning. In: Third International Conference on Modeling Decisions for Artificial Intelligence, pp. 7–18 (2006)

[6] Greco, S., Matarazzo, B., Słowiński, R.: Fuzzy set extensions of the dominance-based rough set approach. In: Fuzzy Sets and Their Extensions: Representation, Aggregation and Models, pp. 239–261 (2008)

[7] Greco, S., Słowiński, R., Yao, Y. Y.: Bayesian decision theory for dominance-based rough set approach. In: The Second International Conference on Rough Sets and Knowledge Technology, pp. 143–141 (2007)

[8] Błaszczyński, J., Greco, S., Słowiński, R.: On variable consistency dominance-based rough set approaches. In: The Fifth International Conference on Rough Sets and Current Trends in Computing, pp. 191–202 (2006)

[9] Błaszczyński, J., Greco, S., Słowiński, R.: Monotonic variable consistency rough set approaches. In: Second International Conference on Rough Sets and Knowledge Technology, pp. 126–133 (2007)

[10] Błaszczyński, J., Greco, S., Słowiński, R.: Multi-criteria classification–A new scheme for application of dominance-based decision rules. Eur. J. Oper. Res. **181**, 1030–1044 (2007)

[11] Błaszczyński, J., Greco, S., Słowiński, R., Szeląg, M.: Monotonic variable consistency rough set approaches. Int. J. Approx. Reason. **50**, 979–999 (2009)

[12] Dembczyński, K., Greco, S., Słowiński, R.: Rough set approach to multiple criteria classification with imprecise evaluations and assignments. Eur. J. Oper. Res. **198**, 626–636 (2009)

[13] Dembczyński, K., Pindur R.: Susmaga, R.: Dominance-based rough set classifier without induction of decision rules. Electronic Notes in Theoretical Computer Science **82** 84–95 (2003)

[14] Qian, Y. H., Liang, J. Y., Dang, C. Y.: Interval ordered information systems. Comput. Math. Appl. **56**, 1994–2009 (2008)

[15] Yang, X. B., Yu, D. J., Yang, J. Y., Wei, L. H. : Dominance-based rough set approach to incomplete interval-valued information system. Data Knowl. Eng. **68**, 1331–1347 (2009)

[16] Fan, T. F., Liu, D. R., Tzeng, G. H.: Rough set-based logics for multicriteria decision analysis. Eur. J. Oper. Res. **182**, 340–355 (2007)

[17] Hu, Q. H., Yu, D. R.: Variable precision dominance based rough set model and reduction algorithm for preference-ordered data. In: Proceedings of the 3rd International Conference on Machine Learning and Cybernetics, pp. 26–29 (2004)

[18] Hu, Q. H., Yu, D. R., Guo, M. Z.: Fuzzy preference based rough sets. Inform. Sci. **180**, 2003–2022 (2010)

[19] Hu, Q. H., Zhang, L., Chen, D. G., Pedrycz, W., Yu, D. R.: Gaussian kernel based fuzzy rough sets: Model, uncertainty measures and applications. Int. J. Approx. Reason. **51**, 453–471 (2010)

[20] Inuiguchi, M., Yoshioka, Y., Kusunoki, Y.: Variable-precision dominance-based rough set approach and attribute reduction. Int. J. Approx. Reason. **20**, 1199–1214 (2009)

[21] Kotłwski, W., Dembczyński, K., Greco, S., Słowiński, R.: Stochastic dominance-based rough set model for ordinal classification. Inform. Sci. **178**, 4019–4037 (2008)

[22] Shao, M. W., Zhang, W. X.: Dominance relation and rules in an incomplete ordered information system. Int. J. Intell. Syst. **20**, 13–27 (2005)

[23] Wei, L. H., Tang, Z. M., Wang, R. Y., Yang, X. B.: Extensions of dominance-based rough set approach in incomplete information system. Autom. Control Comp. Sci. **42**, 255-263 (2008)

[24] Yang, X. B., Xie, J., Song X. N., Yang, J. Y.: Credible rules in incomplete decision system based on descriptors. Knowl.-Based Syst. **22**, 8–17 (2009)

[25] Luo G. Z., Yang, X. B.: Limited dominance-based rough set model and knowledge reductions in incomplete decision system. J. Inform. Sci. Eng. **26**, 2199–2211 (2010)

Chapter 5
Dominance-based Rough Sets in "?" Incomplete Information System

Abstract In this chapter, the dominance-based rough set approach is introduced into the incomplete information system, in which the unknown values cannot be compared with any other values in the domains of the corresponding attributes. The similarity dominance-based rough sets are then constructed in crisp and fuzzy system, respectively. The similarity dominance relation is a generalization of the similarity relation and the dominance relation. It should also be noticed that different from several dominance relations, which have been presented in Chapter 4, similarity dominance relation has two different forms, one is the similarity increasing preference and the other is the similarity decreasing preference.

In Chapter 4, several dominance-based rough set approaches [1–7] for dealing with the incomplete information systems have been introduced. However, it should be noticed that the expanded dominance relation [8], the valued dominance relation [9] and the limited dominance relation [10] are all based on the assumption that the unknown values can be compared with any other values in the domains of the corresponding attributes.

Moreover, in Chapter 2, we have mentioned that there is another explanation of the unknown values in the incomplete information system, i.e. the unknown values cannot be compared with any other values [11, 12]. Based on such semantic explanation of the unknown values, Yang et al. [13] have proposed a new definition of the dominance relation for dealing the incomplete information systems with preference-ordered domains of the attributes, such dominance relation is referred to as the similarity dominance relation. Presently, work on the similarity dominance-based rough set progressing rapidly. For example, based on the proposed similarity dominance relation, Yang et al. explored the similarity dominance-based rough fuzzy set [14] in the incomplete fuzzy environment. Gao et al. [15] further generalized the similarity dominance relation and then proposed the the limited similarity dominance-based rough set. Zhao et al. [16] introduced the basic idea of the variable precision into similarity dominance-based rough set approach.

5.1 Similarity Dominance Relation

5.1.1 Definition of Similarity Dominance Relation

Definition 5.1. Let I be an incomplete decision system in which $A \subseteq AT$, the similarity dominance relations are defined as:

$$SDOM^{\geqslant}(A) = \{(x, y) \in U^2 : \forall a \in A, f(x, a) = ? \vee f(x, a) \geqslant f(y, a)\} \tag{5.1}$$

$$SDOM^{\leqslant}(A) = \{(x, y) \in U^2 : \forall a \in A, f(x, a) = ? \vee f(x, a) \leqslant f(y, a)\} \tag{5.2}$$

Obviously, the above two similarity dominance relations are reflexive and transitive, but not necessarily symmetric. $SDOM^{\geqslant}(A)$ indicates the similarity increasing preference while $SDOM^{\leqslant}(A)$ indicates the similarity decreasing preference.

$\forall x, y \in U$, if $(x, y) \in SDOM^{\geqslant}(A)$, then we do not always have $(y, x) \in SDOM^{\leqslant}(A)$. However, if the decision system is a complete one, then only $SDOM^{\geqslant}(A)$ or $SDOM^{\leqslant}(A)$ is necessary because $(x, y) \in SDOM^{\geqslant}(A) \Leftrightarrow (y, x) \in SDOM^{\leqslant}(A)$. Here, $SDOM^{\geqslant}(A)$ is just the dominance relation $DOM(A)$ that was showed in Section 4.1.

$\forall x \in U$, let us denote by

- $SDOM_A^{\geqslant -}(x) = \{y \in U : (x, y) \in SDOM^{\geqslant}(A)\}$, $\forall y \in SDOM_A^{\geqslant -}(x)$, x is similar to y and x dominates y;
- $SDOM_A^{\geqslant +}(x) = \{y \in U : (y, x) \in SDOM^{\geqslant}(A)\}$, $\forall y \in SDOM_A^{\geqslant +}(x)$, y is similar to x and y dominates x;
- $SDOM_A^{\leqslant -}(x) = \{y \in U : (x, y) \in SDOM^{\leqslant}(A)\}$, $\forall y \in SDOM_A^{\leqslant -}(x)$, x is similar to y and x is dominated by y;
- $SDOM_A^{\leqslant +}(x) = \{y \in U : (y, x) \in SDOM^{\leqslant}(A)\}$, $\forall y \in SDOM_A^{\leqslant +}(x)$, y is similar to x and y is dominated by x.

Generally speaking, we have $SDOM_A^{\geqslant -}(x) \neq SDOM_A^{\leqslant +}(x)$ and $SDOM_A^{\leqslant -}(x) \neq SDOM_A^{\geqslant +}(x)$. However, if I is a complete decision system, then $SDOM_A^{\geqslant +}(x) = SDOM_A^{\leqslant -}(x) = DOM_A^{+}(x)$, $SDOM_A^{\geqslant -}(x) = SDOM_A^{\leqslant +}(x) = DOM_A^{-}(x)$. On the other hand, if all of the condition attributes are not criteria, i.e. the regular attributes, then $SDOM_A^{\geqslant -}(x) = SDOM_A^{\leqslant -}(x) = SIM_A^{-1}(x)$, $SDOM_A^{\geqslant +}(x) = SDOM_A^{\leqslant +}(x) = SIM_A(x)$.

From discussions above, we can see that the similarity dominance relation is a generalization of the similarity relation and the dominance relation.

Proposition 5.1. Let I be an incomplete decision system in which $A \subseteq AT$, we have

$$SDOM^{\geqslant}(AT) \subseteq SDOM^{\geqslant}(A), SDOM^{\leqslant}(AT) \subseteq SDOM^{\leqslant}(A) \tag{5.3}$$

$$y \in SDOM_A^{\geqslant -}(x) \Leftrightarrow SDOM_A^{\geqslant -}(y) \subseteq SDOM_A^{\geqslant -}(x) \tag{5.4}$$

$$y \in SDOM_A^{\leqslant -}(x) \Leftrightarrow SDOM_A^{\leqslant -}(y) \subseteq SDOM_A^{\leqslant -}(x) \tag{5.5}$$

$$y \in SDOM_A^{\geqslant +}(x) \Leftrightarrow SDOM_A^{\geqslant +}(y) \subseteq SDOM_A^{\geqslant +}(x) \tag{5.6}$$

$$y \in SDOM_A^{\leqslant +}(x) \Leftrightarrow SDOM_A^{\leqslant +}(y) \subseteq SDOM_A^{\leqslant +}(x) \tag{5.7}$$

$$SDOM_A^{\geqslant -}(x) = \cup\{SDOM_{AT}^{\geqslant -}(y) : y \in SDOM_A^{\geqslant -}(x)\} \tag{5.8}$$

$$SDOM_A^{\leqslant -}(x) = \cup\{SDOM_{AT}^{\leqslant -}(y) : y \in SDOM_A^{\leqslant -}(x)\} \tag{5.9}$$

$$SDOM_A^{\geqslant+}(x) = \cup\{SDOM_{AT}^{\geqslant+}(y) : y \in SDOM_A^{\geqslant+}(x)\} \tag{5.10}$$

$$SDOM_A^{\leqslant+}(x) = \cup\{SDOM_{AT}^{\leqslant+}(y) : y \in SDOM_A^{\leqslant+}(x)\} \tag{5.11}$$

where $x, y \in U$.

Proof. Since the similarity dominance relation is reflexive and transitive, then the proof of Proposition 5.1 is similar to the proof of Proposition 2.4. □

5.1.2 Reducts of Similarity Dominance Relations

Definition 5.2. Let I be an incomplete decision system in which $A \subseteq AT$,

1. if $SDOM^{\geqslant}(A) = SDOM^{\geqslant}(AT)$, then A is referred to as a $\geqslant$-consistent attributes set of the similarity dominance relation $SDOM^{\geqslant}(AT)$ in I, if A is a $\geqslant$-consistent attributes set of the similarity dominance relation $SDOM^{\geqslant}(AT)$ in I and no proper subset of A is the $\geqslant$-consistent attributes set of the similarity dominance relation $SDOM^{\geqslant}(AT)$ in I, then A is referred to as a $\geqslant$-reduct of the similarity dominance relation $SDOM^{\geqslant}(AT)$ I;
2. if $SDOM^{\leqslant}(A) = SDOM^{\leqslant}(AT)$, then A is referred to as a $\leqslant$-consistent attributes set of the similarity dominance relation $SDOM^{\leqslant}(AT)$ in I, if A is a $\leqslant$-consistent attributes set of the similarity dominance relation $SDOM^{\leqslant}(AT)$ in I and no proper subset of A is the $\leqslant$-consistent attributes set of the similarity dominance relation $SDOM^{\leqslant}(AT)$ in I, then A is referred to as a $\leqslant$-reduct of the similarity dominance relation $SDOM^{\leqslant}(AT)$ in I.

By Definition 5.2, we can see that a $\geqslant$-consistent attributes set of the similarity dominance relation $SDOM^{\geqslant}(AT)$ in I is a subset of AT, which preserves the similarity dominance relation $SDOM^{\geqslant}(AT)$; a $\geqslant$-reduct of the similarity dominance relation $SDOM^{\geqslant}(AT)$ in I is a minimal subset of AT, which preserves the similarity dominance relation $SDOM^{\geqslant}(AT)$; a $\leqslant$-consistent attributes set of the similarity dominance relation $SDOM^{\leqslant}(AT)$ in I is a subset of AT, which preserves the similarity dominance relation $SDOM^{\leqslant}(AT)$; a $\leqslant$-reduct of the similarity dominance relation $SDOM^{\leqslant}(AT)$ in I is a *minimal* subset of AT, which preserves the similarity dominance relation $SDOM^{\leqslant}(AT)$.

To present the practical approach to compute the above two reducts of the similarity dominance relations, $\forall x, y \in U$, let us denote by

$$DIS^{\geqslant}(x,y) = \begin{cases} \{a \in AT : (x,y) \notin SDOM^{\geqslant}(\{a\}), & (x,y) \notin SDOM^{\geqslant}(AT) \\ \varnothing & \text{otherwise} \end{cases}$$

$$DIS^{\leqslant}(x,y) = \begin{cases} \{a \in AT : (x,y) \notin SDOM^{\leqslant}(\{a\}) & (x,y) \notin SDOM^{\leqslant}(AT) \\ \varnothing & \text{otherwise} \end{cases}$$

then $DIS^{\geqslant}(x,y)$, $DIS^{\leqslant}(x,y)$ are referred to as the $\geqslant$ and $\leqslant$-discernibility attributes sets for pair of the objects (x,y), respectively,

$$\boldsymbol{M}_{AT}^{\geqslant} = \{DIS^{\geqslant}(x,y) : (x,y) \notin SDOM^{\geqslant}(AT)\} \tag{5.12}$$

$$\boldsymbol{M}_{AT}^{\leqslant} = \{DIS^{\leqslant}(x,y) : (x,y) \notin SDOM^{\leqslant}(AT)\} \tag{5.13}$$

are referred to as the $\geqslant$ and $\leqslant$-discernibility matrixes of the similarity dominance relations, respectively.

Theorem 5.1. *Let I be an incomplete decision system in which $A \subseteq AT$, we have*

1. $SDOM^{\geqslant}(A) = SDOM^{\geqslant}(AT) \Leftrightarrow A \cap DIS^{\geqslant}(x,y) \neq \emptyset (\forall DIS^{\geqslant}(x,y) \neq \emptyset)$;
2. $SDOM^{\leqslant}(A) = SDOM^{\leqslant}(AT) \Leftrightarrow A \cap DIS^{\leqslant}(x,y) \neq \emptyset (\forall DIS^{\leqslant}(x,y) \neq \emptyset)$.

Proof.
1. "$\Rightarrow$": Since $SDOM^{\geqslant}(A) = SDOM^{\geqslant}(AT)$, then we have $SDOM_A^{\geqslant -}(x) = SDOM_{AT}^{\geqslant -}(x)$ for each $x \in U$. $\forall DIS^{\geqslant}(x,y) \neq \emptyset$, we have $y \notin SDOM_{AT}^{\geqslant -}(x)$, it follows that $y \notin SDOM_A^{\geqslant -}(x)$ since $SDOM_A^{\geqslant -}(x) = SDOM_{AT}^{\geqslant -}(x)$. Therefore, there must be $a \in A \subseteq AT$ such that $(x,y) \notin SDOM^{\geqslant}(\{a\})$, i.e. $A \cap DIS^{\geqslant}(x,y) \neq \emptyset$.
   "$\Leftarrow$": Since $A \subseteq AT$, then by Eq. (5.3), we have $SDOM^{\geqslant}(AT) \subseteq SDOM^{\geqslant}(A)$, thus, it must be proved that $SDOM^{\geqslant}(AT) \supseteq SDOM^{\geqslant}(A)$. $\forall (x,y) \notin SDOM^{\geqslant}(AT)$, there must be $DIS^{\geqslant}(x,y) \neq \emptyset$. Since by assumption $A \cap DIS^{\geqslant}(x,y) \neq \emptyset$, there must be $(x,y) \notin SDOM^{\geqslant}(A)$. It follows that $(x,y) \notin SDOM^{\geqslant}(AT) \Rightarrow (x,y) \notin SDOM^{\geqslant}(A)$, i.e. $SDOM^{\geqslant}(A) \subseteq SDOM^{\geqslant}(AT)$.
2. The proof of 2 is similar to the proof of 1. □

Definition 5.3. Let I be an incomplete decision system in which $A \subseteq AT$, let

$$\Delta^{\geqslant} = \bigwedge_{(x,y) \notin SDOM^{\geqslant}(AT)} \vee DIS^{\geqslant}(x,y) \tag{5.14}$$

$$\Delta^{\leqslant} = \bigwedge_{(x,y) \notin SDOM^{\leqslant}(AT)} \vee DIS^{\leqslant}(x,y) \tag{5.15}$$

$\Delta^{\geqslant}(\Delta^{\leqslant})$ are referred to as the $\geqslant(\leqslant)$-discernibility functions of I.

By Boolean Reasoning technique, it is not difficult to obtain the following theorem by Theorem 5.1.

Theorem 5.2. *Let I be an incomplete decision system in which $A \subseteq AT$, then*

1. *A is a $\geqslant$-reduct of the similarity dominance relation $SDOM^{\geqslant}(AT)$ in I if and only if $\wedge A$ is the prime implicant of the $\geqslant$-discernibility function, i.e. $\Delta^{\geqslant}$;*
2. *A is a $\leqslant$-reduct of the similarity dominance relation $SDOM^{\leqslant}(AT)$ in I if and only if $\wedge A$ is the prime implicant of the $\leqslant$-discernibility function, i.e. $\Delta^{\leqslant}$.*

Example 5.1. Let us consider Table 5.1, which is an incomplete information system. In Table 5.1,

- $U = \{x_1, x_2, \cdots, x_{12}\}$ is the universe;
- $AT = \{a, b, c\}$ is the set of the attributes where $V_a = V_b = V_c = \{1, 2, 3\}$.

By Eq. (5.12), we obtain the ⩾-distribution discernibility matrix as Table 5.2 shows. By Definition 5.2,

$$\begin{aligned}\Delta^{\geqslant} &= AT \wedge (a \vee b) \wedge (a \vee c) \wedge (b \vee c) \wedge a \wedge b \wedge \\ &= a \wedge b \wedge c\end{aligned}$$

Table 5.1 An example of incomplete information system

U	a	b	c
x_1	3	2	2
x_2	2	1	1
x_3	2	1	?
x_4	1	1	1
x_5	1	?	1
x_6	3	1	1
x_7	2	1	1
x_8	3	2	2
x_9	2	1	2
x_{10}	1	1	2
x_{11}	2	1	2
x_{12}	3	?	2

Table 5.2 ⩾-Discernibility Matrix of Table 5.1

	x_1	x_2	x_3	x_4	x_5	x_6	x_7	x_8	x_9	x_{10}	x_{11}	x_{12}
x_1			c		b							b
x_2	AT		c		b	a		AT	c	c	c	AT
x_3	$a \vee b$				b	a		$a \vee b$				$a \vee b$
x_4	AT	a	$a \vee c$		b	a	a	AT	$a \vee c$	c	$a \vee c$	AT
x_5	$a \vee c$	a	$a \vee c$			a	a	$a \vee c$	$a \vee c$	c	$a \vee c$	$a \vee c$
x_6	$b \vee c$		c		b			$b \vee c$	c	c	c	$b \vee c$
x_7	AT		c		b	a		AT	c	c	c	AT
x_8			c		b							b
x_9	$a \vee b$		c		b	a		$a \vee b$				$a \vee b$
x_{10}	$a \vee b$	a	$a \vee c$		b	a	a	$a \vee b$	a		a	$a \vee b$
x_{11}	$a \vee b$		c		b	a		$a \vee b$				$a \vee b$
x_{12}			c									

Such result tells us that no attribute can be eliminated to preserve the original similarity dominance relation $SDOM^{\geqslant}(AT)$ in Table 5.1.

Similarly, we have

$$\Delta^{\leqslant} == a \wedge b \wedge c \tag{5.16}$$

i.e., no attribute can be eliminated to preserve the original similarity dominance relation $SDOM^{\leqslant}(AT)$ in Table 5.1.

5.2 Similarity Dominance-based Rough Set and Approximate Distribution Reducts

5.2.1 Similarity Dominance-based Rough Set

Definition 5.4. Let I be an incomplete decision system in which $A \subseteq AT$, $n \in N$, the lower and upper approximations of $CL_n^{\geqslant}$ in terms of the similarity dominance relation are defined as:

$$\underline{SDOM_A}(CL_n^{\geqslant}) = \{x \in U : SDOM_A^{\leqslant -}(x) \subseteq CL_n^{\geqslant}\} \tag{5.17}$$

$$\overline{SDOM_A}(CL_n^{\geqslant}) = \{x \in U : SDOM_A^{\geqslant -}(x) \cap CL_n^{\geqslant} \neq \varnothing\} \tag{5.18}$$

the lower and upper approximations of $CL_n^{\leqslant}$ in terms of the similarity dominance relation are defined as:

$$\underline{SDOM_A}(CL_n^{\leqslant}) = \{x \in U : SDOM_A^{\geqslant -}(x) \subseteq CL_n^{\leqslant}\} \tag{5.19}$$

$$\overline{SDOM_A}(CL_n^{\leqslant}) = \{x \in U : SDOM_A^{\leqslant -}(x) \cap CL_n^{\leqslant} \neq \varnothing\} \tag{5.20}$$

By Definition 5.4 the boundary regions of the unions of decision classes $CL_n^{\geqslant}$ and $CL_n^{\leqslant}$ are:

$$BN_A^S(CL_n^{\geqslant}) = \overline{SDOM_A}(CL_n^{\geqslant}) - \overline{SOM_A}(CL_n^{\geqslant}) \tag{5.21}$$

$$BN_A^S(CL_n^{\leqslant}) = \overline{SDOM_A}(CL_n^{\leqslant}) - \overline{SOM_A}(CL_n^{\leqslant}) \tag{5.22}$$

Proposition 5.2. Let I be an incomplete decision system in which $A \subseteq AT$, then we have following properties about the similarity dominance-based lower and upper approximations:

1. $\forall n \in N$,

$$\underline{SDOM_A}(CL_n^{\geqslant}) \subseteq CL_n^{\geqslant} \subseteq \overline{SDOM_A}(CL_n^{\geqslant}) \tag{5.23}$$

$$\underline{SDOM_A}(CL_n^{\leqslant}) \subseteq CL_n^{\leqslant} \subseteq \overline{SDOM_A}(CL_n^{\leqslant}) \tag{5.24}$$

2. $\forall n \in N$,

$$\overline{SDOM_A}(CL_n^{\geqslant}) = \bigcup_{x \in CL_n^{\geqslant}} SDOM_A^{\geqslant +}(x) \tag{5.25}$$

$$\overline{SDOM_A}(CL_n^{\leqslant}) = \bigcup_{x \in CL_n^{\leqslant}} SDOM_A^{\leqslant +}(x) \tag{5.26}$$

3. a. $\underline{SDOM_A}(CL_n^{\geqslant}) = U - \overline{SDOM_A}(CL_{n-1}^{\leqslant})$ $(n = 2, \cdots, m)$,

   b. $\underline{SDOM_A}(CL_n^{\leqslant}) = U - \overline{SDOM_A}(CL_{n+1}^{\geqslant})$ $(n = 1, \cdots, m-1)$,

   c. $\overline{SDOM_A}(CL_n^{\geqslant}) = U - \underline{SDOM_A}(CL_{n-1}^{\leqslant})$ $(n = 2, \cdots, m)$,

d. $\overline{SDOM_A}(CL_n^{\leqslant}) = U - \underline{SDOM_A}(CL_{n+1}^{\geqslant})$ $(n = 1, \cdots, m-1)$,

e. $BN_A^S(CL_n^{\geqslant}) = BN_A^S(CL_{n-1}^{\leqslant})$ $(n = 2, \cdots, m)$,

f. $BN_A^S(CL_n^{\leqslant}) = BN_A^S(CL_{n+1}^{\geqslant})$ $(n = 1, \cdots, m-1)$;

4. $\forall n \in N$,

a. $\underline{SDOM_A}(CL_n^{\geqslant}) \subseteq \underline{SDOM_{AT}}(CL_n^{\geqslant})$,

b. $\overline{SDOM_A}(CL_n^{\geqslant}) \supseteq \overline{SDOM_{AT}}(CL_n^{\geqslant})$,

c. $\underline{SDOM_A}(CL_n^{\leqslant}) \subseteq \underline{SDOM_{AT}}(CL_n^{\leqslant})$,

d. $\overline{SDOM_A}(CL_n^{\leqslant}) \supseteq \overline{SDOM_{AT}}(CL_n^{\leqslant})$,

e. $BN_A^S(CL_n^{\geqslant}) \supseteq BN_{AT}^S(CL_n^{\geqslant})$,

f. $BN_A^S(CL_n^{\leqslant}) \supseteq BN_{AT}^S(CL_n^{\leqslant})$.

Proof. The proof of Proposition 5.2 is similar to the proof of Proposition 4.2. □

Similar to the classical dominance-based rough set, by the similarity dominance-based rough set model, we may also generate the "at least" and "at most" rules from the incomplete decision system such that:

- certain "at least" rule: if $f(y, a_1) \geqslant f(x, a_1)$ and $f(y, a_2) \geqslant f(x, a_2)$ and $\cdots$ and $f(y, a_k) \geqslant f(x, a_k)$, then $y \in CL_n^{\geqslant}$ // supported by $x \in \underline{SDOM_A}(CL_n^{\geqslant})$;
- possible "at least" rule: if $f(y, a_1) \geqslant f(x, a_1)$ and $f(y, a_2) \geqslant f(x, a_2)$ and $\cdots$ and $f(y, a_k) \geqslant f(x, a_k)$, then y may belong to $CL_n^{\geqslant}$ // supported by $x \in \overline{SDOM_A}(CL_n^{\geqslant})$;
- certain "at most" rule: if $f(y, a_1) \leqslant f(x, a_1)$ and $f(y, a_2) \leqslant f(x, a_2)$ and $\cdots$ and $f(y, a_k) \leqslant f(x, a_k)$, then $y \in CL_n^{\leqslant}$ // supported by $x \in \underline{SDOM_A}(CL_n^{\leqslant})$;
- possible "at most" rule: if $f(y, a_1) \leqslant f(x, a_1)$ and $f(y, a_2) \leqslant f(x, a_2)$ and $\cdots$ and $f(y, a_k) \leqslant f(x, a_k)$, then y may belong to $CL_n^{\leqslant}$ // supported by $x \in \overline{SDOM_A}(CL_n^{\leqslant})$;

where $A = \{a_1, a_2, \cdots, a_k\}$ and $f(x, a_i)$ is known $\forall i = 1, 2, \cdots, k$.

Example 5.2. Let us consider Table 5.3, which is an expansion of Table 5.1 by adding the decision attribute d.

Table 5.3 An example of incomplete decision system

U	a_1	a_2	a_3	d
x_1	3	2	2	2
x_2	2	1	1	1
x_3	2	1	?	2
x_4	1	1	1	1
x_5	1	?	1	1
x_6	3	1	1	2
x_7	2	1	1	2
x_8	3	2	2	2
x_9	2	1	2	2
x_{10}	1	1	2	1
x_{11}	2	1	2	2
x_{12}	3	?	2	2

Since the decision attribute make a partition on the universe such that $\boldsymbol{CL} = \{CL_1, CL_2\} = \Big\{\{x_2, x_4, x_5, x_{10}\}, \{x_1, x_3, x_6, x_7, x_8, x_9, x_{11}, x_{12}\}\Big\}$, then we can obtain the following lower and upper approximations:

$$\underline{SDOM_{AT}}(CL_1^{\geqslant}) = U$$
$$\underline{SDOM_{AT}}(CL_2^{\geqslant}) = \{x_1, x_6, x_8, x_9, x_{11}, x_{12}\}$$
$$\overline{SDOM_{AT}}(CL_1^{\geqslant}) = U$$
$$\overline{SDOM_{AT}}(CL_2^{\geqslant}) = \{x_1, x_2, x_3, x_6, x_7, x_8, x_9, x_{11}, x_{12}\}$$
$$\underline{SDOM_{AT}}(CL_1^{\leqslant}) = \{x_4, x_5, x_{10}\}$$
$$\underline{SDOM_{AT}}(CL_2^{\leqslant}) = U$$
$$\overline{SDOM_{AT}}(CL_1^{\leqslant}) = \{x_2, x_3, x_4, x_5, x_7, x_{10}\}$$
$$\overline{SDOM_{AT}}(CL_2^{\leqslant}) = U$$

By the above results, we may generate "at least" and "at most" decision rules from Table 5.3. We only present the certain "at least" and "at most" decision rules in the following.

Certain "at least" rules:

1. $f(y, a_1) \geqslant 3 \wedge f(y, a_2) \geqslant 2 \wedge f(y, a_3) \geqslant 2 \rightarrow y \in CL_2^{\geqslant}$ // supported by $x_1 \in \underline{SDOM_{AT}}(CL_2^{\geqslant})$;
2. $f(y, a_1) \geqslant 3 \wedge f(y, a_2) \geqslant 1 \wedge f(y, a_3) \geqslant 1 \rightarrow y \in CL_2^{\geqslant}$ // supported by $x_6 \in \underline{SDOM_{AT}}(CL_2^{\geqslant})$;
3. $f(y, a_1) \geqslant 3 \wedge f(y, a_2) \geqslant 2 \wedge f(y, a_3) \geqslant 2 \rightarrow y \in CL_2^{\geqslant}$ // supported by $x_8 \in \underline{SDOM_{AT}}(CL_2^{\geqslant})$;
4. $f(y, a_1) \geqslant 2 \wedge f(y, a_2) \geqslant 1 \wedge f(y, a_3) \geqslant 2 \rightarrow y \in CL_2^{\geqslant}$ // supported by $x_9 \in \underline{SDOM_{AT}}(CL_2^{\geqslant})$;
5. $f(y, a_1) \geqslant 2 \wedge f(y, a_2) \geqslant 1 \wedge f(y, a_3) \geqslant 2 \rightarrow y \in CL_2^{\geqslant}$ // supported by $x_{11} \in \underline{SDOM_{AT}}(CL_2^{\geqslant})$;
6. $f(y, a_1) \geqslant 3 \wedge f(y, a_3) \geqslant 2 \rightarrow y \in CL_2^{\geqslant}$ // supported by $x_{12} \in \underline{SDOM_{AT}}(CL_2^{\geqslant})$.

Certain "at most" rules:

1. $f(y, a_1) \leqslant 1 \wedge f(y, a_2) \leqslant 1 \wedge f(y, a_3) \leqslant 1 \rightarrow y \in CL_1^{\leqslant}$ // supported by $x_4 \in \underline{SDOM_{AT}}(CL_1^{\leqslant})$;
2. $f(y, a_1) \leqslant 1 \wedge f(y, a_3) \leqslant 1 \rightarrow y \in CL_1^{\leqslant}$ // supported by $x_5 \in \underline{SDOM_{AT}}(CL_1^{\leqslant})$;
3. $f(y, a_1) \leqslant 1 \wedge f(y, a_2) \leqslant 1 \wedge f(y, a_3) \leqslant 2 \rightarrow y \in CL_1^{\leqslant}$ // supported by $x_{10} \in \underline{SDOM_{AT}}(CL_1^{\leqslant})$.

5.2.2 Approximate Distribute Reducts in Similarity Dominance-based Rough Set

Definition 5.5. Let I be an incomplete decision system in which $A \subseteq AT$, let us denote by

$$\underline{SDOM_A}(\boldsymbol{CL}^{\geqslant}) = \left\{\underline{SDOM_A}(CL_1^{\geqslant}), \underline{SDOM_A}(CL_2^{\geqslant}), \cdots, \underline{SDOM_A}(CL_m^{\geqslant})\right\} \tag{5.27}$$

$$\underline{SDOM_A}(\boldsymbol{CL}^{\leqslant}) = \left\{\underline{SDOM_A}(CL_1^{\leqslant}), \underline{SDOM_A}(CL_2^{\leqslant}), \cdots, \underline{SDOM_A}(CL_m^{\leqslant})\right\} \tag{5.28}$$

$$\overline{SDOM_A}(\boldsymbol{CL}^{\geqslant}) = \left\{\overline{SDOM_A}(CL_1^{\geqslant}), \overline{SDOM_A}(CL_2^{\geqslant}), \cdots, \overline{SDOM_A}(CL_m^{\geqslant})\right\} \tag{5.29}$$

$$\overline{SDOM_A}(\boldsymbol{CL}^{\leqslant}) = \left\{\overline{SDOM_A}(CL_1^{\leqslant}), \overline{SDOM_A}(CL_2^{\leqslant}), \cdots, \overline{SDOM_A}(CL_m^{\leqslant})\right\} \tag{5.30}$$

1. If $\underline{SDOM_A}(\boldsymbol{CL}^{\geqslant}) = \underline{SDOM_{AT}}(\boldsymbol{CL}^{\geqslant})$, then A is referred to as a similarity dominance-based $\geqslant$-lower approximate distribution consistent attributes set in I; if A is a similarity dominance-based $\geqslant$-lower approximate distribution consistent attributes set in I and $\forall B \subset A$, B is not the similarity dominance-based $\geqslant$-lower approximate distribution consistent attributes set in I, then A is referred to as a similarity dominance-based $\geqslant$-lower approximate distribution reduct in I.
2. If $\underline{SDOM_A}(\boldsymbol{CL}^{\leqslant}) = \underline{SDOM_{AT}}(\boldsymbol{CL}^{\leqslant})$, then A is referred to as a similarity dominance-based $\leqslant$-lower approximate distribution consistent attributes set in I; if A is a similarity dominance-based $\leqslant$-lower approximate distribution consistent attributes set in I and $\forall B \subset A$, B is not the similarity dominance-based $\leqslant$-lower approximate distribution consistent attributes set in I, then A is referred to as a similarity dominance-based $\leqslant$-lower approximate distribution reduct in I.
3. If $\overline{SDOM_A}(\boldsymbol{CL}^{\geqslant}) = \overline{SDOM_{AT}}(\boldsymbol{CL}^{\geqslant})$, then A is referred to as a similarity dominance-based $\geqslant$-upper approximate distribution consistent attributes set in I; if A is a similarity dominance-based $\geqslant$-upper approximate distribution consistent attributes set in I and $\forall B \subset A$, B is not the similarity dominance-based $\geqslant$-upper approximate distribution consistent attributes set in I, then A is referred to as a similarity dominance-based $\geqslant$-upper approximate distribution reduct in I.
4. If $\overline{SDOM_A}(\boldsymbol{CL}^{\leqslant}) = \overline{SDOM_{AT}}(\boldsymbol{CL}^{\leqslant})$, then A is referred to as a similarity dominance-based $\leqslant$-upper approximate distribution consistent attributes set in I; if A is a similarity dominance-based $\leqslant$-upper approximate distribution consistent attributes set in I and $\forall B \subset A$, B is not the similarity dominance-based $\leqslant$-upper approximate distribution consistent attributes set in I, then A is referred to as a similarity dominance-based $\leqslant$-upper approximate distribution reduct in I.

By Definition 5.5, we know that

1. a similarity dominance-based $\geqslant$-lower approximate distribution consistent attributes set in I is a subset of the attributes, which preserves all of the lower approximations of the upward unions of the decision classes, a similarity dominance-based $\geqslant$-lower approximate distribution reduct in I is a minimal subset of the attributes, which preserves all of the lower approximations of the upward unions of the decision classes;

2. a similarity dominance-based $\leqslant$-lower approximate distribution consistent attributes set in I is a subset of the attributes, which preserves all of the lower approximations of the downward unions of the decision classes, a similarity dominance-based $\leqslant$-lower approximate distribution reduct in I is a minimal subset of the attributes, which preserves all of the lower approximations of the downward unions of the decision classes;
3. a similarity dominance-based $\geqslant$-upper approximate distribution consistent attributes set in I is a subset of the attributes, which preserves all of the upper approximations of the upward unions of the decision classes, a similarity dominance-based $\geqslant$-upper approximate distribution reduct in I is a minimal subset of the attributes, which preserves all of the upper approximations of the upward unions of the decision classes;
4. a similarity dominance-based $\leqslant$-upper approximate distribution consistent attributes set in I is a subset of the attributes, which preserves all of the upper approximations of the downward unions of the decision classes, a similarity dominance-based $\leqslant$-upper approximate distribution reduct in I is a *minimal* subset of the attributes, which preserves all of the upper approximations of the downward unions of the decision classes.

Theorem 5.3. *Let I be an incomplete decision system in which $A \subseteq AT$, we have*

1. *A is a similarity dominance-based $\geqslant$-lower approximate distribution consistent attributes set $\Leftrightarrow$ A is a similarity dominance-based $\leqslant$-upper approximate distribution consistent attributes set;*
2. *A is a similarity dominance-based $\leqslant$-lower approximate distribution consistent attributes set $\Leftrightarrow$ A is a similarity dominance-based $\geqslant$-upper approximate distribution consistent attributes set.*

Proof. We only prove 1, the proof of 2 is similar to the proof of 1.

$\Rightarrow$: Since A is a similarity dominance-based $\geqslant$-lower approximate distribution consistent attributes set, then we have $\underline{SDOM_A}(\boldsymbol{CL}^{\geqslant}) = \underline{SDOM_{AT}}(\boldsymbol{CL}^{\geqslant})$, i.e. $\underline{SDOM_A}(CL_n^{\geqslant}) = \underline{SDOM_{AT}}(CL_n^{\geqslant})$ for each $n = 1, 2, \cdots, m$. By Proposition 5.2 we have $\underline{SDOM_A}(CL_n^{\geqslant}) = U - \overline{SDOM_A}(CL_{n-1}^{\leqslant})$ and $\underline{SDOM_{AT}}(CL_n^{\geqslant}) = U - \overline{SDOM_{AT}}(CL_{n-1}^{\leqslant})$ for $n = 2, \cdots, m$, i.e. $\overline{SDOM_A}(CL_{n-1}^{\leqslant}) = \overline{SDOM_{AT}}(CL_{n-1}^{\leqslant})$ for $n = 2, \cdots, m$. Since $CL_m^{\leqslant} = U$, we have $\overline{SDOM_A}(CL_m^{\leqslant}) = \overline{SDOM_{AT}}(CL_m^{\leqslant}) = U$, it follows that $\overline{SDOM_A}(CL_n^{\leqslant}) = \overline{SDOM_{AT}}(CL_n^{\leqslant})$ for $n = 1, 2, \cdots, m$, i.e. $\overline{SDOM_A}(\boldsymbol{CL}^{\leqslant}) = \overline{SDOM_{AT}}(\boldsymbol{CL}^{\leqslant})$.

$\Leftarrow$: A is a similarity dominance-based $\leqslant$-upper approximate distribution consistent attributes set, then we have $\overline{SDOM_A}(\boldsymbol{CL}^{\geqslant}) = \overline{SDOM_{AT}}(\boldsymbol{CL}^{\geqslant})$, i.e. $\overline{SDOM_A}(CL_n^{\leqslant}) = \overline{SDOM_{AT}}(CL_n^{\leqslant})$ for each $n = 1, 2, \cdots, m$. By Proposition 5.2 we have $\overline{SDOM_A}(CL_n^{\leqslant}) = U - \underline{SDOM_A}(CL_{n+1}^{\geqslant})$ and $\overline{SDOM_{AT}}(CL_n^{\leqslant}) = U - \underline{SDOM_{AT}}(CL_{n+1}^{\geqslant})$ for $n = 1, \cdots, m-1$, i.e. $\underline{SDOM_A}(CL_{n+1}^{\geqslant}) = \underline{SDOM_{AT}}(CL_{n+1}^{\geqslant})$ for $n = 1, \cdots, m-1$. Since $CL_1^{\geqslant} = U$, we have $\underline{SDOM_A}(CL_1^{\geqslant}) = \underline{SDOM_{AT}}(CL_1^{\geqslant}) = U$, it follows that $\underline{SDOM_A}(CL_n^{\geqslant}) = \underline{SDOM_{AT}}(CL_n^{\geqslant})$ for $n = 1, 2, \cdots, m$, i.e. $\underline{SDOM_A}(\boldsymbol{CL}^{\geqslant}) = \underline{SDOM_{AT}}(\boldsymbol{CL}^{\geqslant})$. $\square$

Theorem 5.4. *Let I be an incomplete decision system in which $A \subseteq AT$, we have*

1. *A is a similarity dominance-based $\geqslant$-lower approximate distribution reduct $\Leftrightarrow$ A is a similarity dominance-based $\leqslant$-upper approximate distribution reduct;*
2. *A is a similarity dominance-based $\leqslant$-lower approximate distribution reduct $\Leftrightarrow$ A is a similarity dominance-based $\geqslant$-upper approximate distribution reduct.*

Proof. We only prove 1, the proof of 2 is similar to the proof of 1.

$\Rightarrow$: Since A is a similarity dominance-based $\geqslant$-lower approximate distribution reduct, then by Theorem 5.3, we know that A is a similarity dominance-based $\leqslant$-upper approximate distribution consistent attributes set. Suppose that A is not a similarity dominance-based $\leqslant$-upper approximate distribution reduct, then there must be $B \subset A$ such that B is a similarity dominance-based $\leqslant$-upper approximate distribution reduct. By Theorem 5.3 we know that B is a similarity dominance-based $\geqslant$-lower approximate distribution consistent attributes set, which contradicts that A is a similarity dominance-based $\geqslant$-lower approximate distribution reduct, from which we can conclude that A is a similarity dominance-based $\leqslant$-upper approximate distribution reduct.

$\Leftarrow$: Since A is a similarity dominance-based $\leqslant$-upper approximate distribution reduct, then by Theorem 5.3, we know that A is a similarity dominance-based $\geqslant$ -lower approximate distribution consistent attributes set. Suppose that A is not a similarity dominance-based $\geqslant$-lower approximate distribution reduct, then there must be $B \subset A$ such that B is a similarity dominance-based $\geqslant$-lower approximate distribution reduct. By Theorem 5.3 we know that B is a similarity dominance-based $\leqslant$ -upper approximate distribution consistent attributes set, which contradicts that A is a similarity dominance-based $\leqslant$-upper approximate distribution reduct, from which we can conclude that A is a similarity dominance-based $\geqslant$-lower approximate distribution reduct. □

In the following, we will provide the practical approaches to compute the similarity dominance-based $\geqslant$ ($\leqslant$)-lower and upper approximate distribution reducts in incomplete decision system.

Theorem 5.5. *Let I be an incomplete decision system in which $A \subseteq AT$, $\forall x \in U$, we denote*

$$L^{\geqslant}_{AT}(x) = \{CL^{\geqslant}_n : x \in \underline{SDOM_{AT}}(CL^{\geqslant}_n)\} \tag{5.31}$$

$$L^{\leqslant}_{AT}(x) = \{CL^{\leqslant}_n : x \in \underline{SDOM_{AT}}(CL^{\leqslant}_n)\} \tag{5.32}$$

$$U^{\geqslant}_{AT}(x) = \{CL^{\geqslant}_n : x \in \overline{SDOM_{AT}}(CL^{\geqslant}_n)\} \tag{5.33}$$

$$U^{\leqslant}_{AT}(x) = \{CL^{\leqslant}_n : x \in \overline{SDOM_{AT}}(CL^{\leqslant}_n)\} \tag{5.34}$$

then:

1. *A is a similarity dominance-based $\geqslant$-lower approximate distribution consistent attributes set $\Leftrightarrow L^{\geqslant}_A(x) = L^{\geqslant}_{AT}(x)$ for each $x \in U$;*
2. *A is a similarity dominance-based $\leqslant$-lower approximate distribution consistent attributes set $\Leftrightarrow L^{\leqslant}_A(x) = L^{\leqslant}_{AT}(x)$ for each $x \in U$;*
3. *A is a similarity dominance-based $\geqslant$-upper approximate distribution consistent attributes set $\Leftrightarrow U^{\geqslant}_A(x) = U^{\geqslant}_{AT}(x)$ for each $x \in U$;*

4. *A is a similarity dominance-based $\leqslant$-upper approximate distribution consistent attributes set* $\Leftrightarrow U_A^{\leqslant}(x) = U_{AT}^{\leqslant}(x)$ *for each* $x \in U$.

Proof. It can be derived directly from Definition 5.5. □

Theorem 5.6. *Let I be an incomplete decision system in which* $A \subseteq AT$, *then*

1. *A is a similarity dominance-based $\geqslant$-lower approximate distribution consistent attributes set* $\Leftrightarrow \forall x, y \in U$, *if* $L_{AT}^{\geqslant}(x) \nsubseteq L_{AT}^{\geqslant}(y)$, *then* $SDOM_A^{\leqslant -}(y) \nsubseteq SDOM_A^{\leqslant -}(x)$;
2. *A is a similarity dominance-based $\leqslant$-lower approximate distribution consistent attributes set* $\Leftrightarrow \forall x, y \in U$, *if* $L_{AT}^{\leqslant}(x) \nsubseteq L_{AT}^{\leqslant}(y)$, *then* $SDOM_A^{\geqslant -}(y) \nsubseteq SDOM_A^{\geqslant -}(x)$;
3. *A is a similarity dominance-based $\geqslant$-upper approximate distribution consistent attributes set* $\Leftrightarrow \forall x, y \in U$, *if* $U_{AT}^{\geqslant}(y) \nsubseteq U_{AT}^{\geqslant}(x)$, *then* $SDOM_A^{\geqslant -}(y) \nsubseteq SDOM_A^{\geqslant -}(x)$;
4. *A is a similarity dominance-based $\leqslant$-upper approximate distribution consistent attributes set* $\Leftrightarrow \forall x, y \in U$, *if* $U_{AT}^{\leqslant}(y) \nsubseteq U_{AT}^{\leqslant}(x)$, *then* $SDOM_A^{\leqslant -}(y) \nsubseteq SDOM_A^{\leqslant -}(x)$.

Proof.

1. $\Rightarrow$: $\forall x, y \in U$, if $SDOM_A^{\leqslant -}(y) \subseteq SDOM_A^{\leqslant -}(x)$, then $L_A^{\geqslant}(x) \subseteq L_A^{\geqslant}(y)$. Since A is a similarity dominance-based $\geqslant$-lower approximate distribution consistent attributes set, then $L_A^{\geqslant}(x) = L_{AT}^{\geqslant}(x)$ and $L_A^{\geqslant}(y) = L_{AT}^{\geqslant}(y)$, it follows that $L_{AT}^{\geqslant}(x) \subseteq L_{AT}^{\geqslant}(y)$.

   $\Leftarrow$: By condition we have $SDOM_A^{\leqslant -}(y) \subseteq SDOM_A^{\leqslant -}(x) \Rightarrow L_{AT}^{\geqslant}(x) \subseteq L_{AT}^{\geqslant}(y)$. Since $A \subseteq AT$, then $L_{AT}^{\geqslant}(x) \supseteq L_A^{\geqslant}(x)$ holds obviously. Consequently, it must be proved that $L_{AT}^{\geqslant}(x) \subseteq L_A^{\geqslant}(x)$. $\forall CL_n^{\geqslant} \in L_{AT}^{\geqslant}(x)$ $(1 \leqslant n \leqslant m)$, we have $SDOM_{AT}^{\leqslant -}(x) \subseteq CL_n^{\geqslant}$. $\forall y \in U$, if $SDOM_A^{\leqslant -}(y) \subseteq SDOM_A^{\leqslant -}(x)$, then by Eqs. (5.5) and (5.9), we have $SDOM_A^{\leqslant -}(x) = \cup\{SDOM_{AT}^{\leqslant -}(y) : y \in SDOM_A^{\leqslant -}(x)\}$. Moreover, by $SDOM_A^{\leqslant -}(y) \subseteq SDOM_A^{\leqslant -}(x) \Rightarrow L_{AT}^{\geqslant}(x) \subseteq L_{AT}^{\geqslant}(y)$ we know $CL_n^{\geqslant} \in L_{AT}^{\geqslant}(y)$, i.e. $SDOM_{AT}^{\leqslant -}(y) \subseteq CL_n^{\geqslant}$. From discussions above, we have $SDOM_A^{\leqslant -}(x) = \cup\{SDOM_{AT}^{\leqslant -}(y) : y \in SDOM_A^{\leqslant -}(x)\} \subseteq CL_n^{\geqslant}$, i.e. $L_{AT}^{\geqslant}(x) \subseteq L_A^{\geqslant}(x)$. That completes the proof.
2. The proof of 2 is similar to the proof of 1.
3. $\Rightarrow$: $\forall x, y \in U$, if $SDOM_A^{\geqslant -}(y) \subseteq SDOM_A^{\geqslant -}(x)$, then $U_A^{\geqslant}(y) \subseteq U_A^{\geqslant}(x)$. Since A is a similarity dominance-based $\geqslant$-lower approximate distribution consistent attributes set, then $U_A^{\geqslant}(x) = U_{AT}^{\geqslant}(x)$ and $U_A^{\geqslant}(y) = U_{AT}^{\geqslant}(y)$, it follows that $U_{AT}^{\geqslant}(y) \subseteq U_{AT}^{\geqslant}(x)$.

   $\Leftarrow$: By condition we have $SDOM_A^{\geqslant -}(y) \subseteq SDOM_A^{\geqslant -}(x) \Rightarrow U_{AT}^{\geqslant}(y) \subseteq U_{AT}^{\geqslant}(x)$. Since $A \subseteq AT$, then $U_{AT}^{\geqslant}(x) \subseteq U_A^{\geqslant}(x)$ holds obvisouly. Consequently, it must be proved that $U_A^{\geqslant}(x) \subseteq U_{AT}^{\geqslant}(x)$. $\forall CL_n^{\geqslant} \in U_A^{\geqslant}(x)$, we have $SDOM_A^{\geqslant -}(x) \cap CL_n^{\geqslant} \neq \emptyset$. $\forall y \in SDOM_A^{\geqslant -}(x) \cap CL_n^{\geqslant}$, $y \in SDOM_A^{\geqslant -}(x)$ and $y \in CL_n^{\geqslant}$, then we have $SDOM_A^{\geqslant -}(y) \subseteq SDOM_A^{\geqslant -}(x)$ by Eq. (5.4). Since $y \in SDOM_{AT}^{\geqslant -}(y) \cap CL_n^{\geqslant}$, then $CL_n^{\geqslant} \in U_{AT}^{\geqslant}(y)$, i.e. $CL_n^{\geqslant} \in U_{AT}^{\geqslant}(x)$, $U_A^{\geqslant}(x) \subseteq U_{AT}^{\geqslant}(x)$.
4. The proof of 4 is similar to the proof of 3. □

Let us denote

$$DIS_L^{\geqslant}(x,y) = \begin{cases} \{a \in AT : (x,y) \notin SDOM^{\leqslant}(\{a\})\}, & L_{AT}^{\geqslant}(x) \nsubseteq L_{AT}^{\geqslant}(y) \\ \emptyset, & \text{otherwise} \end{cases}$$

$$DIS_L^{\leqslant}(x,y) = \begin{cases} \{a \in AT : (x,y) \notin SDOM^{\geqslant}(\{a\})\}, & L_{AT}^{\leqslant}(x) \nsubseteq L_{AT}^{\leqslant}(y) \\ \emptyset, & \text{otherwise} \end{cases}$$

$$DIS_U^{\geqslant}(y,x) = \begin{cases} \{a \in AT : (x,y) \notin SDOM^{\geqslant}(\{a\})\}, & U_{AT}^{\geqslant}(y) \nsubseteq U_{AT}^{\geqslant}(x) \\ \emptyset, & \text{otherwise} \end{cases}$$

$$DIS_U^{\leqslant}(y,x) = \begin{cases} \{a \in AT : (x,y) \notin SDOM^{\leqslant}(\{a\})\}, & U_{AT}^{\leqslant}(y) \nsubseteq U_{AT}^{\leqslant}(x) \\ \emptyset, & \text{otherwise} \end{cases}$$

then

$$\boldsymbol{M}_L^{\geqslant} = \{DIS_L^{\geqslant}(x,y) : L_{AT}^{\geqslant}(x) \nsubseteq L_{AT}^{\geqslant}(y)\} \tag{5.35}$$

$$\boldsymbol{M}_L^{\leqslant} = \{DIS_L^{\leqslant}(x,y) : L_{AT}^{\leqslant}(x) \nsubseteq L_{AT}^{\leqslant}(y)\} \tag{5.36}$$

$$\boldsymbol{M}_U^{\geqslant} = \{DIS_U^{\geqslant}(y,x) : U_{AT}^{\geqslant}(y) \nsubseteq U_{AT}^{\geqslant}(x)\} \tag{5.37}$$

$$\boldsymbol{M}_U^{\leqslant} = \{DIS_U^{\leqslant}(y,x) : U_{AT}^{\leqslant}(y) \nsubseteq U_{AT}^{\leqslant}(x)\} \tag{5.38}$$

are referred to as the similarity dominance-based $\geqslant$-lower approximate distribution discernibility matrix, the similarity dominance-based $\leqslant$-lower approximate distribution discernibility matrix, the similarity dominance-based $\geqslant$-upper approximate distribution discernibility matrix, the similarity dominance-based $\leqslant$-upper approximate distribution discernibility matrix, respectively.

Theorem 5.7. *Let I be an incomplete decision system in which $A \subseteq AT$, then*

1. *A is a similarity dominance-based $\geqslant$-lower approximate distribution consistent attributes set $\Leftrightarrow A \cap DIS_L^{\geqslant}(x,y) \neq \emptyset$ for each $DIS_L^{\geqslant}(x,y) \in \boldsymbol{M}_L^{\geqslant}$;*
2. *A is a similarity dominance-based $\leqslant$-lower approximate distribution consistent attributes set $\Leftrightarrow A \cap DIS_L^{\leqslant}(x,y) \neq \emptyset$ for each $DIS_L^{\leqslant}(x,y) \in \boldsymbol{M}_L^{\leqslant}$;*
3. *A is a similarity dominance-based $\geqslant$-upper approximate distribution consistent attributes set $\Leftrightarrow A \cap DIS_U^{\geqslant}(y,x) \neq \emptyset$ for each $DIS_U^{\geqslant}(y,x) \in \boldsymbol{M}_U^{\geqslant}$;*
4. *A is a similarity dominance-based $\leqslant$-upper approximate distribution consistent attributes set $\Leftrightarrow A \cap DIS_U^{\leqslant}(y,x) \neq \emptyset$ for each $DIS_U^{\leqslant}(y,x) \in \boldsymbol{M}_U^{\leqslant}$.*

Proof. We only prove 1, others can be proved analogously.

$\Rightarrow$: Since A is a similarity dominance-based $\geqslant$-lower approximate distribution consistent attributes set, then by Theorem 5.6, $\forall x,y \in U$, if $L_{AT}^{\geqslant}(x) \nsubseteq L_{AT}^{\geqslant}(y)$, then $SDOM_A^{\leqslant-}(y) \nsubseteq SDOM_A^{\leqslant-}(x)$, that is, there must be $z \in U$ such that $z \in SDOM_A^{\leqslant-}(y)$ and $z \notin SDOM_A^{\leqslant-}(x)$. Suppose that $(x,y) \in SDOM^{\leqslant}(A)$, then we have $(x,z) \in SDOM^{\leqslant}(A)$ because $(y,z) \in SDOM^{\leqslant}(A)$ and the transitivity of $SDOM^{\leqslant}(A)$. Such result is contradictive to $z \notin SDOM_A^{\leqslant-}(x)$, which indicates that $(x,y) \notin SDOM^{\leqslant}(A)$, from which we can conclude that there must be $a \in A$ such that $(x,y) \notin SDOM^{\leqslant}(a)$, i.e. $A \cap DIS_L^{\geqslant}(x,y) \neq \emptyset$.

$\Leftarrow$: Since $DIS_L^{\geqslant}(x,y) \in \boldsymbol{M}_L^{\geqslant}$, then $L_{AT}^{\geqslant}(x) \not\subseteq L_{AT}^{\geqslant}(y)$. By condition we have $A \cap DIS_L^{\geqslant}(x,y) \neq \emptyset$, then there must be $a \in A$ such that $(x,y) \notin SDOM^{\leqslant}(\{a\})$, $y \notin SDOM_A^{\leqslant-}(x)$. Due to the reflexive of the similarity dominance relation, $y \in SDOM_A^{\leqslant-}(y)$, thus $SDOM_A^{\leqslant-}(y) \not\subseteq SDOM_A^{\leqslant-}(x)$, i.e. $L_{AT}^{\geqslant}(x) \not\subseteq L_{AT}^{\geqslant}(y) \Rightarrow SDOM_A^{\leqslant-}(y) \not\subseteq SDOM_A^{\leqslant-}(x)$, by Theorem 5.6, we know that A is a imitability dominance-based $\geqslant$-lower approximate distribution consistent attributes set. □

Definition 5.6. Let I be an incomplete decision system, then define

$$\Delta_L^{\geqslant} = \bigwedge_{DIS_L^{\geqslant}(x,y) \in \boldsymbol{M}_L^{\geqslant}} \vee DIS_L^{\geqslant}(x,y) \tag{5.39}$$

$$\Delta_L^{\leqslant} = \bigwedge_{DIS_L^{\leqslant}(x,y) \in \boldsymbol{M}_L^{\leqslant}} \vee DIS_L^{\leqslant}(x,y) \tag{5.40}$$

$$\Delta_U^{\geqslant} = \bigwedge_{DIS_U^{\geqslant}(y,x) \in \boldsymbol{M}_U^{\geqslant}} \vee DIS_U^{\geqslant}(y,x) \tag{5.41}$$

$$\Delta_U^{\leqslant} = \bigwedge_{DIS_U^{\leqslant}(y,x) \in \boldsymbol{M}_U^{\leqslant}} \vee DIS_U^{\leqslant}(y,x) \tag{5.42}$$

are referred to as the similarity dominance-based $\geqslant$-lower approximate distribution discernibility function, the similarity dominance-based $\leqslant$-lower approximate distribution discernibility function, the similarity dominance-based $\geqslant$-upper approximate distribution discernibility function, the similarity dominance-based $\leqslant$-upper approximate distribution discernibility function, respectively.

By Boolean Reasoning technique, it is not difficult to obtain the following theorem by Theorem 5.7.

Theorem 5.8. *Let I be an incomplete decision system in which $A \subseteq AT$, then*

1. *A is a similarity dominance-based $\geqslant$-lower approximate distribution reduct if and only if $\wedge A$ is the prime implicant of the similarity dominance-based $\geqslant$-lower approximate distribution discernibility function, i.e. $\Delta_L^{\geqslant}$;*
2. *A is a similarity dominance-based $\leqslant$-lower approximate distribution reduct if and only if $\wedge A$ is the prime implicant of the similarity dominance-based $\leqslant$-lower approximate distribution discernibility function, i.e. $\Delta_L^{\leqslant}$;*
3. *A is a similarity dominance-based $\geqslant$-upper approximate distribution reduct if and only if $\wedge A$ is the prime implicant of the similarity dominance-based $\geqslant$-upper approximate distribution discernibility function, i.e. $\Delta_U^{\geqslant}$;*
4. *A is a similarity dominance-based $\leqslant$-upper approximate distribution reduct if and only if $\wedge A$ is the prime implicant of the similarity dominance-based $\leqslant$-upper approximate distribution discernibility function, i.e. $\Delta_U^{\leqslant}$.*

Example 5.3. Let us consider Table 4.7, which is a summary of reviewers' reports for ten papers submitted to a journal. If all unknown values are replaced by "?", then by Definition 5.4, we have

$$\underline{SDOM_{AT}}(CL_1^{\geqslant}) = U$$
$$\underline{SDOM_{AT}}(CL_2^{\geqslant}) = \{x_1, x_2, x_3, x_4, x_7, x_8, x_{10}\}$$
$$\underline{SDOM_{AT}}(CL_3^{\geqslant}) = \{x_1, x_2, x_{10}\}$$
$$\underline{SDOM_{AT}}(CL_4^{\geqslant}) = \{x_1\}$$
$$\underline{SDOM_{AT}}(CL_1^{\leqslant}) = \varnothing$$
$$\underline{SDOM_{AT}}(CL_2^{\leqslant}) = \{x_4, x_5, x_6, x_7, x_8\}$$
$$\underline{SDOM_{AT}}(CL_3^{\leqslant}) = \{x_2, x_3, x_4, x_5, x_6, x_7, x_8, x_9, x_{10}\}$$
$$\underline{SDOM_{AT}}(CL_4^{\leqslant}) = U$$
$$\overline{SDOM_{AT}}(CL_1^{\geqslant}) = U$$
$$\overline{SDOM_{AT}}(CL_2^{\geqslant}) = U$$
$$\overline{SDOM_{AT}}(CL_3^{\geqslant}) = \{x_1, x_2, x_3, x_9, x_{10}\}$$
$$\overline{SDOM_{AT}}(CL_4^{\geqslant}) = \{x_1\}$$
$$\overline{SDOM_{AT}}(CL_1^{\leqslant}) = \{x_5, x_6, x_9\}$$
$$\overline{SDOM_{AT}}(CL_2^{\leqslant}) = \{x_3, x_4, x_5, x_6, x_7, x_8, x_9\}$$
$$\overline{SDOM_{AT}}(CL_3^{\leqslant}) = \{x_2, x_3, x_4, x_5, x_6, x_7, x_8, x_9, x_{10}\}$$
$$\overline{SDOM_{AT}}(CL_4^{\leqslant}) = U$$

By Eq. (5.35), we obtain the similarity dominance-based $\geqslant$-lower approximate distribution discernibility matrix as Table 5.4 shows.

Table 5.4 Similarity dominance-based $\geqslant$-lower approximate distribution discernibility matrix of Table 4.7

	x_1	x_2	x_3	x_4	x_5	x_6	x_7	x_8	x_9	x_{10}
x_1		a_1, a_2	a_2, a_3	AT	AT	AT	AT	a_2, a_3	AT	a_2
x_2			a_2, a_3	a_1, a_3	AT	a_2, a_3	a_1, a_3	a_2, a_3	a_2, a_3	
x_3					a_1	a_1			a_1	
x_4					a_2	a_2			a_2	
x_5										
x_6										
x_7					a_2	a_2			a_2	
x_8					a_1	a_1			a_1, a_2	
x_9										
x_{10}			a_2, a_3	AT	AT	AT	AT	a_2, a_3	AT	

By Definition 5.6,

$$\begin{aligned}\Delta_L^{\geqslant} &= \bigwedge_{DIS_L^{\geqslant}(x,y)\in \boldsymbol{M}_L^{\geqslant}} \vee DIS_L^{\geqslant}(x,y)\\ &= a_1 \wedge a_2 \wedge (a_1 \vee a_2) \wedge (a_2 \vee a_3) \wedge (a_1 \vee a_3) \wedge AT\\ &= a_1 \wedge a_2\end{aligned}$$

By Theorem 5.8, we know that $\{a_1, a_2\}$ is the similarity dominance-based $\geqslant$-lower approximate distribution reduct of Table 4.7.

Similarly, it is not difficult to obtain that AT is the similarity dominance-based $\geqslant$-upper approximate distribution reduct of Table 4.7, i.e., no attribute can be eliminated for preserving the upper approximations of all of the upward unions of the decision classes.

Moreover, by Theorem 5.4, we know that $\{a_1, a_2\}$ is the similarity dominance-based $\leqslant$-upper approximate distribution reduct of Table 4.7, AT is the similarity dominance-based $\leqslant$-lower approximate distribution reduct of Table 4.7.

5.3 Similarity Dominance-based Rough Sets in Fuzzy Decision System

5.3.1 Similarity Dominance-based Rough Fuzzy Set

An incomplete fuzzy decision system is an incomplete decision system, in which $\forall a \in AT \cup \{d\}$, if $f(x, a)$ is known, then $f(x, a) \in [0, 1]$.

Example 5.4. To demonstrate the incomplete fuzzy information system, let us consider data in Table 5.5, which describes a small training set with fuzzy objects. In Table 5.5,

- $U = \{x_1, x_2, \cdots, x_{10}\}$ is the universe of discourse;
- $AT = \{a_1, a_2, a_3, a_4\}$ is the set of the condition attributes while d is the decision attribute.

Table 5.5 An example of incomplete fuzzy information system

	a_1	a_2	a_3	a_4	d
x_1	0.9	?	0.2	0.7	0.7
x_2	0.9	0.2	0.2	0.1	0.8
x_3	0.1	0.1	0.1	0.9	0.0
x_4	0.0	0.9	?	0.8	0.5
x_5	0.1	0.1	1.0	0.8	0.4
x_6	?	0.2	0.9	0.1	0.3
x_7	0.0	0.1	0.9	0.2	0.0
x_8	0.9	0.9	0.1	1.0	0.6
x_9	0.8	0.4	1.0	1.0	0.9
x_{10}	0.0	1.0	1.0	?	0.0

Obviously, in Table 5.5, the unknown values are represented by the symbol "?" and then we should use the similarity dominance relation for classification analysis. Moreover, in an incomplete fuzzy decision system, the set to be approximated is a fuzzy set instead of a crisp subset of the universe. For example, in Table 5.4, by the decision attribute d, we may consider a fuzzy set such that $\tilde{d} = \{f(x, d)/x : x \in U\}$. From these two point of views, it is not difficult to construct similarity dominance-based rough fuzzy set as Definition 5.7 shows.

Definition 5.7. Let I be an incomplete fuzzy decision system in which $A \subseteq AT$, the lower and upper approximations of $\widetilde{d}$ are denoted by $\underline{SDOM_A}(\widetilde{d})$ and $\overline{SDOM_A}(\widetilde{d})$, respectively, whose membership functions for each $x \in U$, are defined as:

$$\mu_{\underline{SDOM_A}(\widetilde{d})}(x) = \wedge\{f(y,d) : y \in SDOM_A^{\leqslant -}(x)\} \tag{5.43}$$

$$\mu_{\overline{SDOM_A}(\widetilde{d})}(x) = \vee\{f(y,d) : y \in SDOM_A^{\geqslant -}(x)\} \tag{5.44}$$

Example 5.5. Following Example 5.4, since the fuzzy set to be approximated is

$$\widetilde{d} = \{0.7/x_1, 0.8/x_2, 0.0/x_3, 0.5/x_4, 0.4/x_5, 0.3/x_6, 0.0/x_7, 0.6/x_8, 0.9/x_9, 0.0/x_{10}\}$$

then we may obtain the lower and upper approximate memberships for each object in Table 5.5 as Table 5.6 shows.

Table 5.6 Lower and upper approximate memberships in Table 5.5

	x_1	x_2	x_3	x_4	x_5	x_6	x_7	x_8	x_9	x_{10}
$\mu_{\underline{SDOM_{AT}}(\widetilde{d})}(x)$	0.7	0.8	0.0	0.5	0.4	0.3	0.0	0.6	0.9	0.0
$\mu_{\overline{SDOM_{AT}}(\widetilde{d})}(x)$	0.8	0.8	0.0	0.5	0.4	0.8	0.0	0.6	0.9	0.3

Proposition 5.3. *Let I be an incomplete fuzzy decision system in which $A \subseteq AT$, then we have following properties about the similarity dominance-based rough fuzzy approximation:*

1.
$$\underline{SDOM_A}(\widetilde{d}) \subseteq \widetilde{d} \subseteq \overline{SDOM_A}(\widetilde{d}) \tag{5.45}$$

2.
$$\widetilde{d_1} \subseteq \widetilde{d_2} \Rightarrow \quad \underline{SDOM_A}(\widetilde{d_1}) \subseteq \underline{SDOM_A}(\widetilde{d_2}) \tag{5.46}$$

$$\overline{SDOM_A}(\widetilde{d_1}) \subseteq \overline{SDOM_A}(\widetilde{d_2}) \tag{5.47}$$

3.
$$\underline{SDOM_A}(\widetilde{d}) \subseteq \underline{SDOM_{AT}}(\widetilde{d}) \tag{5.48}$$

$$\overline{SDOM_A}(\widetilde{d}) \supseteq \overline{SDOM_{AT}}(\widetilde{d}) \tag{5.49}$$

4.
$$\underline{SDOM_A}(\widetilde{d_1} \cap \widetilde{d_2}) \subseteq \underline{SDOM_A}(\widetilde{d_1}) \cap \underline{SDOM_A}(\widetilde{d_2}) \tag{5.50}$$

$$\underline{SDOM_A}(\widetilde{d_1} \cup \widetilde{d_2}) \supseteq \underline{SDOM_A}(\widetilde{d_1}) \cup \underline{SDOM_A}(\widetilde{d_2}) \tag{5.51}$$

$$\overline{SDOM_A}(\widetilde{d_1} \cap \widetilde{d_2}) \subseteq \overline{SDOM_A}(\widetilde{d_1}) \cap \overline{SDOM_A}(\widetilde{d_2}) \tag{5.52}$$

$$\overline{SDOM_A}(\widetilde{d_1} \cup \widetilde{d_2}) \supseteq \overline{SDOM_A}(\widetilde{d_1}) \cup \overline{SDOM_A}(\widetilde{d_2}) \tag{5.53}$$

Proof.

1. $\forall x \in U$, since the similarity dominance relation is reflexive, then $x \in SDOM_A^{\leqslant -}(x)$, we then obtain that
$$\mu_{\underline{SDOM_A(\widetilde{d})}}(x) = \wedge\{f(y,d) : y \in SDOM_A^{\leqslant -}(x)\} \leqslant f(x,d), \text{i.e. } \underline{SDOM_A(\widetilde{d})} \subseteq \widetilde{d}.$$
Similarly, it is not difficult to prove that $\widetilde{d} \subseteq \overline{SDOM_A(\widetilde{d})}$.
2. Since $\widetilde{d_1} \subseteq \widetilde{d_2}$, then $\forall y \in SDOM_A^{\leqslant -}(x)$, we have $f(y,d_1) \leqslant f(y,d_2)$. By Eq. (5.43), we can conclude that
$$\mu_{\underline{SDOM_A(\widetilde{d_1})}}(x) \leqslant \mu_{\underline{SDOM_A(\widetilde{d_2})}}(x), \text{i.e. } \underline{SDOM_A(\widetilde{d_1})} \subseteq \underline{SDOM_A(\widetilde{d_2})}.$$
Similarly, it is not difficult to prove that $\overline{SDOM_A(\widetilde{d_1})} \subseteq \overline{SDOM_A(\widetilde{d_2})}$.
3. Since $A \subseteq AT$, then $\forall x \in U$, we have $SDOM_{AT}^{\leqslant -}(x) \subseteq SDOM_A^{\leqslant -}(x)$. By Eq. (5.43), $\mu_{\underline{SDOM_{AT}(\widetilde{d})}}(x) = \wedge\{f(y,d) : y \in SDOM_{AT}^{\leqslant -}(x)\} \geqslant \mu_{\underline{SDOM_A(\widetilde{d})}}(x) = \wedge\{f(y,d) : y \in SDOM_A^{\leqslant -}(x)\}$, i.e. $\underline{SDOM_A(\widetilde{d})} \subseteq \underline{SDOM_{AT}(\widetilde{d})}$.
Similarly, it is not difficult to prove $\overline{SDOM_A(\widetilde{d})} \supseteq \overline{SDOM_{AT}(\widetilde{d})}$.
4. By Eq. (5.46), since $\widetilde{d_1} \cap \widetilde{d_2} \subseteq \widetilde{d_1}, \widetilde{d_2}$, then we have $\underline{SDOM_A(\widetilde{d_1} \cap \widetilde{d_2})} \subseteq \underline{SDOM_A(\widetilde{d_1})}$, $\underline{SDOM_A(\widetilde{d_1} \cap \widetilde{d_2})} \subseteq \underline{SDOM_A(\widetilde{d_2})}$, i.e. $\underline{SDOM_A(\widetilde{d_1} \cap \widetilde{d_2})} \subseteq \underline{SDOM_A(\widetilde{d_1})} \cap \underline{SDOM_A(\widetilde{d_2})}$.
Similarly, it is not difficult to prove other formulas. □

Similar to the classical dominance-based rough set, we may also generate the "at least" and "at most" rules from the incomplete fuzzy decision system such that:

- "at least" rule: if $f(y,a_1) \geqslant f(x,a_1)$ and $f(y,a_2) \geqslant f(x,a_2)$ and $\cdots$ and $f(y,a_k) \geqslant f(x,a_k)$, then $f(y,d) \geqslant \mu_{\underline{SDOM_A(\widetilde{d})}}(x)$;
- "at most" rule: if $f(y,a_1) \leqslant f(x,a_1)$ and $f(y,a_2) \leqslant f(x,a_2)$ and $\cdots$ and $f(y,a_k) \leqslant f(x,a_k)$, then $f(y,d) \leqslant \mu_{\overline{SDOM_A(\widetilde{d})}}(x)$;

where $A = \{a_1, a_2, \cdots, a_k\}$ and $f(x,a_i)$ is known $\forall i = 1, 2, \cdots, k$.

Example 5.6. Take for instance Table 5.5, by the result of Table 5.6, we may obtain the following decision rules:

"at least" rules:

1. $f(y,a_1) \geqslant 0.9 \wedge f(y,a_3) \geqslant 0.2 \wedge f(y,a_4) \geqslant 0.7 \to f(y,d) \geqslant 0.7$;
2. $f(y,a_1) \geqslant 0.9 \wedge f(y,a_2) \geqslant 0.2 \wedge f(y,a_3) \geqslant 0.2 \wedge f(y,a_4) \geqslant 0.1 \to f(y,d) \geqslant 0.8$;
3. $f(y,a_1) \geqslant 0.1 \wedge f(y,a_2) \geqslant 0.1 \wedge f(y,a_3) \geqslant 0.1 \wedge f(y,a_4) \geqslant 0.9 \to f(y,d) \geqslant 0.0$;
4. $f(y,a_1) \geqslant 0.0 \wedge f(y,a_2) \geqslant 0.9 \wedge f(y,a_4) \geqslant 0.8 \to f(y,d) \geqslant 0.5$;
5. $f(y,a_1) \geqslant 0.1 \wedge f(y,a_2) \geqslant 0.1 \wedge f(y,a_3) \geqslant 1.0 \wedge f(y,a_4) \geqslant 0.8 \to f(y,d) \geqslant 0.4$;
6. $f(y,a_2) \geqslant 0.2 \wedge f(y,a_3) \geqslant 0.9 \wedge f(y,a_4) \geqslant 0.1 \to f(y,d) \geqslant 0.3$;
7. $f(y,a_1) \geqslant 0.0 \wedge f(y,a_2) \geqslant 0.1 \wedge f(y,a_3) \geqslant 0.9 \wedge f(y,a_4) \geqslant 0.2 \to f(y,d) \geqslant 0.0$;
8. $f(y,a_1) \geqslant 0.9 \wedge f(y,a_2) \geqslant 0.9 \wedge f(y,a_3) \geqslant 0.1 \wedge f(y,a_4) \geqslant 1.0 \to f(y,d) \geqslant 0.6$;
9. $f(y,a_1) \geqslant 0.8 \wedge f(y,a_2) \geqslant 0.4 \wedge f(y,a_3) \geqslant 1.0 \wedge f(y,a_4) \geqslant 1.0 \to f(y,d) \geqslant 0.9$;

10. $f(y, a_1) \geqslant 0.0 \wedge f(y, a_2) \geqslant 1.0 \wedge f(y, a_3) \geqslant 1.0 \rightarrow f(y, d) \geqslant 0.0$.

"at most" rules:

1. $f(y, a_1) \leqslant 0.9 \wedge f(y, a_3) \leqslant 0.2 \wedge f(y, a_4) \leqslant 0.7 \rightarrow f(y, d) \leqslant 0.8$;
2. $f(y, a_1) \leqslant 0.9 \wedge f(y, a_2) \leqslant 0.2 \wedge f(y, a_3) \leqslant 0.2 \wedge f(y, a_4) \leqslant 0.1 \rightarrow f(y, d) \leqslant 0.8$;
3. $f(y, a_1) \leqslant 0.1 \wedge f(y, a_2) \leqslant 0.1 \wedge f(y, a_3) \leqslant 0.1 \wedge f(y, a4) \leqslant 0.9 \rightarrow f(y, d) \leqslant 0.0$;
4. $f(y, a_1) \leqslant 0.0 \wedge f(y, a_2) \leqslant 0.9 \wedge f(y, a_4) \leqslant 0.8 \rightarrow f(y, d) \leqslant 0.5$;
5. $f(y, a_1) \leqslant 0.1 \wedge f(y, a_2) \leqslant 0.1 \wedge f(y, a_3) \leqslant 1.0 \wedge f(y, a_4) \leqslant 0.8 \rightarrow f(y, d) \leqslant 0.4$;
6. $f(y, a_2) \leqslant 0.2 \wedge f(y, a_3) \leqslant 0.9 \wedge f(y, a_4) \leqslant 0.1 \rightarrow f(y, d) \leqslant 0.8$;
7. $f(y, a_1) \leqslant 0.0 \wedge f(y, a_2) \leqslant 0.1 \wedge f(y, a_3) \leqslant 0.9 \wedge f(y, a_4) \leqslant 0.2 \rightarrow f(y, d) \leqslant 0.0$;
8. $f(y, a_1) \leqslant 0.9 \wedge f(y, a_2) \leqslant 0.9 \wedge f(y, a_3) \leqslant 0.1 \wedge f(y, a_4) \leqslant 1.0 \rightarrow f(y, d) \leqslant 0.6$;
9. $f(y, a_1) \leqslant 0.8 \wedge f(y, a_2) \leqslant 0.4 \wedge f(y, a_3) \leqslant 1.0 \wedge f(y, a_4) \leqslant 1.0 \rightarrow f(y, d) \leqslant 0.9$;
10. $f(y, a_1) \leqslant 0.0 \wedge f(y, a_2) \leqslant 1.0 \wedge f(y, a_3) \leqslant 1.0 \rightarrow f(y, d) \leqslant 0.3$.

5.3.2 Relative Approximate Distribution Reducts of Similarity Dominance-based Rough Fuzzy Set

Definition 5.8. Let I be an incomplete fuzzy decision system in which $A \subseteq AT$, $\forall x \in U$,

1. if $\mu_{\underline{SDOM_A(\widetilde{d})}}(x) = \mu_{\underline{SDOM_{AT}(\widetilde{d})}}(x)$, then A is referred to as a relative similarity dominance-based lower approximate distribution consistent attributes set of x in I; if A is a relative similarity dominance-based lower approximate distribution consistent attributes set of x in I and $\forall B \subset A$, B is not the relative similarity dominance-based lower approximate distribution consistent attributes set of x in I, then A is referred to as a relative similarity dominance-based lower approximate distribution reduct of x in I.
2. if $\mu_{\overline{SDOM_A(\widetilde{d})}}(x) = \mu_{\overline{SDOM_{AT}(\widetilde{d})}}(x)$, then A is referred to as a relative similarity dominance-based upper approximate distribution consistent attributes set of x in I; if A is a relative similarity dominance-based upper approximate distribution consistent attributes set of x in I and $\forall B \subset A$, B is not the relative similarity dominance-based upper approximate distribution consistent attributes set of x in I, then A is referred to as a relative similarity dominance-based upper approximate distribution reduct of x in I.

By Definition 5.8, we can see that

- A relative similarity dominance-based lower approximate distribution consistent attributes set of x in I is the subset of the condition attributes, which preserves the similarity dominance-based lower membership of x in I; a relative similarity dominance-based lower approximate distribution reduct of x in I is the minimal subset of the condition attributes, which preserves the similarity dominance-based lower membership of x in I.

- A relative similarity dominance-based upper approximate distribution consistent attributes set of x in I is the subset of the condition attributes, which preserves the similarity dominance-based upper membership of x in I; a relative similarity dominance-based upper approximate distribution reduct of x in I is the minimal subset of the condition attributes, which preserves the similarity dominance-based upper membership of x in I.

Let us denote

$$DIS_L^{SD}(x,y) = \begin{cases} \{a \in AT : (x,y) \notin SDOM^{\leqslant}(\{a\})\}, & \mu_{\underline{SDOM_A(\tilde{d})}}(x) \geqslant f(y,d) \\ \varnothing, & \text{otherwise} \end{cases}$$

$$DIS_U^{SD}(x,y) = \begin{cases} \{a \in AT : (x,y) \notin SDOM^{\geqslant}(\{a\})\}, & \mu_{\overline{SDOM_A(\tilde{d})}}(x) \leqslant f(y,d) \\ \varnothing, & \text{otherwise} \end{cases}$$

then

$$\boldsymbol{M}_L^{SD}(x) = \{DIS_L^{SD}(x,y) : \mu_{\underline{SDOM_{AT}(\tilde{d})}}(x) \geqslant f(y,d)\} \tag{5.54}$$

$$\boldsymbol{M}_U^{SD}(x) = \{DIS_U^{SD}(x,y) : \mu_{\overline{SDOM_{AT}(\tilde{d})}}(x) \leqslant f(y,d)\} \tag{5.55}$$

are referred to as the relative similarity dominance-based lower and upper approximate distribution discernibility matrixes of x in I, respectively.

Theorem 5.9. *Let I be an incomplete fuzzy decision system in which $A \subseteq AT$, then*

1. *A is a similarity dominance-based lower approximate distribution consistent attributes set of x in $I \Leftrightarrow A \cap DIS_L^{SD}(x,y) \neq \varnothing$ for each $DIS_L^{SD}(x,y) \in \boldsymbol{M}_L^{SD}(x)$;*
2. *A is a similarity dominance-based upper approximate distribution consistent attributes set of x in $I \Leftrightarrow A \cap DIS_U^{SD}(x,y) \neq \varnothing$ for each $DIS_U^{SD}(x,y) \in \boldsymbol{M}_U^{SD}(x)$.*

Proof. We only prove 1, the proof of 2 is similar to the proof of 1.

$\Rightarrow$: Suppose $\exists DIS_L^{SD}(x,y) \in \boldsymbol{M}_L^{SD}(x)$ such that $A \cap DIS_L^{SD}(x,y) = \varnothing$, then we have $(x,y) \in SDOM^{\leqslant}(A)$, i.e. $y \in SDOM_A^{\leqslant-}(x)$. By Definition 5.7, we have $\mu_{\underline{SDOM_A(\tilde{d})}}(x) \leqslant f(y,d)$. Since A is a similarity dominance-based lower approximate distribution consistent attributes set of x in I, then we have $\mu_{\underline{SDOM_A(\tilde{d})}}(x) = \mu_{\underline{SDOM_{AT}(\tilde{d})}}(x) \leqslant f(y,d)$, which is contradictive to the assumption because $DIS_L^{SD}(x,y) \in \boldsymbol{M}_L^{SD} \Rightarrow \mu_{\underline{SDOM_{AT}(\tilde{d})}}(x) \geqslant f(y,d)$.

$\Leftarrow$: Since $A \subseteq AT$, then by 3 of Proposition 5.3, we have $\mu_{\underline{SDOM_A(\tilde{d})}}(x) \leqslant \mu_{\underline{SDOM_{AT}(\tilde{d})}}(x)$. Suppose that $\mu_{\underline{SDOM_A(\tilde{d})}}(x) \neq \mu_{\underline{SDOM_{AT}(\tilde{d})}}(x)$, then we have $\mu_{\underline{SDOM_A(\tilde{d})}}(x) < \mu_{\underline{SDOM_{AT}(\tilde{d})}}(x)$, from which we can conclude that there must be $y \in SDOM_A^{\leqslant-}(x)$ such that $f(y,d) < \mu_{\underline{SDOM_{AT}(\tilde{d})}}(x)$, it follows that $\exists DIS_L^{SD}(x,y) \in \boldsymbol{M}_L^{SD}(x)$ such that $A \cap DIS_L^{SD}(x,y) = \varnothing$. Therefore, if $A \cap DIS_L^{SD}(x,y) \neq \varnothing$ for each $DIS_L^{SD}(x,y) \in \boldsymbol{M}_L^{SD}(x)$, then $\mu_{\underline{SDOM_A(\tilde{d})}}(x) = \mu_{\underline{SDOM_{AT}(\tilde{d})}}(x)$, i.e. A is a similarity dominance-based lower approximate distribution consistent attributes set of x in I. □

Definition 5.9. Let I be an incomplete fuzzy decision system in which $A \subseteq AT$, then define

$$\Delta_L^{SD}(x) = \bigwedge_{DIS_L^{SD}(x,y)\in \boldsymbol{M}_L^{SD}} \vee DIS_L^{SD}(x,y) \tag{5.56}$$

$$\Delta_U^{SD}(x) = \bigwedge_{DIS_U^{SD}(x,y)\in \boldsymbol{M}_U^{SD}} \vee DIS_U^{SD}(x,y) \tag{5.57}$$

$\Delta_L^{SD}(x)$ and $\Delta_U^{SD}(x)$ are referred to as the relative similarity dominance-based lower and upper approximate distribution discernibility functions of x in I, respectively.

By Boolean Reasoning technique, it is not difficult to obtain the following theorem by Theorem 5.9.

Theorem 5.10. *Let I be an incomplete fuzzy decision system in which $A \subseteq AT$, then*

1. *A is a similarity dominance-based lower approximate distribution reduct of x in I if and only if $\wedge A$ is the prime implicant of the similarity dominance-based lower approximate distribution discernibility function of x in I, i.e. $\Delta_L^{SD}(x)$;*
2. *A is a similarity dominance-based upper approximate distribution reduct of x in I if and only if $\wedge A$ is the prime implicant of the similarity dominance-based upper approximate distribution discernibility function of x in I, i.e. $\Delta_U^{SD}(x)$.*

Example 5.7. Take for instance Table 5.5, since

$$\mu_{\underline{SDOM_{AT}(\widetilde{d})}}(x_1) \geqslant f(x_3,d), f(x_4,d), f(x_5,d), \\ f(x_6,d), f(x_7,d), f(x_8,d), f(x_{10},d)$$

then by Definition 5.9, the relative similarity dominance lower approximate distribution discernibility function of x_1 is

$$\begin{aligned}\Delta_L^{SD}(x) &= \bigwedge_{DIS_L^{SD}(x,y)\in \boldsymbol{M}_L^{SD}} \vee DIS_L^{SD}(x,y) \\ &= (a_1 \vee a_3) \wedge (a_1 \vee a_4) \wedge a_1 \wedge a_3 \wedge a_4 \\ &= a_1 \wedge a_3 \wedge a_4\end{aligned}$$

By Theorem 5.10, we know that $\{a_1, a_3, a_4\}$ is the similarity dominance-based lower approximate distribution reduct of x_1.

Similarly, we obtain the following results:

- $\{a_3\}, \{a_4\}$ are the similarity dominance-based upper approximate distribution reducts of x_1;
- $\{a_1, a_2, a_3\}$ is the similarity dominance-based lower approximate distribution reduct of x_2, $\{a_2\}, \{a_3\}, \{a_4\}$ are the similarity dominance-based upper approximate distribution reducts of x_2;

- $\{a_1\}, \{a_2\}, \{a_3\}, \{a_4\}$ are the similarity dominance-based lower approximate distribution reducts of x_3, $\{a_3\}$ is the similarity dominance-based upper approximate distribution reduct of x_3;
- $\{a_2, a_4\}$ is the similarity dominance-based lower approximate distribution reduct of x_4, $\{a_1\}$ is the similarity dominance-based upper approximate distribution reduct of x_4;
- $\{a_1, a_3\}, \{a_3, a_4\}$ are the similarity dominance-based lower approximate distribution reducts of x_5, $\{a_1, a_3\}, \{a_2\}$ are the similarity dominance-based upper approximate distribution reducts of x_5;
- $\{a_2, a_4\}$ is the similarity dominance-based lower approximate distribution reduct of x_6, $\{a_2, \{a_3\}, \{a_4\}$ are the similarity dominance-based upper approximate distribution reducts of x_6;
- $\{a_1\}, \{a_2\}, \{a_3\}, \{a_4\}$ are the similarity dominance-based lower approximate distribution reducts of x_7, $\{a_1, a_2, a_3\}, \{a_1, a_2, a_4\}$ are the similarity dominance-based upper approximate distribution reducts of x_7;
- $\{a_1\}, \{a_4\}$ are the similarity dominance-based lower approximate distribution reducts of x_8, $\{a_1\}$ is the similarity dominance-based upper approximate distribution reduct of x_8;
- $\{a_1, a_2, a_3\}, \{a_2, a_3, a_4\}$ are the similarity dominance-based lower approximate distribution reducts of x_9, $\{a_1\}, \{a_2\}, \{a_3\}, \{a_4\}$ are the similarity dominance-based upper approximate distribution reducts of x_9;
- $\{a_1\}, \{a_2\}, \{a_3\}, \{a_4\}$ are the similarity dominance-based lower approximate distribution reducts of x_{10}, $\{a_1\}$ is the similarity dominance-based upper approximate distribution reduct of x_{10}.

5.4 Conclusions

In this chapter, the dominance-based rough set approach is introduced into the incomplete information system with "?" unknown values:

- firstly, the concept of the similarity dominance relation is presented, different from the expanded dominance relation and the limited dominance relation showed in Chapter 4, similarity dominance relation has two different forms, i.e. $SDOM^{\geqslant}(A)$ and $SDOM^{\leqslant}(A)$, the discernibility matrixes approach to compute the reducts of similarity dominance relation is also presented;
- secondly, by the similarity dominance relation, the similarity dominance-based rough set model is constructed, such model is a generalization of the classical dominance-based rough set in the incomplete information system, the concept of the approximate distribution reduct is introduced into the similarity dominance-based rough set and then four types of the similarity dominance-based approximate distribution reducts are proposed, not only the relationships among these reducts are discussed, but also the discernibility matrixes approach to compute these reducts is presented;

- finally, the similarity dominance-based rough set approach is also generalized into the incomplete fuzzy decision system and then the rough fuzzy set in terms of the similarity dominance relation is constructed, we also show how to compute the similarity dominance-based rough fuzzy approximate distribution reducts.

References

[1] Greco, S., Inuiguchi, M., Słowiński, R.: Fuzzy rough sets and multiple-premise gradual decision rules. Int. J. Approx. Reason. **41**, 179–211 (2006)

[2] Greco, S., Matarazzo, B., Słowiński, R.: Handing missing values in rough set analysis of mutiattribute and muti-criteria decision problems. In: 7th International Workshop on New Directions in Rough Sets, Data Mining and Granular-Soft Computing, pp. 146–157 (1999)

[3] Greco, S., Matarazzo, B., Słowiński, R.: Rough approximation by dominance relations. Int. J. Intell. Syst. **17**, 153–171 (2002)

[4] Greco, S., Matarazzo, B., Słowiński, R.: Rough sets theory for multicriteria decision analysis. Eur. J. Oper. Res. **129**, 1–47 (2002)

[5] Greco, S., Matarazzo, B., Słowiński, R.: Dominance-based rough set approach to case-Based reasoning. In: Third International Conference on Modeling Decisions for Artificial Intelligence, pp. 7–18 (2006)

[6] Greco, S., Matarazzo, B., Słowiński, R.: Fuzzy set extensions of the dominance-based rough set approach. In: Fuzzy Sets and Their Extensions: Representation, Aggregation and Models, pp. 239–261 (2008)

[7] Greco, S., Słowiński, R., Yao, Y. Y.: Bayesian decision theory for dominance-based rough set approach. In: The Second International Conference on Rough Sets and Knowledge Technology, pp. 143–141 (2007)

[8] Shao, M. W., Zhang, W. X.: Dominance relation and rules in an incomplete ordered information system. Int. J. Intell. Syst. **20**, 13–27 (2005)

[9] Wei, L. H., Tang, Z. M., Wang, R. Y., Yang, X. B.: Extensions of dominance-based rough set approach in incomplete information system. Autom. Control Comp. Sci. **42**, 255–263 (2008)

[10] Luo G. Z., Yang, X. B.: Limited dominance-based rough set model and knowledge reductions in incomplete decision system. J. Inform. Sci. Eng. **26**, 2199–2211 (2010)

[11] Stefanowski, J., Tsoukiàs, A.: On the extension of rough sets under incomplete information. In: Proceeding of New directions in rough sets, data mining and granular-soft computing, pp. 73–82 (1999)

[12] Stefanowski, J., Tsoukiàs, A.: Incomplete information tables and rough classification. Comput. Intell. **17**, 545–566 (2001)

[13] Yang, X. B., Yang, J. Y., Wu, C., Yu, D. J.: Dominance-based rough set approach and knowledge reductions in incomplete ordered information system. Inform. Sci. **178**, 1219–1234 (2008)

[14] Yang, X. B., Wei, L. H., Yu, D. J., Yang, J. Y.: Rough fuzzy set in incomplete fuzzy information system based on similarity dominance relation. Recent Patents on Computer Science **2**, 68–74 (2009)

[15] Gao, Y. Q., Ha, M. H., Zhao, X., Li, C. J.: Rough set model based on limited similarity dominance relation. In: Fifth International Conference on Fuzzy Systems and Knowledge Discovery, pp. 285–289 (2008)

[16] Zhao, X., Ha, M. H., Zhang, A. Q.: Variable precision rough set model in incomplete ordered decision system. In: 3rd International Conference on Innovative Computing Information and Control, pp. 498–498 (2008)

Part IV
Incomplete Information Systems and Multigranulation Rough Sets

Chapter 6

Multigranulation Rough Sets in Incomplete Information System

Abstract Since multigranulation rough set is an important expansion of Pawlak's rough set and then it is an interesting issue to explore multigranulation rough set approach to incomplete information system. In this chapter, by considering two different semantic explanations of the unknown values, the tolerance relations, similarity relations are employed to construct multigranulation rough sets in incomplete information system, respectively. Following Qian's multigranulation rough set theory, the optimistic and pessimistic cases are also considered in this chapter. The rough set models defined in this chapter provide a new direction for the investigation of rough set theory in incomplete information system.

In Section 1.3, we have briefly introduced the concept of the multigranulation rough set [1–4], which was firstly proposed by Qian et al., multigranulation rough set is different from Pawlak's rough set since Qian et al. use a family of the indiscernibility relations instead of a single indiscernibility relation to construct approximation. In this chapter, the generalizations of multigranulation rough set models will be studied in incomplete information system. Two important binary relations, i.e., tolerance relation [5–8] and similarity relation [9, 10] will be used to construct multigranulation rough sets in the incomplete information system, respectively.

6.1 Tolerance Relations Based Multigranulation Rough Sets

6.1.1 Optimistic and Pessimistic Tolerance Relations Based Multigranulation Rough Sets

In Section 2.1, the tolerance based rough set, which was proposed by Kryszkiewicz has been presented. Following the basic idea of multigranulation rough set, it is not difficult to construct multigranulation rough sets, which are based on the tolerance relations, i.e. a family of the tolerance relations instead of a single tolerance relation are used to construct lower and upper approximations.

Definition 6.1. Let I be an incomplete decision system in which $A_1, A_2, \cdots, A_m \subseteq AT$, then $\forall X \subseteq U$, the optimistic multigranulation lower approximation and optimistic multigranulation upper approximation in terms of the tolerance relations are denoted by $\underline{\sum_{i=1}^{m} A_i}_{TOL}^{OPT}(X)$ and $\overline{\sum_{i=1}^{m} A_i}_{TOL}^{OPT}(X)$, respectively, where

$$\underline{\sum_{i=1}^{m} A_i}_{TOL}^{OPT}(X) = \{x \in U : TOL_{A_1}(x) \subseteq X \vee \cdots \vee TOL_{A_m}(x) \subseteq X\} \tag{6.1}$$

$$\overline{\sum_{i=1}^{m} A_i}_{TOL}^{OPT}(X) = \sim \underline{\sum_{i=1}^{m} A_i}_{TOL}^{OPT}(\sim X) \tag{6.2}$$

Definition 6.2. Let I be an incomplete decision system in which $A_1, A_2, \cdots, A_m \subseteq AT$, then $\forall X \subseteq U$, the pessimistic multigranulation lower approximation and pessimistic multigranulation upper approximation in terms of the tolerance relations are denoted by $\underline{\sum_{i=1}^{m} A_i}_{TOL}^{PES}(X)$ and $\overline{\sum_{i=1}^{m} A_i}_{TOL}^{PES}(X)$, respectively, where

$$\underline{\sum_{i=1}^{m} A_i}_{TOL}^{PES}(X) = \{x \in U : TOL_{A_1}(x) \subseteq X \wedge \cdots \wedge TOL_{A_m}(x) \subseteq X\} \tag{6.3}$$

$$\overline{\sum_{i=1}^{m} A_i}_{TOL}^{PES}(X) = \sim \underline{\sum_{i=1}^{m} A_i}_{TOL}^{PES}(\sim X) \tag{6.4}$$

The pair $\left[\underline{\sum_{i=1}^{m} A_i}_{TOL}^{OPT}(X), \overline{\sum_{i=1}^{m} A_i}_{TOL}^{OPT}(X)\right]$ is referred to as an optimistic multigranulation rough set of X in terms of the set of the tolerance relations $\{TOL(A_1), TOL(A_2), \cdots, TOL(A_m)\}$; the pair $\left[\underline{\sum_{i=1}^{m} A_i}_{TOL}^{PES}(X), \overline{\sum_{i=1}^{m} A_i}_{TOL}^{PES}(X)\right]$ is referred to as a pessimistic multigranulation rough set of X in terms of the set of the tolerance relations $\{TOL(A_1), TOL(A_2), \cdots, TOL(A_m)\}$.

Theorem 6.1. *Let I be an incomplete decision system in which $A_1, A_2, \cdots, A_m \subseteq AT$, then $\forall X \subseteq U$, we have*

1. $\overline{\sum_{i=1}^{m} A_i}_{TOL}^{OPT}(X) = \{x \in U : TOL_{A_1}(x) \cap X \neq \varnothing \wedge \cdots \wedge TOL_{A_m}(x) \cap X \neq \varnothing\}$;
2. $\overline{\sum_{i=1}^{m} A_i}_{TOL}^{PES}(X) = \{x \in U : TOL_{A_1}(x) \cap X \neq \varnothing \vee \cdots \vee TOL_{A_m}(x) \cap X \neq \varnothing\}$;

3. $\overline{\sum_{i=1}^{m} A_i}_{TOL}^{OPT}(X) = \bigcap_{i=1}^{m} \overline{TOL_{A_i}}(X)$;

4. $\overline{\sum_{i=1}^{m} A_i}_{TOL}^{PES}(X) = \bigcup_{i=1}^{m} \overline{TOL_{A_i}}(X)$.

Proof.

1. $\forall x \in U$, by Definition 6.1, we have

$$\begin{aligned} x \in \overline{\sum_{i=1}^{m} A_i}_{TOL}^{OPT}(X) &\Leftrightarrow x \notin \underline{\sum_{i=1}^{m} A_i}_{TOL}^{OPT}(\sim X) \\ &\Leftrightarrow TOL_{A_i}(x) \nsubseteq (\sim X), (\forall i = 1, 2, \cdots, m) \\ &\Leftrightarrow TOL_{A_i}(x) \cap X \neq \varnothing, (\forall i = 1, 2, \cdots, m) \end{aligned}$$

2. The proof of 2 is similar to the proof of 1.

3. $\forall x \in \overline{\sum_{i=1}^{m} A_i}_{TOL}^{OPT}(X)$, by the proof of 1, we have $TOL_{A_1}(x) \cap X \neq \varnothing \wedge TOL_{A_2}(x) \cap X \neq \varnothing \wedge \cdots \wedge TOL_{A_m}(x) \cap X \neq \varnothing$, it follows that $x \in \overline{TOL_{A_1}}(X) \wedge \overline{TOL_{A_2}}(X) \wedge \cdots \wedge \overline{TOL_{A_m}}(X)$, i.e. $x \in \bigcap_{i=1}^{m} \overline{TOL_{A_i}}(X)$.

   Conversely, $\forall x \in \bigcap_{i=1}^{m} \overline{TOL_{A_i}}(X)$, we have $x \in \overline{TOL_{A_1}}(X) \wedge \overline{TOL_{A_2}}(X) \wedge \cdots \wedge \overline{TOL_{A_m}}(X)$, then by the proof of 1, $x \in \overline{\sum_{i=1}^{m} A_i}_{TOL}^{OPT}(X)$.

   From discussions above, we can conclude that $\overline{\sum_{i=1}^{m} A_i}_{TOL}^{OPT}(X) = \bigcap_{i=1}^{m} \overline{TOL_{A_i}}(X)$.

4. The proof of 4 is similar to the proof of 3. □

In the above theorem, results of 1 says that the optimistic multigranulation upper approximation is a set of objects, in which each x has an intersection between the tolerance class and the target in terms of all of the tolerance relations; 2 says that the pessimistic multigranulation upper approximation is a set of objects, in which each x has an intersection between the tolerance class and the target in terms of at least one of the tolerance relations. Results of 3 and 4 show that the optimistic (pessimistic) multigranulation upper approximations can also be considered as the intersection (union) of all the single granulation upper approximations.

Example 6.1. Let us consider Table 6.1, which is an incomplete evaluation table about venture investment.

In Table 6.1,

- $U = \{x_1, x_2, \cdots, x_{10}\}$ is the set of ten investment projects, i.e. universe;
- $AT = \{a_1, a_2, \cdots, a_5\}$ is the set of five evaluation experts, i.e. condition attributes; d is the decision attribute.

Table 6.1 An incomplete evaluation table about venture investment

U	a_1	a_2	a_3	a_4	a_5	d
x_1	2	3	3	2	2	High
x_2	1	3	3	2	2	High
x_3	1	1	1	1	1	Low
x_4	1	1	1	1	1	Low
x_5	1	1	1	1	1	Low
x_6	*	2	1	2	3	High
x_7	2	2	2	2	2	High
x_8	3	2	2	3	3	High
x_9	2	3	2	3	1	High
x_{10}	1	1	*	1	*	High

Since the decision attribute partitions the universe into subsets such that

$$U/IND(\{d\}) = \{\text{high}, \text{low}\} = \left\{\{x_1, x_2, x_6, x_7, x_8, x_9, x_{10}\}, \{x_3, x_4, x_5\}\right\}$$

then by Definition 6.1 and Definition 6.2, we obtain the following multigranulation lower and upper approximations:

$$\underline{\sum_{i=1}^{5} a_i}_{TOL}^{OPT}(\text{high}) = \{x_1, x_2, x_6, x_7, x_8, x_9\}$$

$$\overline{\sum_{i=1}^{5} a_i}_{TOL}^{OPT}(\text{high}) = U$$

$$\underline{\sum_{i=1}^{5} a_i}_{TOL}^{OPT}(\text{low}) = \varnothing$$

$$\overline{\sum_{i=1}^{5} a_i}_{TOL}^{OPT}(\text{low}) = \{x_3, x_4, x_5, x_{10}\}$$

$$\underline{\sum_{i=1}^{5} a_i}_{TOL}^{PES}(\text{high}) = \{x_1, x_7, x_8\}$$

$$\overline{\sum_{i=1}^{5} a_i}_{TOL}^{PES}(\text{high}) = U$$

$$\underline{\sum_{i=1}^{5} a_i}_{TOL}^{PES}(\text{low}) = \emptyset$$

$$\overline{\sum_{i=1}^{5} a_i}_{TOL}^{PES}(\text{low}) = \{x_2, x_3, x_4, x_5, x_6, x_9, x_{10}\}$$

6.1.2 Properties of Multigranulation Rough Sets Based on Tolerance Relations

Proposition 6.1. *Let I be an incomplete decision system in which $A_1, A_2, \cdots, A_m \subseteq AT$, then $\forall X \subseteq U$, the optimistic multigranulation rough set in terms of the tolerance relations has the following properties:*

1. $\underline{\sum_{i=1}^{m} A_i}_{TOL}^{OPT}(X) \subseteq X \subseteq \overline{\sum_{i=1}^{m} A_i}_{TOL}^{OPT}(X)$;
2. $\underline{\sum_{i=1}^{m} A_i}_{TOL}^{OPT}(\emptyset) = \overline{\sum_{i=1}^{m} A_i}_{TOL}^{OPT}(\emptyset) = \emptyset$, $\underline{\sum_{i=1}^{m} A_i}_{TOL}^{OPT}(U) = \overline{\sum_{i=1}^{m} A_i}_{TOL}^{OPT}(U) = U$;
3. $\underline{\sum_{i=1}^{m} A_i}_{TOL}^{OPT}(\sim X) = \sim \overline{\sum_{i=1}^{m} A_i}_{TOL}^{OPT}(X)$, $\overline{\sum_{i=1}^{m} A_i}_{TOL}^{OPT}(\sim X) = \sim \underline{\sum_{i=1}^{m} A_i}_{TOL}^{OPT}(X)$.

Proof.

1. $\forall x \in \underline{\sum_{i=1}^{m} A_i}_{TOL}^{OPT}(X)$, there must be $i \in \{1, 2, \cdots, m\}$ such that $TOL_{A_i}(x) \subseteq X$. Since the tolerance relation is reflexive, then we have $x \in X$, it follows that $\underline{\sum_{i=1}^{m} A_i}_{TOL}^{OPT}(X) \subseteq X$.

   Similarly, it is not difficult to prove that $X \subseteq \overline{\sum_{i=1}^{m} A_i}_{TOL}^{OPT}(X)$.

2. By the proof of 1 we know that $\underline{\sum_{i=1}^{m} A_i}_{TOL}^{OPT}(\emptyset) \subseteq \emptyset$. Moreover, $\emptyset \subseteq \underline{\sum_{i=1}^{m} A_i}_{TOL}^{OPT}(\emptyset)$ holds obviously since empty set is included in every set, thus, $\underline{\sum_{i=1}^{m} A_i}_{TOL}^{OPT}(\emptyset) = \emptyset$.

   Since $\emptyset \subseteq \overline{\sum_{i=1}^{m} A_i}_{TOL}^{OPT}(\emptyset)$, then it must be proved that $\overline{\sum_{i=1}^{m} A_i}_{TOL}^{OPT}(\emptyset) \subseteq \emptyset$. $\forall x \notin \emptyset$, we have $x \in U$, since tolerance relation is reflexive, then $TOL_{A_i}(x) \neq \varnothing$ for each $i = 1, 2, \cdots, m$, it follows that $TOL_{A_i}(x) \cap \varnothing = \varnothing$ for each $i = 1, 2, \cdots, m$. By 1 of Theorem 6.1, we know that $x \notin \overline{\sum_{i=1}^{m} A_i}^{OPT}_{TOL}(\varnothing)$, from which we can conclude that $\overline{\sum_{i=1}^{m} A_i}^{OPT}_{TOL}(\varnothing) \subseteq \varnothing$. From discussions above, we have $\underline{\sum_{i=1}^{m} A_i}^{OPT}_{TOL}(\varnothing) = \overline{\sum_{i=1}^{m} A_i}^{OPT}_{TOL}(\varnothing) = \varnothing$.

Similarly, it is not difficult to prove that $\underline{\sum_{i=1}^{m} A_i}^{OPT}_{TOL}(U) = \overline{\sum_{i=1}^{m} A_i}^{OPT}_{TOL}(U) = U$.

3. $\forall x \in U$,

$$
\begin{aligned}
x \in \underline{\sum_{i=1}^{m} A_i}^{OPT}_{TOL}(\sim X) &\Leftrightarrow \exists i \in \{1, 2, \cdots, m\} \text{ s.t. } TOL_{A_i}(x) \subseteq (\sim X)\\
&\Leftrightarrow \exists i \in \{1, 2, \cdots, m\} \text{ s.t. } TOL_{A_i}(x) \cap X = \varnothing\\
&\Leftrightarrow x \in \sim \overline{\sum_{i=1}^{m} A_i}^{OPT}_{TOL}(X)
\end{aligned}
$$

Similarly, it is not difficult to prove that $\overline{\sum_{i=1}^{m} A_i}^{OPT}_{TOL}(\sim X) = \sim \underline{\sum_{i=1}^{m} A_i}^{OPT}_{TOL}(X)$. □

Proposition 6.2. *Let I be an incomplete decision system in which $A_1, A_2, \cdots, A_m \subseteq AT$, then $\forall X \subseteq U$, the pesimistic multigranulation rough set in terms of the tolerance relations has the following properties:*

1. $\underline{\sum_{i=1}^{m} A_i}^{PES}_{TOL}(X) \subseteq X \subseteq \overline{\sum_{i=1}^{m} A_i}^{PES}_{TOL}(X)$;
2. $\underline{\sum_{i=1}^{m} A_i}^{PES}_{TOL}(\varnothing) = \overline{\sum_{i=1}^{m} A_i}^{PES}_{TOL}(\varnothing) = \varnothing$, $\underline{\sum_{i=1}^{m} A_i}^{PES}_{TOL}(U) = \overline{\sum_{i=1}^{m} A_i}^{PES}_{TOL}(U) = U$;
3. $\underline{\sum_{i=1}^{m} A_i}^{PES}_{TOL}(\sim X) = \sim \overline{\sum_{i=1}^{m} A_i}^{PES}_{TOL}(X)$, $\overline{\sum_{i=1}^{m} A_i}^{PES}_{TOL}(\sim X) = \sim \underline{\sum_{i=1}^{m} A_i}^{PES}_{TOL}(X)$.

Proof. The proof of Proposition 6.2 is similar to the proof of Proposition 6.1. □

6.1.3 Comparisons Among Several Rough Sets

Theorem 6.2. *Let I be an incomplete decision system in which $A_1, A_2, \cdots, A_m \subseteq AT$, then $\forall X \subseteq U$, we have*

1. $\underline{\sum_{i=1}^{m} A_i}_{TOL}^{PES}(X) \subseteq \underline{\sum_{i=1}^{m} A_i}_{TOL}^{OPT}(X)$;
2. $\overline{\sum_{i=1}^{m} A_i}_{TOL}^{OPT}(X) \subseteq \overline{\sum_{i=1}^{m} A_i}_{TOL}^{PES}(X)$.

Proof. $\forall x \in \underline{\sum_{i=1}^{m} A_i}_{TOL}^{PES}(X)$, then we have $TOL_{A_i}(x) \subseteq X$ for each $i = 1, 2, \cdots, m$. By Definition 6.1, $x \in \underline{\sum_{i=1}^{m} A_i}_{TOL}^{OPT}(X)$ holds obviously.

Similarly, it is not difficult to prove $\overline{\sum_{i=1}^{m} A_i}_{TOL}^{OPT}(X) \subseteq \overline{\sum_{i=1}^{m} A_i}_{TOL}^{PES}(X)$. □

Theorem 6.2 shows the relationships between optimistic and pessimistic multigranulatioin rough sets, which are both based on the tolerance relations. The details are: the pessimistic multigranulation lower approximation is included into the optimistic multigranulation lower approximation while the optimistic multigranulation upper approximation is included into the pessimistic multigranulation upper approximation. From this point of view, the optimistic multigranulation rough set can generate higher accuracy of approximation than pessimistic multigranulation rough set can do.

Theorem 6.3. *Let I be an incomplete decision system in which $A_1, A_2, \cdots, A_m \subseteq AT$, suppose that $A = A_1 \cup A_2 \cup \cdots \cup A_m$, then $\forall X \subseteq U$, we have*

1. $\underline{\sum_{i=1}^{m} A_i}_{TOL}^{OPT}(X) \subseteq \underline{TOL_A}(X)$;
2. $\overline{\sum_{i=1}^{m} A_i}_{TOL}^{OPT}(X) \supseteq \overline{TOL_A}(X)$;
3. $\underline{\sum_{i=1}^{m} A_i}_{TOL}^{PES}(X) \subseteq \underline{TOL_A}(X)$;
4. $\overline{\sum_{i=1}^{m} A_i}_{TOL}^{PES}(X) \supseteq \overline{TOL_A}(X)$.

Proof. We only prove 1, others can be proved analogously.

$\forall x \in \underline{\sum_{i=1}^{m} A_i}_{TOL}^{OPT}(X)$, then there must be $i \in \{1, 2, \cdots, m\}$ such that $TOL_{A_i}(x) \subseteq X$.

Moreover, since $A = A_1 \cup A_2 \cup \cdots \cup A_m$, then $TOL(A) = \bigcap_{i=1}^{m} TOL(A_i)$, it follows that $TOL_A(x) \subseteq TOL_{A_i}(x)$, i.e. $TOL_A(x) \subseteq X$, $x \in \underline{TOL_A}(X)$. □

Theorem 6.3 shows the relationships between multigranulation rough sets and single granulation rough set, which are both based on the tolerance relations. The details are: both the optimistic and pessimistic multigranulation lower approximations are included into the single granulation lower approximation, while both the optimistic and pessimistic multigranulation upper approximations include the single granulation upper approximation.

6.1.4 Approximation Distribution Reducts in Tolerance Relations Based Multigranulation Rough Sets

Definition 6.3. Let I be an incomplete decision system in which $A = \{a_1, a_2, \cdots, a_m\} \subseteq AT = \{a_1, a_2, \cdots, a_n\}$, then define

$$\underline{AT}_{TOL}^{OPT}(d) = \left\{ \underline{\sum_{i=1}^{n} a_i}_{TOL}^{OPT}(X_1), \underline{\sum_{i=1}^{n} a_i}_{TOL}^{OPT}(X_2), \cdots, \underline{\sum_{i=1}^{n} a_i}_{TOL}^{OPT}(X_l) \right\} \tag{6.5}$$

$$\overline{AT}_{TOL}^{OPT}(d) = \left\{ \overline{\sum_{i=1}^{n} a_i}_{TOL}^{OPT}(X_1), \overline{\sum_{i=1}^{n} a_i}_{TOL}^{OPT}(X_2), \cdots, \overline{\sum_{i=1}^{n} a_i}_{TOL}^{OPT}(X_l) \right\} \tag{6.6}$$

$$\underline{AT}_{TOL}^{PES}(d) = \left\{ \underline{\sum_{i=1}^{n} a_i}_{TOL}^{PES}(X_1), \underline{\sum_{i=1}^{n} a_i}_{TOL}^{PES}(X_2), \cdots, \underline{\sum_{i=1}^{n} a_i}_{TOL}^{PES}(X_l) \right\} \tag{6.7}$$

$$\overline{AT}_{TOL}^{PES}(d) = \left\{ \overline{\sum_{i=1}^{n} a_i}_{TOL}^{PES}(X_1), \overline{\sum_{i=1}^{n} a_i}_{TOL}^{PES}(X_2), \cdots, \overline{\sum_{i=1}^{n} a_i}_{TOL}^{PES}(X_l) \right\} \tag{6.8}$$

1. If $\underline{A}_{TOL}^{OPT}(d) = \underline{AT}_{TOL}^{OPT}(d)$, then A is referred to as a tolerance relations optimistic multigranulation lower approximate distribution consistent attributes set in I; if A is a tolerance relations optimistic multigranulation lower approximate distribution consistent attributes set in I and $\forall B \subset A$, B is not the tolerance relations optimistic multigranulation lower approximate distribution consistent attributes set in I, then A is referred to as a tolerance relations optimistic multigranulation lower approximate distribution reduct in I.
2. If $\overline{A}_{TOL}^{OPT}(d) = \overline{AT}_{TOL}^{OPT}(d)$, then A is referred to as a tolerance relations optimistic multigranulation upper approximate distribution consistent attributes set in I; if A is a tolerance relations optimistic multigranulation upper approximate distribution

consistent attributes set in I and $\forall B \subset A$, B is not the tolerance relations optimistic multigranulation upper approximate distribution consistent attributes set in I, then A is referred to as a tolerance relations optimistic multigranulation upper approximate distribution reduct in I.

3. If $\underline{A}_{TOL}^{OPT}(d) = \underline{AT}_{TOL}^{OPT}(d)$, then A is referred to as a tolerance relations pessimistic multigranulation lower approximate distribution consistent attributes set in I; if A is a tolerance relations pessimistic multigranulation lower approximate distribution consistent attributes set in I and $\forall B \subset A$, B is not the tolerance relations pessimistic multigranulation lower approximate distribution consistent attributes set in I, then A is referred to as a tolerance relations pessimistic multigranulation lower approximate distribution reduct in I.
4. If $\overline{A}_{TOL}^{OPT}(d) = \overline{AT}_{TOL}^{OPT}(d)$, then A is referred to as a tolerance relations pessimistic multigranulation upper approximate distribution consistent attributes set in I; if A is a tolerance relations pessimistic multigranulation upper approximate distribution consistent attributes set in I and $\forall B \subset A$, B is not the tolerance relations pessimistic multigranulation upper approximate distribution consistent attributes set in I, then A is referred to as a tolerance relations pessimistic multigranulation upper approximate distribution reduct in I.

By Definition 6.3, we can see that

- A tolerance relations optimistic multigranulation lower approximate distribution consistent attributes set in I is a subset of the attributes, which preserves the optimistic multigranulation lower approximations of all the decision classes; a tolerance relations optimistic multigranulation lower approximate distribution reduct in I is a minimal subset of the attributes, which preserves the optimistic multigranulation lower approximations of all the decision classes.
- A tolerance relations optimistic multigranulation upper approximate distribution consistent attributes set in I is a subset of the attributes, which preserves the optimistic multigranulation upper approximations of all the decision classes; a tolerance relations optimistic multigranulation upper approximate distribution reduct in I is a minimal subset of the attributes, which preserves the optimistic multigranulation upper approximations of all the decision classes.
- A tolerance relations pessimistic multigranulation lower approximate distribution consistent attributes set in I is a subset of the attributes, which preserves the pessimistic multigranulation lower approximations of all the decision classes; a tolerance relations pessimistic multigranulation lower approximate distribution reduct in I is a minimal subset of the attributes, which preserves the pessimistic multigranulation lower approximations of all the decision classes.
- A tolerance relations pessimistic multigranulation upper approximate distribution consistent attributes set in I is a subset of the attributes, which preserves the pessimistic multigranulation upper approximations of all the decision classes; a tolerance relations pessimistic multigranulation upper approximate distribution reduct

in I is a minimal subset of the attributes, which preserves the pessimistic multigranulation upper approximations of all the decision classes.

Theorem 6.4. *Let I be an incomplete decision system in which $A = \{a_1, a_2, \cdots, a_m\} \subseteq AT = \{a_1, a_2, \cdots, a_n\}$, then $\forall x \in U$, denote*

$$OL_{AT}^{MT}(x) = \left\{ X_k \in U/IND(\{d\}) : x \in \underline{\sum_{i=1}^{n} a_i}_{TOL}^{OPT}(X_k) \right\} \tag{6.9}$$

$$OU_{AT}^{MT}(x) = \left\{ X_k \in U/IND(\{d\}) : x \in \overline{\sum_{i=1}^{n} a_i}_{TOL}^{OPT}(X_k) \right\} \tag{6.10}$$

$$PL_{AT}^{MT}(x) = \left\{ X_k \in U/IND(\{d\}) : x \in \underline{\sum_{i=1}^{n} a_i}_{TOL}^{PES}(X_k) \right\} \tag{6.11}$$

$$PU_{AT}^{MT}(x) = \left\{ X_k \in U/IND(\{d\}) : x \in \overline{\sum_{i=1}^{n} a_i}_{TOL}^{PES}(X_k) \right\} \tag{6.12}$$

then

1. *A is a tolerance relations optimistic multigranulation lower approximate distribution consistent attributes set in $I \Leftrightarrow OL_{A}^{MT}(x) = OL_{AT}^{MT}(x)$ for each $x \in U$;*
2. *A is a tolerance relations optimistic multigranulation upper approximate distribution consistent attributes set in $I \Leftrightarrow OU_{A}^{MT}(x) = OU_{AT}^{MT}(x)$ for each $x \in U$;*
3. *A is a tolerance relations pessimistic multigranulation lower approximate distribution consistent attributes set in $I \Leftrightarrow PL_{A}^{MT}(x) = PL_{AT}^{MT}(x)$ for each $x \in U$;*
4. *A is a tolerance relations pessimistic multigranulation upper approximate distribution consistent attributes set in $I \Leftrightarrow PU_{A}^{MT}(x) = PU_{AT}^{MT}(x)$ for each $x \in U$.*

Proof. We only prove 1, others can be proved analogously.

$\Rightarrow$: Since A is a tolerance relations optimistic multigranulation lower approximate distribution consistent attributes set in I, then $\underline{A}_{TOL}^{OPT}(d) = \underline{AT}_{TOL}^{OPT}(d)$. By Eq. (6.5), we have $\underline{\sum_{i=1}^{m} a_i}_{TOL}^{OPT}(X_k) = \underline{\sum_{i=1}^{n} a_i}_{TOL}^{OPT}(X_k)$ for each $X_k \in U/IND(\{d\})$, it follows that $x \in \underline{\sum_{i=1}^{m} a_i}_{TOL}^{OPT}(X_k) \Leftrightarrow x \in \underline{\sum_{i=1}^{n} a_i}_{TOL}^{OPT}(X_k)$ and $x \notin \underline{\sum_{i=1}^{m} a_i}_{TOL}^{OPT}(X_k) \Leftrightarrow x \notin \underline{\sum_{i=1}^{n} a_i}_{TOL}^{OPT}(X_k)$ for each $x \in U$, i.e. $OL_{A}^{MT}(x) = OL_{AT}^{MT}(x)$ for each $x \in U$.

$\Leftarrow$: Since $OL_{A}^{MT}(x) = OL_{AT}^{MT}(x)$ for each $x \in U$, then $x \in \underline{\sum_{i=1}^{m} a_i}_{TOL}^{OPT}(X_k) \Leftrightarrow x \in$

$\underline{\sum_{i=1}^{n} a_i}_{TOL}^{OPT}(X_k)$ and $x \notin \underline{\sum_{i=1}^{m} a_i}_{TOL}^{OPT}(X_k) \Leftrightarrow x \notin \underline{\sum_{i=1}^{n} a_i}_{TOL}^{OPT}(X_k)$ for each $x \in U$, it follows that $\underline{\sum_{i=1}^{m} a_i}_{TOL}^{OPT}(X_k) = \underline{\sum_{i=1}^{n} a_i}_{TOL}^{OPT}(X_k)$ for each $X_k \in U/IND(\{d\})$, i.e. $\underline{A}_{TOL}^{OPT}(d) = \underline{AT}_{TOL}^{OPT}(d)$ and then A is a tolerance relations optimistic multigranulation lower approximate distribution consistent attributes set in I. □

Given an incomplete decision system I in which $A = \{a_1, a_2, \cdots, a_m\} \subseteq AT = \{a_1, a_2, \cdots, a_n\}$, let us denoted

$$DIS_{OL}^{MT} = \{(x,y) \in U^2 : X_k \in OL_{AT}^{MT}(x), y \notin X_k\}$$
$$DIS_{OU}^{MT} = \{(x,y) \in U^2 : X_k \notin OU_{AT}^{MT}(y), x \in X_k\}$$
$$DIS_{PL}^{MT} = \{(x,y) \in U^2 : X_k \in PL_{AT}^{MT}(x), y \notin X_k\}$$
$$DIS_{PU}^{MT} = \{(x,y) \in U^2 : X_k \notin PU_{AT}^{MT}(y), x \in X_k\}$$

$$DIS_{OL}^{MT}(x,y) = \begin{cases} \{a \in AT : (x,y) \notin TOL(\{a\})\}, & (x,y) \in DIS_{OL}^{MT} \\ \emptyset, & \text{otherwise} \end{cases}$$

$$DIS_{OU}^{MT}(x,y) = \begin{cases} \{a \in AT : (x,y) \notin TOL(\{a\})\}, & (x,y) \in DIS_{OU}^{MT} \\ \emptyset, & \text{otherwise} \end{cases}$$

$$DIS_{PL}^{MT}(x,y) = \begin{cases} \{a \in AT : (x,y) \notin TOL(\{a\})\}, & (x,y) \in DIS_{PL}^{MT} \\ \emptyset, & \text{otherwise} \end{cases}$$

$$DIS_{PU}^{MT}(x,y) = \begin{cases} \{a \in AT : (x,y) \notin TOL(\{a\})\}, & (x,y) \in DIS_{PU}^{MT} \\ \emptyset, & \text{otherwise} \end{cases}$$

then

$$\boldsymbol{M}_{OL}^{MT} = \{DIS_{OL}^{MT}(x,y) : (x,y) \in DIS_{OL}^{MT}\} \tag{6.13}$$
$$\boldsymbol{M}_{OU}^{MT} = \{DIS_{OU}^{MT}(x,y) : (x,y) \in DIS_{OU}^{MT}\} \tag{6.14}$$
$$\boldsymbol{M}_{PL}^{MT} = \{DIS_{PL}^{MT}(x,y) : (x,y) \in DIS_{PL}^{MT}\} \tag{6.15}$$
$$\boldsymbol{M}_{PU}^{MT} = \{DIS_{PU}^{MT}(x,y) : (x,y) \in DIS_{PU}^{MT}\} \tag{6.16}$$

are referred to as the tolerance relations optimistic multigranulation lower approximate distribution discernibility matrix, tolerance relations optimistic multigranulation upper approximate distribution discernibility matrix, tolerance relations pessimistic multigranulation lower approximate distribution discernibility matrix and tolerance relations pessimistic multigranulation upper approximate distribution discernibility matrix, respectively.

Theorem 6.5. *Let I be an incomplete decision system in which $A = \{a_1, a_2, \cdots, a_m\} \subseteq AT = \{a_1, a_2, \cdots, a_n\}$, then*

1. *A is a tolerance relations optimistic multigranulation lower approximate distribution consistent attributes set in $I \Leftrightarrow A \cap DIS_{OL}^{MT}(x,y) \neq \emptyset$ for each $DIS_{OL}^{MT}(x,y) \in \boldsymbol{M}_{OL}^{MT}$;*
2. *A is a tolerance relations optimistic multigranulation upper approximate distribution consistent attributes set in $I \Leftrightarrow A \cap DIS_{OU}^{MT}(x,y) \neq \emptyset$ for each $DIS_{OU}^{MT}(x,y) \in \boldsymbol{M}_{OU}^{MT}$;*
3. *A is a tolerance relations pessimistic multigranulation lower approximate distribution consistent attributes set in $I \Leftrightarrow A \cap DIS_{PL}^{MT}(x,y) \neq \emptyset$ for each $DIS_{PL}^{MT}(x,y) \in \boldsymbol{M}_{PL}^{MT}$;*
4. *A is a tolerance relations pessimistic multigranulation upper approximate distribution consistent attributes set in $I \Leftrightarrow A \cap DIS_{PU}^{MT}(x,y) \neq \emptyset$ for each $DIS_{PU}^{MT}(x,y) \in \boldsymbol{M}_{PU}^{MT}$.*

Proof.

1. $\Rightarrow$: Suppose $\exists DIS_{OL}^{MT}(x,y) \in \boldsymbol{M}_{OL}^{MT}$ such that $A \cap DIS_{OL}^{MT}(x,y) = \emptyset$, then we have $(x,y) \in TOL(A)$, i.e. $y \in TOL_A(x)$. Since $DIS_{OL}^{MT}(x,y) \in \boldsymbol{M}_{OL}^{MT}$, then there must be $X_k \in U/IND(\{d\})$ such that $x \in \underline{\sum_{i=1}^{n} a_i}_{TOL}^{OPT}(X_k)$ and $y \notin X_k$.

   By condition we also know that A is a tolerance relations optimistic multigranulation lower approximate distribution consistent attributes set in I, then $x \in \underline{\sum_{i=1}^{m} a_i}_{TOL}^{OPT}(X_k)$, i.e. $\exists a_j \in A$ such that $TOL_{\{a_j\}}(x) \subseteq X_k$. By the basic property of the tolerance relation, we have $TOL_A(x) \subseteq TOL_{\{a_j\}}(x) \subseteq X_k$, from which we can conclude that $y \in X_k$, which is contradictive to the assumption $DIS_{OL}^{MT}(x,y) \in \boldsymbol{M}_{OL}^{MT}$ because $DIS_{OL}^{MT}(x,y) \in \boldsymbol{M}_{OL}^{MT} \Rightarrow y \notin X_k$.

   $\Leftarrow$: Since $A \subseteq AT$, then we have $OL_A^{MT}(x) \subseteq OL_{AT}^{MT}(x)$ for each $x \in U$ obviously. Therefore, it must be proved that $OL_{AT}^{MT}(x) \subseteq OL_A^{MT}(x)$ for each $x \in U$.

   Suppose $\exists DIS_{OL}^{MT}(x,y) \in \boldsymbol{M}_{OL}^{MT} \left(X_k \in OL_{AT}^{MT}(x), y \notin X_k\right)$ such that $A \cap DIS_{OL}^{MT}(x,y) = \emptyset$, then we have $(x,y) \in TOL(A)$, i.e. $y \in TOL_A(x)$. By the basic property of the tolerance relation, we have $y \in TOL_{\{a\}}(x)$ for each $a \in A$, i.e. $TOL_{\{a\}}(x) \nsubseteq X_k$ for each $a \in A$, it follows that $x \notin \underline{\sum_{i=1}^{m} a_i}_{TOL}^{OPT}(X_k)$, $X_k \in OL_A^{MT}(x)$. From discussions above, we can conclude that if $A \cap DIS_{OL}^{MT}(x,y) \neq \emptyset$ for each $DIS_{OL}^{MT}(x,y) \in \boldsymbol{M}_{OL}^{MT}$, then we have $x \in \underline{\sum_{i=1}^{m} a_i}_{TOL}^{OPT}(X_k)\left(X_k \in OL_A^{MT}(x)\right)$ where $X_k \in OL_{AT}^{MT}(x)$, i.e. $OL_{AT}^{MT}(x) \subseteq OL_A^{MT}(x)$.

2. $\Rightarrow$: Suppose $\exists DIS_{OU}^{MT}(x,y) \in \boldsymbol{M}_{OU}^{MT}$ such that $A \cap DIS_{OU}^{MT}(x,y) = \emptyset$, then we have $(y,x) \in TOL(A)$, i.e. $x \in TOL_A(y)$. Since $DIS_{OU}^{MT}(x,y) \in \boldsymbol{M}_{OU}^{MT}$, then there must be $X_k \in U/IND(\{d\})$ such that $y \notin \overline{\sum_{i=1}^{n} a_i}_{TOL}^{OPT}(X_k)$ and $x \in X_k$.

By condition we also know that A is a tolerance relations optimistic multigranulation upper approximate distribution consistent attributes set in I, then $y \notin \overline{\sum_{i=1}^{m} a_i}_{TOL}^{OPT}(X_k)$, by 1 of Theorem 6.1, $\exists a_j \in A$ such that $TOL_{\{a_j\}}(y) \cap X_k = \emptyset$. By the basic property of the tolerance relation, we have $TOL_A(y) \subseteq TOL_{\{a_j\}}(y)$ and then $TOL_A(y) \cap X_k = \emptyset$, from which we can conclude that $x \notin X_k$, which is contradictive to the assumption $DIS_{OU}^{MT}(x,y) \in \boldsymbol{M}_{OU}^{MT}$ because $DIS_{OU}^{MT}(x,y) \in \boldsymbol{M}_{OU}^{MT} \Rightarrow x \in X_k$.
$\Leftarrow$: Since $A \subseteq AT$, then we have $OU_{AT}^{MT}(y) \subseteq OL_{A}^{MT}(y)$ for each $y \in U$ obviously. Therefore, it must be proved that $OU_{A}^{MT}(y) \subseteq OU_{AT}^{MT}(y)$ for each $y \in U$.

Suppose $\exists DIS_{OU}^{MT}(x,y) \in \boldsymbol{M}_{OU}^{MT}\left(X_k \notin OU_{AT}^{MT}(y), x \in X_k\right)$ such that $A \cap DIS_{OU}^{MT}(x,y) = \emptyset$, then we have $(x,y) \in TOL(A)$, i.e. $x \in TOL_A(y)$. By the basic property of the tolerance relation, we have $x \in TOL_{\{a\}}(y)$ for each $a \in A$, i.e. $TOL_{\{a\}}(y) \cap X_k \neq \emptyset$ for each $a \in A$, it follows that $y \in \overline{\sum_{i=1}^{m} a_i}_{TOL}^{OPT}(X_k)$, $X_k \in OU_{A}^{MT}(y)$. From discussions above, we can conclude that if $A \cap DIS_{OL}^{MT}(x,y) \neq \emptyset$ for each $DIS_{OU}^{MT}(x,y) \in \boldsymbol{M}_{OU}^{MT}$, then we have $y \notin \overline{\sum_{i=1}^{m} a_i}_{TOL}^{OPT}(X_k)\left(X_k \notin OU_{A}^{MT}(y)\right)$ where $X_k \notin OU_{AT}^{MT}(y)$, i.e. $OU_{A}^{MT}(y) \subseteq OU_{AT}^{MT}(y)$.

3. The proof of 3 is similar to the proof of 1.
4. The proof of 4 is similar to the proof of 2. □

Definition 6.4. Let I be an incomplete decision system in which $A = \{a_1, a_2, \cdots, a_m\} \subseteq AT = \{a_1, a_2, \cdots, a_n\}$, define

$$\Delta_{OL}^{MT} = \bigwedge_{DIS_{OL}^{MT}(x,y) \in \boldsymbol{M}_{OL}^{MT}} \vee DIS_{OL}^{MT}(x,y) \tag{6.17}$$

$$\Delta_{OU}^{MT} = \bigwedge_{DIS_{OU}^{MT}(x,y) \in \boldsymbol{M}_{OU}^{MT}} \vee DIS_{OU}^{MT}(x,y) \tag{6.18}$$

$$\Delta_{PL}^{MT} = \bigwedge_{DIS_{PL}^{MT}(x,y) \in \boldsymbol{M}_{PL}^{MT}} \vee DIS_{PL}^{MT}(x,y) \tag{6.19}$$

$$\Delta_{PU}^{MT} = \bigwedge_{DIS_{PU}^{MT}(x,y) \in \boldsymbol{M}_{PU}^{MT}} \vee DIS_{PU}^{MT}(x,y) \tag{6.20}$$

Δ_{OL}^{MT} is referred to as the tolerance relations optimistic multigranulation lower approximate distribution discernibility function, Δ_{OU}^{MT} is referred to as the tolerance relations optimistic multigranulation upper approximate distribution discernibility function, Δ_{PL}^{MT} is referred to as the tolerance relations pessimistic multigranulation lower approximate distribution discernibility function, Δ_{PU}^{MT} is referred to as the tolerance relations pessimistic multigranulation upper approximate distribution discernibility function.

By Boolean Reasoning technique, it is not difficult to obtain the following theorem by Theorem 6.5.

Theorem 6.6. *Let I be an incomplete decision system in which* $A = \{a_1, a_2, \cdots, a_m\} \subseteq AT = \{a_1, a_2, \cdots, a_n\}$, *then*

1. *A is the tolerance relations optimistic multigranulation lower approximate distribution reduct if and only if* $\wedge A$ *is the prime implicant of the tolerance relations optimistic multigranulation lower approximate distribution discernibility function, i.e.* Δ_{OL}^{MT};
2. *A is the tolerance relations optimistic multigranulation upper approximate distribution reduct if and only if* $\wedge A$ *is the prime implicant of the tolerance relations optimistic multigranulation upper approximate distribution discernibility function, i.e.* Δ_{OU}^{MT};
3. *A is the tolerance relations pessimistic multigranulation lower approximate distribution reduct if and only if* $\wedge A$ *is the prime implicant of the tolerance relations pessimistic multigranulation lower approximate distribution discernibility function, i.e.* Δ_{PL}^{MT};
4. *A is the tolerance relations pessimistic multigranulation upper approximate distribution reduct if and only if* $\wedge A$ *is the prime implicant of the tolerance relations pessimistic multigranulation upper approximate distribution discernibility function, i.e.* Δ_{PU}^{MT}.

Example 6.2. Take for instance Table 6.1, by Eqs.(6.13) and (6.14), we obtain the tolerance relations optimistic multigranulation lower approximate distribution discernibility matrix and tolerance relations optimistic multigranulation upper approximate distribution discernibility matrix as Table 6.2 and Table 6.3 show, respectively.

Table 6.2 Tolerance relations optimistic multigranulation lower approximate distribution discernibility matrix $\boldsymbol{M}_{OL}^{MT}$ in Table 6.1

	x_1	x_2	x_3	x_4	x_5	x_6	x_7	x_8	x_9	x_{10}
x_1			AT	AT	AT					
x_2			a_2, a_3, a_4, a_5	a_2, a_3, a_4, a_5	a_2, a_3, a_4, a_5					
x_3										
x_4										
x_5										
x_6			a_2, a_4, a_5	a_2, a_4, a_5	a_2, a_4, a_5					
x_7			AT	AT	AT					
x_8			AT	AT	AT					
x_9			a_1, a_2, a_3, a_4	a_1, a_2, a_3, a_4	a_1, a_2, a_3, a_4					
x_{10}										

Table 6.3 Tolerance relations optimistic multigranulation upper approximate distribution discernibility matrix $\boldsymbol{M}_{OU}^{MT}$ in Table 6.1

x_1	x_2	x_3	x_4	x_5	x_6	x_7	x_8	x_9	x_{10}
x_1									
x_2									
x_3	AT	a_2, a_3, a_4, a_5				a_2, a_4, a_5	AT	AT	a_1, a_2, a_3, a_4
x_4	AT	a_2, a_3, a_4, a_5				a_2, a_4, a_5	AT	AT	a_1, a_2, a_3, a_4
x_5	AT	a_2, a_3, a_4, a_5				a_2, a_4, a_5	AT	AT	a_1, a_2, a_3, a_4
x_6									
x_7									
x_8									
x_9									
x_{10}									

Then by Eqs. (6.17) and (6.18), the tolerance relations optimistic multigranulation lower approximate distribution discernibility function and the tolerance relations optimistic multigranulation upper approximate distribution discernibility function are:

$$\begin{aligned} \Delta_{OL}^{MT} &= \Delta_{OU}^{MT} \\ &= \left(a_2 \vee a_3 \vee a_4 \vee a_5\right) \wedge \left(a_2 \vee a_4 \vee a_5\right) \wedge \left(a_1 \vee a_2 \vee a_3 \vee a_4\right) \wedge AT \\ &= \left(a_2 \wedge a_4\right) \vee \left(a_1 \wedge a_5\right) \vee \left(a_3 \wedge a_5\right) \end{aligned}$$

By Theorem 6.6, we know that the sets of the attributes $\{a_2, a_4\}$, $\{a_1, a_5\}$ and $\{a_3, a_5\}$ are both the tolerance relations optimistic multigranulation lower approximate distribution reducts and the tolerance relations optimistic multigranulation upper approximate distribution reducts in Table 6.1.

Similarly, it is not difficult to obtain that the tolerance relations pessimistic multigranulation lower approximate distribution reduct and the tolerance relations pessimistic multigranulation upper approximate distribution reducts in Table 6.1 is the set of the attributes $AT = \{a_1, a_2, a_3, a_4, a_5\}$, i.e. no attribute can be deleted.

6.2 Similarity Relations Based Multigranulation Rough Sets

6.2.1 Optimistic and Pessimistic Similarity Relations Based Multigranulation Rough Sets

Definition 6.5. Let I be an incomplete decision system in which $A_1, A_2, \cdots, A_m \subseteq AT$, then $\forall X \subseteq U$, the optimistic multigranulation lower approximation and optimistic multigranulation upper approximation in terms of the similarity relations are denoted by $\underline{\sum_{i=1}^{m} A_i}_{SIM}^{OPT}(X)$ and $\overline{\sum_{i=1}^{m} A_i}_{SIM}^{OPT}(X)$, respectively, where

$$\underline{\sum_{i=1}^{m} A_i}_{SIM}^{OPT}(X) = \{x \in U : SIM_{A_1}^{-1}(x) \subseteq X \vee \cdots \vee SIM_{A_m}^{-1}(x) \subseteq X\} \tag{6.21}$$

$$\overline{\sum_{i=1}^{m} A_i}_{SIM}^{OPT}(X) = \sim \underline{\sum_{i=1}^{m} A_i}_{SIM}^{OPT}(\sim X) \tag{6.22}$$

Definition 6.6. Let I be an incomplete decision system in which $A_1, A_2, \cdots, A_m \subseteq AT$, then $\forall X \subseteq U$, the pessimistic multigranulation lower approximation and pessimistic multigranulation upper approximation in terms of the similarity relations are denoted by $\underline{\sum_{i=1}^{m} A_i}_{SIM}^{PES}(X)$ and $\overline{\sum_{i=1}^{m} A_i}_{SIM}^{PES}(X)$, respectively, where

$$\underline{\sum_{i=1}^{m} A_i}_{SIM}^{PES}(X) = \{x \in U : SIM_{A_1}^{-1}(x) \subseteq X \wedge \cdots \wedge SIM_{A_m}^{-1}(x) \subseteq X\} \tag{6.23}$$

$$\overline{\sum_{i=1}^{m} A_i}_{SIM}^{PES}(X) = \sim \underline{\sum_{i=1}^{m} A_i}_{SIM}^{PES}(\sim X) \tag{6.24}$$

The pair $\left[\underline{\sum_{i=1}^{m} A_i}_{SIM}^{OPT}(X), \overline{\sum_{i=1}^{m} A_i}_{SIM}^{OPT}(X)\right]$ is referred to as an optimistic multigranulation rough set of X in terms of the set of the similarity relations $\{SIM(A_1), SIM(A_2), \cdots, SIM(A_m)\}$; the pair $\left[\underline{\sum_{i=1}^{m} A_i}_{SIM}^{PES}(X), \overline{\sum_{i=1}^{m} A_i}_{SIM}^{PES}(X)\right]$ is referred to as a pessimistic multigranulation rough set of X in terms of the set of the similarity relations $\{SIM(A_1), SIM(A_2), \cdots, SIM(A_m)\}$.

Theorem 6.7. *Let I be an incomplete decision system in which $A_1, A_2, \cdots, A_m \subseteq AT$, then $\forall X \subseteq U$, we have*

1. $\overline{\sum_{i=1}^{m} A_i}_{SIM}^{OPT}(X) = \{x \in U : SIM_{A_1}^{-1}(x) \cap X \neq \varnothing \wedge \cdots \wedge SIM_{A_m}^{-1}(x) \cap X \neq \varnothing\}$;
2. $\overline{\sum_{i=1}^{m} A_i}_{SIM}^{PES}(X) = \{x \in U : SIM_{A_1}^{-1}(x) \cap X \neq \varnothing \vee \cdots \vee SIM_{A_m}^{-1}(x) \cap X \neq \varnothing\}$;
3. $\overline{\sum_{i=1}^{m} A_i}_{SIM}^{OPT}(X) = \bigcap_{i=1}^{m} \overline{SIM_{A_i}}(X)$;

4. $\overline{\sum_{i=1}^{m} A_i}^{PES}_{SIM}(X) = \bigcup_{i=1}^{m} \overline{SIM_{A_i}}(X)$.

Proof. The proof of Theorem 6.7 is similar to the proof of Theorem 6.1. □

Similar to Theorem 6.1, in Theorem 6.7, 1 says that the optimistic multigranulation upper approximations is the set of the objects, in which each x has an intersection between $SIM^{-1}_{A_i}(x)$ and the target in terms of all of the similarity relations, 2 says that the pessimistic multigranulation upper approximations is the set of the objects, in which each x has an intersection between $SIM^{-1}_{A_i}(x)$ and the target in terms of at least one of the similarity relations. Results of 3 and 4 show that the optimistic (pessimistic) multigranulation upper approximations can also be considered as the intersection (union) of all single granulation upper approximations.

Example 6.3. Let us consider Table 2.11, which is an incomplete decision system with lost unknown values. Since the decision attribute determine the universe into partition such that

$$U/IND(\{d\}) = \{\Phi, \Psi\} = \Big\{\{x_1, x_2, x_4, x_7, x_{10}, x_{12}\}, \{x_3, x_5, x_6, x_8, x_9, x_{11}\}\Big\}$$

then by Definition 6.5 and Definition 6.6, we obtain the following multigranulation lower and upper approximations:

$$\underline{\sum_{i=1}^{4} a_i}^{OPT}_{SIM}(\Phi) = \{x_{10}\}$$

$$\overline{\sum_{i=1}^{4} a_i}^{OPT}_{SIM}(\Phi) = \{x_1, x_2, x_3, x_4, x_5, x_6, x_7, x_9, x_{10}, x_{11}, x_{12}\}$$

$$\underline{\sum_{i=1}^{4} a_i}^{OPT}_{SIM}(\Psi) = \{x_8\}$$

$$\overline{\sum_{i=1}^{4} a_i}^{OPT}_{SIM}(\Psi) = \{x_1, x_2, x_3, x_4, x_5, x_6, x_7, x_8, x_9, x_{11}, x_{12}\}$$

$$\underline{\sum_{i=1}^{4} a_i}^{PES}_{SIM}(\Phi) = \emptyset$$

$$\overline{\sum_{i=1}^{4} a_i}^{PES}_{SIM}(\Phi) = U$$

$$\underline{\sum_{i=1}^{4} a_i}^{PES}_{SIM}(\Psi) = \emptyset$$

$$\overline{\sum_{i=1}^{4} a_i}^{PES}_{SIM}(\Psi) = U$$

6.2.2 Properties of Multigranulation Rough Sets Based on Similarity Relations

Proposition 6.3. *Let I be an incomplete decision system in which $A_1, A_2, \cdots, A_m \subseteq AT$, then $\forall X \subseteq U$, the optimistic multigranulation rough set in terms of the similarity relations has the following properties*:

1. $\underline{\sum_{i=1}^{m} A_i}^{OPT}_{SIM}(X) \subseteq X \subseteq \overline{\sum_{i=1}^{m} A_i}^{OPT}_{SIM}(X)$;
2. $\underline{\sum_{i=1}^{m} A_i}^{OPT}_{SIM}(\varnothing) = \overline{\sum_{i=1}^{m} A_i}^{OPT}_{SIM}(\varnothing) = \varnothing$, $\underline{\sum_{i=1}^{m} A_i}^{OPT}_{SIM}(U) = \overline{\sum_{i=1}^{m} A_i}^{OPT}_{SIM}(U) = U$;
3. $\underline{\sum_{i=1}^{m} A_i}^{OPT}_{SIM}(\sim X) = \sim \overline{\sum_{i=1}^{m} A_i}^{OPT}_{SIM}(X)$, $\overline{\sum_{i=1}^{m} A_i}^{OPT}_{SIM}(\sim X) = \sim \underline{\sum_{i=1}^{m} A_i}^{OPT}_{SIM}(X)$.

Proof. The proof of Proposition 6.3 is similar to the proof of Proposition 6.1. □

Proposition 6.4. *Let I be an incomplete decision system in which $A_1, A_2, \cdots, A_m \subseteq AT$, then $\forall X \subseteq U$, the pessimistic multigranulation rough set in terms of the similarity relations has the following properties*:

1. $\underline{\sum_{i=1}^{m} A_i}^{PES}_{SIM}(X) \subseteq X \subseteq \overline{\sum_{i=1}^{m} A_i}^{PES}_{SIM}(X)$;
2. $\underline{\sum_{i=1}^{m} A_i}^{PES}_{SIM}(\varnothing) = \overline{\sum_{i=1}^{m} A_i}^{PES}_{SIM}(\varnothing) = \varnothing$, $\underline{\sum_{i=1}^{m} A_i}^{PES}_{SIM}(U) = \overline{\sum_{i=1}^{m} A_i}^{PES}_{SIM}(U) = U$;
3. $\underline{\sum_{i=1}^{m} A_i}^{PES}_{SIM}(\sim X) = \sim \overline{\sum_{i=1}^{m} A_i}^{PES}_{SIM}(X)$, $\overline{\sum_{i=1}^{m} A_i}^{PES}_{SIM}(\sim X) = \sim \underline{\sum_{i=1}^{m} A_i}^{PES}_{SIM}(X)$.

Proof. The proof of Proposition 6.4 is similar to the proof of Proposition 6.1. □

6.2.3 Comparisons Among Several Rough Sets

Theorem 6.8. *Let I be an incomplete decision system in which $A_1, A_2, \cdots, A_m \subseteq AT$, then $\forall X \subseteq U$, we have*

1. $\underline{\sum_{i=1}^{m} A_i}^{PES}_{SIM}(X) \subseteq \underline{\sum_{i=1}^{m} A_i}^{OPT}_{SIM}(X)$;

2. $\overline{\sum_{i=1}^{m} A_i}^{OPT}_{SIM}(X) \subseteq \overline{\sum_{i=1}^{m} A_i}^{PES}_{SIM}(X)$.

Proof. The proof of Theorem 6.8 is similar to the proof of Theorem 6.2. □

Similar to Theorem 6.2, Theorem 6.8 shows the relationships between optimistic and pessimistic multigranulatioin rough sets, which are both based on the similarity relations. The details are: the pessimistic multigranulation lower approximation is included into the optimistic multigranulation lower approximation while the optimistic multigranulation upper approximation is included into the pessimistic multigranulation upper approximation. From this point of view, the optimistic multigranulation rough set can generate higher accuracy of approximation than pessimistic multigranulation rough set can do.

Theorem 6.9. *Let I be an incomplete decision system in which $A_1, A_2, \cdots, A_m \subseteq AT$, suppose that $A = A_1 \cup A_2 \cup \cdots \cup A_m$, then $\forall X \subseteq U$, we have*

1. $\underline{\sum_{i=1}^{m} A_i}^{OPT}_{SIM}(X) \subseteq \underline{SIM_A}(X)$;
2. $\overline{\sum_{i=1}^{m} A_i}^{OPT}_{SIM}(X) \supseteq \overline{SIM_A}(X)$;
3. $\underline{\sum_{i=1}^{m} A_i}^{PES}_{SIM}(X) \subseteq \underline{SIM_A}(X)$;
4. $\overline{\sum_{i=1}^{m} A_i}^{PES}_{SIM}(X) \supseteq \overline{SIM_A}(X)$.

Proof. The proof of Theorem 6.9 is similar to the proof of Theorem 6.3. □

Similar to Theorem 6.3, Theorem 6.9 shows the relationships between multigranulation rough sets and classical single granulation rough set, which are both based on the similarity relations. The details are: both the optimistic and pessimistic multigranulation lower approximations are included into the single granulation lower approximation while both the optimistic and pessimistic multigranulation upper approximations include the single granulation upper approximation.

Theorem 6.10. *Let I be an incomplete decision system in which $A_1, A_2, \cdots, A_m \subseteq AT$, then $\forall X \subseteq U$, we have*

1. $\underline{\sum_{i=1}^{m} A_i}^{OPT}_{TOL}(X) \subseteq \underline{\sum_{i=1}^{m} A_i}^{OPT}_{SIM}(X)$;
2. $\overline{\sum_{i=1}^{m} A_i}^{OPT}_{TOL}(X) \supseteq \overline{\sum_{i-1}^{m} A_i}^{OPT}_{SIM}(X)$;

3. $\underline{\sum_{i=1}^{m} A_i}_{TOL}^{PES}(X) \subseteq \underline{\sum_{i=1}^{m} A_i}_{SIM}^{PES}(X)$;

4. $\overline{\sum_{i=1}^{m} A_i}_{TOL}^{PES}(X) \supseteq \overline{\sum_{i=1}^{m} A_i}_{SIM}^{PES}(X)$.

Proof. We only prove 1, others can be proved analogously.

$\forall x \in \underline{\sum_{i=1}^{m} A_i}_{TOL}^{OPT}(X)$, then there must be $i \in \{1, 2, \cdots, m\}$ such that $TOL_{A_i}(x) \subseteq X$.

Since $SIM(A_i) \subseteq TOL(A_i)$, then we have $SIM_{A_i}^{-1} \subseteq TOL_{A_i}(x)$, it follows that $x \in \underline{\sum_{i=1}^{m} A_i}_{SIM}^{OPT}(X)$, i.e. $\underline{\sum_{i=1}^{m} A_i}_{TOL}^{OPT}(X) \subseteq \underline{\sum_{i=1}^{m} A_i}_{SIM}^{OPT}(X)$. □

Theorem 6.10 tells us no matter optimistic or pessimistic method is selected, by using the similarity relations, we can obtain greater lower approximations and smaller upper approximations than the tolerance relation.

6.2.4 Approximate Distribution Reducts in Similarity Relations Based Multigranulation Rough Sets

Definition 6.7. Let I be an incomplete decision system in which $A = \{a_1, a_2, \cdots, a_m\} \subseteq AT = \{a_1, a_2, \cdots, a_n\}$, then define

$$\underline{AT}_{SIM}^{OPT}(d) = \left\{\underline{\sum_{i=1}^{n} a_i}_{SIM}^{OPT}(X_1), \underline{\sum_{i=1}^{n} a_i}_{SIM}^{OPT}(X_2), \cdots, \underline{\sum_{i=1}^{n} a_i}_{SIM}^{OPT}(X_l)\right\} \quad (6.25)$$

$$\overline{AT}_{SIM}^{OPT}(d) = \left\{\overline{\sum_{i=1}^{n} a_i}_{SIM}^{OPT}(X_1), \overline{\sum_{i=1}^{n} a_i}_{SIM}^{OPT}(X_2), \cdots, \overline{\sum_{i=1}^{n} a_i}_{SIM}^{OPT}(X_l)\right\} \quad (6.26)$$

$$\underline{AT}_{SIM}^{PES}(d) = \left\{\underline{\sum_{i=1}^{n} a_i}_{SIM}^{PES}(X_1), \underline{\sum_{i=1}^{n} a_i}_{SIM}^{PES}(X_2), \cdots, \underline{\sum_{i=1}^{n} a_i}_{SIM}^{PES}(X_l)\right\} \quad (6.27)$$

$$\overline{AT}_{SIM}^{PES}(d) = \left\{\overline{\sum_{i=1}^{n} a_i}_{SIM}^{PES}(X_1), \overline{\sum_{i=1}^{n} a_i}_{SIM}^{PES}(X_2), \cdots, \overline{\sum_{i=1}^{n} a_i}_{SIM}^{PES}(X_l)\right\} \quad (6.28)$$

1. If $\underline{A}_{SIM}^{OPT}(d) = \underline{AT}_{SIM}^{OPT}(d)$, then A is referred to as a similarity relations optimistic multigranulation lower approximate distribution consistent attributes set in I; if A is a similarity relations optimistic multigranulation lower approximate distribution consistent attributes set in I and $\forall B \subset A$, B is not the similarity relations optimistic multigranulation lower approximate distribution consistent attributes set in I, then A is referred to as a similarity relations optimistic multigranulation lower approximate distribution reduct in I.

2. If $\overline{A}_{SIM}^{OPT}(d) = \overline{AT}_{SIM}^{OPT}(d)$, then A is referred to as a similarity relations optimistic multigranulation upper approximate distribution consistent attributes set in I; if A is a similarity relations optimistic multigranulation upper approximate distribution consistent attributes set in I and $\forall B \subset A$, B is not the similarity relations optimistic multigranulation upper approximate distribution consistent attributes set in I, then A is referred to as a similarity relations optimistic multigranulation upper approximate distribution reduct in I.
3. If $\underline{A}_{SIM}^{OPT}(d) = \underline{AT}_{SIM}^{OPT}(d)$, then A is referred to as a similarity relations pessimistic multigranulation lower approximate distribution consistent attributes set in I; if A is a similarity relations pessimistic multigranulation lower approximate distribution consistent attributes set in I and $\forall B \subset A$, B is not the similarity relations pessimistic multigranulation lower approximate distribution consistent attributes set in I, then A is referred to as a similarity relations pessimistic multigranulation lower approximate distribution reduct in I.
4. If $\overline{A}_{SIM}^{OPT}(d) = \overline{AT}_{SIM}^{OPT}(d)$, then A is referred to as a similarity relations pessimistic multigranulation upper approximate distribution consistent attributes set in I; if A is a similarity relations pessimistic multigranulation upper approximate distribution consistent attributes set in I and $\forall B \subset A$, B is not the similarity relations pessimistic multigranulation upper approximate distribution consistent attributes set in I, then A is referred to as a similarity relations pessimistic multigranulation upper approximate distribution reduct in I.

By Definition 6.7, we can see that

- A similarity relations optimistic multigranulation lower approximate distribution consistent attributes set in I is a subset of the attributes, which preserves the optimistic multigranulation lower approximations of all the decision classes; a similarity relations optimistic multigranulation lower approximate distribution reduct in I is a minimal subset of the attributes, which preserves the optimistic multigranulation lower approximations of all the decision classes.
- A similarity relations optimistic multigranulation upper approximate distribution consistent attributes set in I is a subset of the attributes, which preserves the optimistic multigranulation upper approximations of all the decision classes; a similarity relations optimistic multigranulation upper approximate distribution reduct in I is a minimal subset of the attributes, which preserves the optimistic multigranulation upper approximations of all the decision classes.
- A similarity relations pessimistic multigranulation lower approximate distribution consistent attributes set in I is a subset of the attributes, which preserves the pessimistic multigranulation lower approximations of all the decision classes; a similarity relations pessimistic multigranulation lower approximate distribution reduct in I is a minimal subset of the attributes, which preserves the pessimistic multigranulation lower approximations of all the decision classes.
- A similarity relations pessimistic multigranulation upper approximate distribution consistent attributes set in I is a subset of the attributes, which preserves the pes-

simistic multigranulation upper approximations of all the decision classes; a similarity relations pessimistic multigranulation upper approximate distribution reduct in I is a minimal subset of the attributes, which preserves the pessimistic multigranulation upper approximations of all the decision classes.

Theorem 6.11. *Let I be an incomplete decision system in which $A = \{a_1, a_2, \cdots, a_m\} \subseteq AT = \{a_1, a_2, \cdots, a_n\}$, then $\forall x \in U$, denote*

$$OL_{AT}^{MS}(x) = \left\{X_k \in U/IND(\{d\}) : x \in \underline{\sum_{i=1}^{n} a_i}_{SIM}^{OPT}(X_k)\right\}$$

$$OU_{AT}^{MS}(x) = \left\{X_k \in U/IND(\{d\}) : x \in \overline{\sum_{i=1}^{n} a_i}_{SIM}^{OPT}(X_k)\right\}$$

$$PL_{AT}^{MS}(x) = \left\{X_k \in U/IND(\{d\}) : x \in \underline{\sum_{i=1}^{n} a_i}_{SIM}^{PES}(X_k)\right\}$$

$$PU_{AT}^{MS}(x) = \left\{X_k \in U/IND(\{d\}) : x \in \overline{\sum_{i=1}^{n} a_i}_{SIM}^{PES}(X_k)\right\}$$

then

1. *A is a similarity relations optimistic multigranulation lower approximate distribution consistent attributes set in $I \Leftrightarrow OL_{A}^{MS}(x) = OL_{AT}^{MS}(x)$ for each $x \in U$;*
2. *A is a similarity relations optimistic multigranulation upper approximate distribution consistent attributes set in $I \Leftrightarrow OU_{A}^{MS}(x) = OU_{AT}^{MS}(x)$ for each $x \in U$;*
3. *A is a similarity relations pessimistic multigranulation lower approximate distribution consistent attributes set in $I \Leftrightarrow PL_{A}^{MS}(x) = PL_{AT}^{MS}(x)$ for each $x \in U$;*
4. *A is a similarity relations pessimistic multigranulation upper approximate distribution consistent attributes set in $I \Leftrightarrow PU_{A}^{MS}(x) = PU_{AT}^{MS}(x)$ for each $x \in U$.*

Proof. The proof of Theorem 6.11 is similar to the proof of Theorem 6.4. □

Theorem 6.12. *Let I be an incomplete decision system in which $A = \{a_1, a_2, \cdots, a_m\} \subseteq AT = \{a_1, a_2, \cdots, a_n\}$, then*

1. *$OL_{A}^{MS}(x) = OL_{AT}^{MS}(x)$ for each $x \in U \Leftrightarrow \exists a_i \in A$, if $OL_{AT}(x) \nsubseteq OL_{AT}(y)$, then $SIM_{a_i}^{-1}(y) \nsubseteq SIM_{a_i}^{-1}(x)$;*
2. *$OU_{A}^{MS}(x) = OU_{AT}^{MS}(x)$ for each $x \in U \Leftrightarrow \exists a_i \in A$, if $OU_{AT}(y) \nsubseteq OU_{AT}(x)$, then $SIM_{a_i}^{-1}(y) \nsubseteq SIM_{a_i}^{-1}(x)$;*
3. *$PL_{A}^{MS}(x) = PL_{AT}^{MS}(x)$ for each $x \in U \Leftrightarrow \exists a_i \in A$, if $PL_{AT}(x) \nsubseteq PL_{AT}(y)$, then $SIM_{a_i}^{-1}(y) \nsubseteq SIM_{a_i}^{-1}(x)$;*

4. $PU_A^{MS}(x) = PU_{AT}^{MS}(x)$ *for each* $x \in U \Leftrightarrow \exists a_i \in A$, *if* $PU_{AT}(y) \nsubseteq PU_{AT}(x)$, *then* $SIM_{a_i}^{-1}(y) \nsubseteq SIM_{a_i}^{-1}(x)$.

Proof.

1. $\Rightarrow$: Suppose that $OL_{AT}^{MS}(x) \nsubseteq OL_{AT}^{MS}(y)$, then there must be $X_k \in IND(\{d\})$ such that $X_k \in OL_{AT}^{MS}(x)$ and $X_k \notin OL_{AT}^{MS}(y)$. By condition we know that $OL_A^{MS}(x) = OL_{AT}^{MS}(x)$ for each $x \in U$, thus, we have $X_k \in OL_A^{MS}(x)$ and $X_k \notin OL_A^{MS}(y)$. Since $X_k \in OL_A^{MS}(x)$, then there must be $a_i \in A$ such that $SIM_{a_i}^{-1}(x) \subseteq X_k$; since $X_k \notin OL_A^{MS}(y)$, then $SIM_a^{-1}(y) \nsubseteq X_k$ for each $a \in A$, i.e. $SIM_{a_i}^{-1}(y) \nsubseteq X_k$, from which we can conclude that $\exists a_i \in A$ such that $SIM_{a_i}^{-1}(y) \nsubseteq SIM_{a_i}^{-1}(x)$.

   $\Leftarrow$: Since $A \subseteq AT$, then by the definition of the optimistic multigranulation lower approximation, $\underline{\sum_{i=1}^{m} a_i}_{SIM}^{OPT}(X_k) \subseteq \underline{\sum_{i=1}^{n} a_i}_{SIM}^{OPT}(X_k)$ holds obviously for each $X_k \in U/IND(\{d\})$, from which we can conclude that $OL_A^{MS}(x) \subseteq OL_{AT}^{MS}(x)$ for each $x \in U$. Thus, it must be proved $OL_A^{MS}(x) \supseteq OL_{AT}^{MS}(x)$ for each $x \in U$.

   By condition we know that $\exists a_i \in A$ such that $SIM_{a_i}^{-1}(y) \subseteq SIM_{a_i}^{-1}(x) \Rightarrow OL_{AT}^{MS}(x) \subseteq OL_{AT}^{MS}(y)$. Since similarity relation is reflexive and transitive, then $SIM_{a_i}^{-1}(y) \subseteq SIM_{a_i}^{-1}(x) \Leftrightarrow y \in SIM_{a_i}^{-1}(x)$ holds obviously, from which we can conclude that $\exists a_i \in A$ where $y \in SIM_{a_i}^{-1}(x) \Rightarrow OL_{AT}^{MS}(x) \subseteq OL_{AT}^{MS}(y)$.

   $\forall y \in SIM_{a_i}^{-1}(x)$ and $\forall X_k \in OL_{AT}^{MS}(x)$, $X_k \in OL_{AT}^{MS}(y)$ holds. Thus, $\exists a_j \in AT$ such that $SIM_{a_j}^{-1}(y) \subseteq X_k$. Since $a_j \in AT$, then we have $SIM_{AT}^{-1}(y) \subseteq SIM_{a_j}^{-1}(x)$, it follows that $SIM_{AT}^{-1}(y) \subseteq X_k$. Since similarity relation is reflexive and transitive, then $SIM_{a_i}^{-1}(x) = \bigcup\{SIM_{AT}^{-1}(y) : y \in SIM_{a_i}^{-1}(x)\}$ holds obviously. It has been proved if $y \in SIM_{a_i}^{-1}(x)$, then $SIM_{AT}^{-1}(y) \subseteq X_k$, from which we can conclude that $SIM_{a_i}^{-1}(x) \subseteq X_k$, i.e. $X_k \in OL_A^{MS}(x)$.
2. $\Rightarrow$: Suppose that $OU_{AT}^{MS}(y) \nsubseteq OU_{AT}^{MS}(x)$, then there must be $X_k \in IND(\{d\})$ such that $X_k \in OU_{AT}^{MS}(y)$ and $X_k \notin OU_{AT}^{MS}(x)$. By condition we know that $OU_A^{MS}(x) = OU_{AT}^{MS}(x)$ for each $x \in U$, thus, we have $X_k \in OU_A^{MS}(y)$ and $X_k \notin OU_A^{MS}(x)$. since $X_k \notin OU_A^{MS}(x)$, then there must be $a_i \in A$ such that $SIM_{a_i}^{-1}(x) \cap X_k = \emptyset$; since $X_k \in OU_A^{MS}(y)$, then $\forall a \in A$ such that $SIM_a^{-1}(y) \cap X_k \neq \emptyset$, i.e. $SIM_{a_i}^{-1}(y) \cap X_k \neq \emptyset$, from which we can conclude that $\exists a_i \in A$ such that $SIM_{a_i}^{-1}(y) \nsubseteq SIM_{a_i}^{-1}(x)$.

   $\Leftarrow$: Since $A \subseteq AT$, then by 1 of Theorem 6.7, $\overline{\sum_{i=1}^{m} a_i}_{SIM}^{OPT}(X_k) \supseteq \overline{\sum_{i=1}^{n} a_i}_{SIM}^{OPT}(X_k)$ holds obviously for each $X_k \in U/IND(\{d\})$, from which we can conclude that $OU_A^{MS}(x) \supseteq OU_{AT}^{MS}(x)$ for each $x \in U$. Thus, it must be proved $OU_A^{MS}(x) \subseteq$

$OU_{AT}^{MS}(x)$ for each $x \in U$.

By condition we know that $\exists a_i \in A$ such that $SIM_{a_i}^{-1}(y) \subseteq SIM_{a_i}^{-1}(x) \Rightarrow OU_{AT}^{MS}(y) \subseteq OU_{AT}^{MS}(x)$. Since similarity relation is reflexive and transitive, then $SIM_{a_i}^{-1}(y) \subseteq SIM_{a_i}^{-1}(x) \Leftrightarrow y \in SIM_{a_i}^{-1}(x)$ holds obviously, from which we can conclude that $\exists a_i \in A$ where $y \in SIM_{a_i}^{-1}(x) \Rightarrow OU_{AT}^{MS}(y) \subseteq OU_{AT}^{MS}(x)$.

$\forall y \in SIM_{a_i}^{-1}(x)$ and $\forall X_k \notin OU_{AT}^{MS}(x)$, $X_k \notin OU_{AT}^{MS}(y)$ holds. Thus, $\exists a_j \in AT$ such that $SIM_{a_j}^{-1}(y) \cap X_k = \varnothing$. Since $a_j \in AT$, then we have $SIM_{AT}^{-1}(y) \subseteq SIM_{a_j}^{-1}(x)$, it follows that $SIM_{AT}^{-1}(y) \cap X_k = \varnothing$. Since similarity relation is reflexive and transitive, then $SIM_{a_i}^{-1}(x) = \bigcup\{SIM_{AT}^{-1}(y) : y \in SIM_{a_i}^{-1}(x)\}$ holds obviously. It has been proved if $y \in SIM_{a_i}^{-1}(x)$, then $SIM_{AT}^{-1}(y) \cap X_k = \varnothing$, from which we can conclude that $SIM_{a_i}^{-1}(x) \cap X_k = \varnothing$, i.e. $X_k \notin OU_{A}^{MS}(x)$.

3. The proof of 3 is similar to the proof of 1.
4. The proof of 4 is similar to the proof of 2. □

Given an incomplete decision system I in which $A = \{a_1, a_2, \cdots, a_m\} \subseteq AT = \{a_1, a_2, \cdots, a_n\}$, let us denoted

$$DIS_{OL}^{MS} = \{(x,y) \in U^2 : OL_{AT}^{MS}(x) \not\subseteq OL_{AT}^{MS}(y)\}$$
$$DIS_{OU}^{MS} = \{(x,y) \in U^2 : OU_{AT}^{MS}(y) \not\subseteq OU_{AT}^{MS}(x)\}$$
$$DIS_{PL}^{MS} = \{(x,y) \in U^2 : PL_{AT}^{MS}(x) \not\subseteq PL_{AT}^{MS}(y)\}$$
$$DIS_{PU}^{MS} = \{(x,y) \in U^2 : PU_{AT}^{MS}(y) \not\subseteq PU_{AT}^{MS}(x)\}$$

$$DIS_{OL}^{MS}(x,y) = \begin{cases} \{a \in AT : (x,y) \notin SIM(\{a\})\}, & (x,y) \in DIS_{OL}^{MS} \\ \varnothing, & \text{otherwise} \end{cases}$$

$$DIS_{OU}^{MS}(x,y) = \begin{cases} \{a \in AT : (x,y) \notin SIM(\{a\})\}, & (x,y) \in DIS_{OU}^{MS} \\ \varnothing, & \text{otherwise} \end{cases}$$

$$DIS_{PL}^{MS}(x,y) = \begin{cases} \{a \in AT : (x,y) \notin SIM(\{a\})\}, & : (x,y) \in DIS_{PL}^{MS} \\ \varnothing, & \text{otherwise} \end{cases}$$

$$DIS_{PU}^{MS}(x,y) = \begin{cases} \{a \in AT : (x,y) \notin SIM(\{a\})\}, & (x,y) \in DIS_{PU}^{MS} \\ \varnothing, & \text{otherwise} \end{cases}$$

then

$$\boldsymbol{M}_{OL}^{MS} = \{DIS_{OL}^{MS}(x,y) : (x,y) \in DIS_{OL}^{MS}\} \tag{6.29}$$
$$\boldsymbol{M}_{OU}^{MS} = \{DIS_{OU}^{MS}(x,y) : (x,y) \in DIS_{OU}^{MS}\} \tag{6.30}$$
$$\boldsymbol{M}_{PL}^{MS} = \{DIS_{PL}^{MS}(x,y) : (x,y) \in DIS_{PL}^{MS}\} \tag{6.31}$$
$$\boldsymbol{M}_{PU}^{MS} = \{DIS_{PU}^{MS}(x,y) : (x,y) \in DIS_{PU}^{MS}\} \tag{6.32}$$

are referred to as the similarity relations optimistic multigranulation lower approximate distribution discernibility matrix, similarity relations optimistic multigranulation upper approximate distribution discernibility matrix, similarity relations pessimistic multigranulation lower approximate distribution discernibility matrix and similarity relations pessimistic multigranulation upper approximate distribution discernibility matrix, respectively.

Theorem 6.13. *Let I be an incomplete decision system in which $A = \{a_1, a_2, \cdots, a_m\} \subseteq AT = \{a_1, a_2, \cdots, a_n\}$, the*

1. *A is a similarity relations optimistic multigranulation lower approximate distribution consistent attributes set in $I \Leftrightarrow A \cap DIS^{MS}_{OL}(x,y) \neq \emptyset$ for each $DIS^{MS}_{OL}(x,y) \neq \emptyset$;*
2. *A is a similarity relations optimistic multigranulation upper approximate distribution consistent attributes set in $I \Leftrightarrow A \cap DIS^{MS}_{OU}(x,y) \neq \emptyset$ for each $DIS^{MS}_{OU}(x,y) \neq \emptyset$;*
3. *A is a similarity relations pessimistic multigranulation lower approximate distribution consistent attributes set in $I \Leftrightarrow A \cap DIS^{MS}_{PL}(x,y) \neq \emptyset$ for each $DIS^{MS}_{PL}(x,y) \neq \emptyset$;*
4. *A is a similarity relations pessimistic multigranulation upper approximate distribution consistent attributes set in $I \Leftrightarrow A \cap DIS^{MS}_{PU}(x,y) \neq \emptyset$ for each $DIS^{MS}_{PU}(x,y) \neq \emptyset$.*

Proof. We only prove 1, others can be proved analogously.

$\Rightarrow$: Since A is a similarity relations optimistic multigranulation lower approximate distribution consistent attributes set in I, then $\underline{A}^{OPT}_{SIM}(d) = \underline{AT}^{OPT}_{SIM}(d)$. By Theorem 6.11, we have $OL^{MS}_{A}(x) = OL^{MS}_{AT}(x)$ for each $x \in U$. $\forall DIS^{MS}_{OL}(x,y) \neq \emptyset$, we have $OL^{MS}_{AT}(x) \nsubseteq OL^{MS}_{AT}(y)$. By Theorem 6.12, we know that $\exists a_i \in A$, if $OL^{MS}_{AT}(x) \nsubseteq OL^{MS}_{AT}(y)$, then $SIM^{-1}_{a_i}(y) \nsubseteq SIM^{-1}_{a_i}(x)$, it follows that $y \notin SIM^{-1}_{a_i}(x)$, i.e. $(x,y) \notin SIM(\{a_i\})$, $A \cap DIS^{MS}_{OL}(x,y) \neq \emptyset$.

$\Leftarrow$: for each $DIS^{MS}_{OL}(x,y) \neq \emptyset$, we have $OL^{MS}_{AT}(x) \nsubseteq OL^{MS}_{AT}(y)$. By condition, we know that $A \cap DIS^{MS}_{OL}(x,y) \neq \emptyset$, it follows that there must be $a_i \in A$ such that $(x,y) \notin SIM(\{a_i\})$. By Theorem 6.12, we can conclude that $OL^{MS}_{A}(x) = OL^{MS}_{AT}(x)$ for each $x \in U$ and then by Theorem 6.11, we obtain $\underline{A}^{OPT}_{SIM}(d) = \underline{AT}^{OPT}_{SIM}(d)$, i.e. A is a similarity relations optimistic multigranulation lower approximate distribution consistent attributes set in I. □

Definition 6.8. Let I be an incomplete decision system in which $A = \{a_1, a_2, \cdots, a_m\} \subseteq AT = \{a_1, a_2, \cdots, a_n\}$, define

$$\Delta^{MS}_{OL} = \bigwedge_{DIS^{MS}_{OL}(x,y)\in \boldsymbol{M}^{MS}_{OL}} \vee DIS^{MS}_{OL}(x,y) \tag{6.33}$$

$$\Delta^{MS}_{OU} = \bigwedge_{DIS^{MS}_{OU}(x,y)\in \boldsymbol{M}^{MS}_{OU}} \vee DIS^{MS}_{OU}(x,y) \tag{6.34}$$

$$\Delta^{MS}_{PL} = \bigwedge_{DIS^{MS}_{PL}(x,y)\in \boldsymbol{M}^{MS}_{PL}} \vee DIS^{MS}_{PL}(x,y) \tag{6.35}$$

$$\Delta_{PU}^{MS} = \bigwedge_{DIS_{PU}^{MS}(x,y) \in \boldsymbol{M}_{PU}^{MS}} \vee DIS_{PU}^{MS}(x,y) \tag{6.36}$$

Δ_{OL}^{MS} is referred to as the similarity relations optimistic multigranulation lower approximate distribution discernibility function, Δ_{OU}^{MS} is referred to as the similarity relations optimistic multigranulation upper approximate distribution discernibility function, Δ_{PL}^{MS} is referred to as the similarity relations pessimistic multigranulation lower approximate distribution discernibility function, Δ_{PU}^{MS} is referred to as the similarity relations pessimistic multigranulation upper approximate distribution discernibility function.

By Boolean Reasoning technique, it is not difficult to obtain the following theorem by Theorem 6.13.

Theorem 6.14. *Let I be an incomplete decision system in which* $A = \{a_1, a_2, \cdots, a_m\} \subseteq AT = \{a_1, a_2, \cdots, a_n\}$, *then*

1. *A is the similarity relations optimistic multigranulation lower approximate distribution reduct if and only if* $\bigwedge A$ *is the similarity relations optimistic multigranulation lower approximate distribution discernibility function, i.e.* Δ_{OL}^{MS};
2. *A is the similarity relations optimistic multigranulation upper approximate distribution reduct if and only if* $\bigwedge A$ *is the similarity relations optimistic multigranulation upper approximate distribution discernibility function, i.e.* Δ_{OU}^{MS};
3. *A is the similarity relations pessimistic multigranulation lower approximate distribution reduct if and only if* $\bigwedge A$ *is the similarity relations pessimistic multigranulation lower approximate distribution discernibility function, i.e.* Δ_{PL}^{MS};
4. *A is the similarity relations pessimistic multigranulation upper approximate distribution reduct if and only if* $\bigwedge A$ *is the similarity relations pessimistic multigranulation upper approximate distribution discernibility function, i.e.* Δ_{PU}^{MS}.

Example 6.4. Take for instance Table 2.11, by Eqs.(6.29) and (6.30), we obtain the similarity relation optimistic multigranulation lower approximate distribution discernibility matrix and similarity relation optimistic multigranulation upper approximate distribution discernibility matrix as Table 6.4 and Table 6.5 show, respectively.

Table 6.4 Similarity relations optimistic multigranulation lower approximate distribution discernibility matrix $\boldsymbol{M}_{OL}^{MS}$ in Table 2.11

	x_1	x_2	x_3	x_4	x_5	x_6	x_7	x_8	x_9	x_{10}	x_{11}	x_{12}
x_1												
x_2												
x_3												
x_4												
x_5												
x_6												
x_7												
x_8	a_2, a_3	a_2, a_3	a_2, a_3	a_2, a_3	a_2, a_3	a_2, a_3	a_2, a_3		a_2, a_3	a_2, a_3	a_2, a_3	a_2, a_3
x_9												
x_{10}	a_1	a_1	a_1	a_1	a_1	a_1	a_1		a_1	a_1	a_1	a_1
x_{11}												
x_{12}												

Table 6.5 Similarity relations optimistic multigranulation upper approximate distribution discernibility matrix $\boldsymbol{M}_{OU}^{MS}$ in Table 2.11

	x_1	x_2	x_3	x_4	x_5	x_6	x_7	x_8	x_9	x_{10}	x_{11}	x_{12}
x_1												
x_2												
x_3												
x_4												
x_5												
x_6												
x_7												
x_8	a_2, a_3	a_2, a_3	a_2, a_3	a_2, a_3	a_2, a_3	a_2, a_3	a_2, a_3		a_2, a_3	a_2, a_3	a_2, a_3	a_2, a_3
x_9												
x_{10}	a_1	a_1	a_1	a_1	a_1	a_1	a_1		a_1	a_1	a_1	a_1
x_{11}												
x_{12}												

Then by Eqs. (6.33) and (6.34), the similarity relations optimistic multigranulation lower approximate distribution discernibility function and the similarity relations optimistic multigranulation upper approximate distribution discernibility function are:

$$\begin{aligned} \Delta_{OL}^{MS} = \Delta_{OU}^{MS} &= \left(a_2 \vee a_3\right) \wedge a_1 \\ &= \left(a_1 \wedge a_2\right) \vee \left(a_1 \wedge a_3\right) \end{aligned}$$

By Theorem 6.14, we know that the sets of the attributes $\{a_1, a_2\}$ and $\{a_1, a_3\}$ are both the similarity relations optimistic multigranulation lower approximate distribution reducts and the similarity relations optimistic multigranulation upper approximate distribution reducts in Table 2.11.

Similarly, it is not difficult to obtain that the similarity relations pessimistic multigranulation lower approximate distribution reduct and the similarity relations pessimistic multigranulation upper approximate distribution reducts in Table 2.11 is the set of the attributes $AT = \{a_1, a_2, a_3, a_4\}$, i.e. no attribute can be deleted.

6.3 Conclusions

In this chapter, the multigranulation rough set approach is introduced into the incomplete information systems. Two widely used binary relation, i.e. the tolerance relation and the similarity relation are used to construct the multigranulation rough set. Similar to the multigranulation rough sets, which are based on the equivalence relations, both the optimistic and pessimistic cases are considered in the incomplete information system for the constructing of the multigranulation rough sets.

Based on these multigranulation rough sets, not only the basic properties are discussed, but also the concept of the distribution reducts are introduced. The practical approaches to compute these distribution reducts are also presented.

Though the tolerance relations and the similarity relations are used to construct multigranulation rough sets in this chapter, it should be noticed that since many other approaches can also be used to deal with the incomplete information system (see details in Section 2), and then these approaches can be generalized into the multigranulation environment with respect to different requirements.

References

[1] Qian, Y. H., Dang, C. Y., Liang, J. Y.: MGRS in incomplete information systems. In: 2007 IEEE International Conference on Granular Computing, pp. 163–168 (2007)

[2] Qian, Y. H., Liang, J. Y., Dang, C. Y.: Incomplete multigranulation rough set, IEEE T. Syst. Man Cy. B. **20**, 420–431 (2010)

[3] Qian, Y. H., Liang, J. Y., Wei, W.: Pessimistic rough decision. In: Second International Workshop on Rough Sets Theory, pp. 440–449 (2010)

[4] Qian, Y. H., Liang, J. Y., Yao, Y. Y., Dang, C. Y.: MGRS: A multi-granulation rough set. Inform. Sci. **180**, 949–970 (2010)

[5] Guan, Y. Y., Wang, H. K.: Set-valued information systems. Inform. Sci. **176**, 2507–2525 (2006)

[6] Kryszkiewicz, M.: Rough set approach to incomplete information systems. Inform. Sci. **112**, 39–49 (1998)

[7] Leung, Y., Li, D. Y.: Maximal consistent block technique for rule acquisition in incomplete information systems. Inform. Sci. **115**, 85–106 (2003)

[8] Qian, Y. H., Liang, J. Y., Li, D. Y., Wang, F., Ma, N. N.: Approximation reduction in inconsistent incomplete decision tables. Knowl.-Based Syst. **23**, 427–433 (2010)

[9] Stefanowski, J., Tsoukiàs, A.: On the extension of rough sets under incomplete information. In: Proceeding of New directions in rough sets, data mining and granular-soft computing, pp. 73–82 (1999)

[10] Stefanowski, J., Tsoukiàs, A.: Incomplete information tables and rough classification. Comput. Intell. **17**, 545–566 (2001)

Glossary

accuracy An accuracy is the inverse of roughness.
accuracy of approximation An accuracy of approximation is the cardinal number of the lower approximation divided by the cardinal number of the upper approximation.
approximate distribution reduct An approximate distribution reduct is a subset of the attributes, which preserves the approximations of all the decision classes.
attribute An attribute is a feature, which can be used to describe the objects in the universe.
characteristic relation A characteristic relation is a relation that is reflexive for the classification analyzing in the generalized incomplete information systems.
core A core is the intersection of all the reducts.
covering A covering of a nonemptyset set is a set of nonemptyset, whose union is the full set.
decision system A decision system is an information system, which contains both condition and decision attributes.
descriptor A descriptor is an atomic property or conjunction of different atomic properties.
difference relation A difference relation is a relation that is not necessarily reflexive, symmetric and transitive, it can be used to induce negative rules from the incomplete information systems with lost unknown values.
discernibility function A discernibility function is a Boolean function, which can be used to delete superfluous attributes in the information system.
discernibility matrix A discernibility matrix is a matrix, in which each element is the set of the attributes to distinguish two different objects.
distribution consistent set A distribution consistent set is a subset of the attributes, which provides the same information for approximate distribution or some purposes as all of the attributes in the information system.
dominance-based rough set A dominance-based rough set is a rough set, which is defined by a dominance relation.
dominance relation A dominance relation is a relation that is reflexive and transitive for the classification analyzing in the preference-ordered domains of the attributes.
equivalence relation An equivalence relation is a relation that is reflexive, symmetric and transitive.
expanded dominance relation An expanded dominance relation is a relation that is only reflexive for the dominance-based classification analyzing in the incomplete information systems with "do not care" unknown values.
fuzzy rough set A fuzzy rough set is a rough set, which is defined through the fuzzy binary relation insetad of an indiscernibility relation.

generalized decision A generalized decision is a function from the universe to the power set of the domain of the decision attribute.
generalized incomplete information system A generalized incomplete information system is an incomplete information system in which "do not care" and lost unknown values are coexisting.
incomplete information system An incomplete information system is an information system with unknown values.
indiscernibility relation An indiscernibility relation is the intersection of the equivalence relations in the knowledge base.
information system An information system is a table representation approach of the knowledge in the knowledge base.
knowledge base A knowledge base consists of a universe and a family of the equivalence relation on this universe.
knowledge dependency Knowledge Q depends on knowledge P if and only if all elementary categories of Q can be defined in terms of some elementary categories of knowledge P.
knowledge granulation A knowledge granulation is the discernibility ability of the knowledge.
knowledge operation A knowledge operation is an operation which is defined on the knowledge.
knowledge reduction A knowledge reduction is the searching of particular subsets of the equivalence relations, which provide the same information for classification or some purposes as all of the equivalence relations in the knowledge base.
limited dominance relation A limited dominance relation is a relation that is reflexive and transitive for the dominance-based classification analyzing in the incomplete information systems with "do not care" unknown values.
limited tolerance relation A limited tolerance relation is a relation that is reflexive and symmetric, whose classification analyzing is between similarity relation and tolerance relation.
lower approximation A lower approximation of X is the collection of objects which can be classified with full certainty as members of X.
maximal consistent block A maximal consistent block is the maximal collection of objects, in which all objects are tolerant to each other.
multigranulation rough set A multigranulation rough set is a rough set, which is defined through a family of the binary relations instead of a single binary relation.
multiple-source information systems A multiple-source information systems is a family of the information systems.
neighborhood A neighborhood is a unit in a neighborhood system.
neighborhood system A neighborhood system is one of the formal models of the granular computing.
partition A partition of a nonemptyset set is a set of nonemptyset, disjoint subsets whose union is the full set.
reduct A reduct is a minimal subset of the attributes, which provide the same information for classification or some purposes as all of the attributes in the information

system.

rough accuracy A rough accuracy is the inverse of the accuracy of approximation.

roughness A roughness is the product of rough accuracy and knowledge granulation.

rough set A rough set is an extension of the set theory for the study of intelligent systems characterized by insufficient and incomplete information.

similarity dominance relation A similarity dominance relation is a relation that is reflexive and transitive for the dominance-based classification analyzing in the incomplete information systems with lost unknown values.

similarity relation A similarity relation is a relation that is reflexive and transitive for the classification analyzing in the incomplete information systems with lost unknown values.

tolerance relation A tolerance relation is a relation that is reflexive and symmetric for the classification analyzing in the incomplete information systems with "do not care" unknown values.

upper approximation An upper approximation of X is the collection of objects which can be classified with possibility as members of X.

valued dominance relation A valued dominance relation is a fuzzy dominance relation that is reflexive for the classification analyzing in the incomplete information systems with "do not care" unknown values.

valued tolerance relation A valued tolerance relation is a fuzzy relation that is reflexive and symmetric for the classification analyzing in the incomplete information systems with "do not care" unknown values.

variable precision rough set A variable precision rough set is a rough set, which is defined on the the concept of inclusion error.

Index

Kolibri 360 GmbH
Ferdinand-Jühlke-Straße 7
99095 Erfurt, Deutschland
produktsicherheit@kolibri360.de